Topology Deduction and Control for Dual-Active-Bridge Converters

Guo Xu · Xiaoying Chen

Topology Deduction and Control for Dual-Active-Bridge Converters

Springer

Guo Xu
School of Automation
Central South University
Changsha, China

Xiaoying Chen
School of Electrical Engineering
and Automation
Fuzhou University
Fuzhou, China

ISBN 978-981-97-4855-6 ISBN 978-981-97-4856-3 (eBook)
https://doi.org/10.1007/978-981-97-4856-3

This Springer imprint is published by the registered company Springer Nature Singapore Pte Ltd.
The registered company address is: 152 Beach Road, #21-01/04 Gateway East, Singapore 189721, Singapore

Preface

The dual-active-bridge (DAB) converters have become interesting topologies for medium/high power applications, especially for bidirectional power applications requiring galvanic isolation, e.g. DC microgrids, solid-state transformers (SSTs), automotive applications, energy storage systems. In comparison with other bidirectional isolation converters, the salient features of DAB converters are as follows. (1) Inherent zero voltage switching (ZVS) capability. (2) Wide voltage conversion range. (3) Natural bidirectional power flow.

This book provides a comprehensive review and classification for popular DAB converters. Based on the unified topology architecture of the DAB converters, the topologies derivation law is studied to develop novel converters. Several novel converters are provided to illustrate different topology modification methods, including modified IO types, modified active bridges, modified transformers, adding auxiliary high-frequency networks, and modular topology structure. Meanwhile, the control optimization and the parameter design of the novel converters are also investigated. By analyzing the converter working principles, power transfer characteristics, and converter losses, the power control strategies are applied to achieve better performance aimed at different control goals.

This book will benefit the researchers and engineers in the field of DC-DC power conversion, and graduate/undergraduate students focused on the bidirectional DC-DC converters. The outcomes will enable effective development and high-quality mass production of novel high-performance DC-DC converters for the aforementioned bidirectional power applications. The main objectives are as follows,

(1) To review and summarize the unified architecture and topologies derivation law of the popular DAB converters.
(2) To derive a novel feasible DAB topology for the application that requires wide voltage gain range, high efficiency, and bidirectional power flow.
(3) To learn how to start analyzing a novel DAB topology and developing effective modulation and control methods to achieve desired goals.
(4) To present the useful methods and principles to develop new topologies and controls for isolated bidirectional DC-DC converters.

This book consists of 13 Chapters based on our several research projects, and covers the aspects of converter topologies and power control strategies for DAB converters. This book starts with a comprehensive review and classification for popular DAB converters in Chap. 1, and the topologies derivation law is also studied. From the perspective of topology modification methods, the remaining content can be divided into 5 parts. The first part focuses on the control and optimization for the converter with the modified IO types which are from Chaps. 2–3. The second part presents the novel topology and control strategy with modified active bridges in Chap. 4. The topologies and control methods in the third part are studied by modifying transformers, which are from Chaps. 5–7. The modification method of adding auxiliary high-frequency networks is presented in the fourth part which are from Chaps. 8–10. The fifth part focuses on the modular structure of the bidirectional DC-DC converters which are from Chaps. 11–13.

The Authors wish to express their sincere thanks to Dr. Liting Li, Prof. Deshang Sha, Prof. Yonglu Liu, Prof. Yao Sun, and Prof. Mei Su for their contribution on the topology's derivation research and other contributions to this book.

The author would also like to acknowledge the contributions of Dr. Jing Guo, Prof. Hua Han, Mr. Zhiqiang Cai, Prof. Hui Wang, Dr. Liang Yuan, and Miss. Yuchen Lin on the current fed DAB converters, Mr. Wenhui Qiu, Dr. Wenjing Xiong on the T-type DAB converters, Mr. Yaohui Qu, Dr. Jingtao Xu, Dr. Dong Liu and Dr. Ziqing Xia on the DAB converters with modified transformer, Dr. Yuefeng Liao, Prof. Tao Peng, Dr. Bin Guo, Prof. Yongheng Yang, and Dr. Zhengmei Lu on the modular DAB converters.

The Author would also like to thank their families who have given tremendous support all the time. Finally, the authors are extremely grateful to Springer and the editorial staff for the opportunity to publish this book and help in all possible manners.

Changsha, China — Guo Xu
Fuzhou, China — Xiaoying Chen

Contents

1 Introduction to Topology Deduction of Dual-Active-Bridge Converters 1
1.1 Basic Cells of the Dual-Active-Bridge Converter 1
1.1.1 The Cells of the Input/Output Type of DAB Converter 2
1.1.2 The Cells of the Active Bridge 2
1.1.3 The Cells of the High-Frequency Transformer 5
1.1.4 The Cells of Auxiliary High-Frequency Networks 6
1.2 Dual-Active-Bridge Converter with Modified I/O Type 7
1.2.1 Voltage-Fed DAB Converter 7
1.2.2 Current-Fed DAB Converter 7
1.3 Dual-Active-Bridge Converter with Modified Active Bridges 9
1.3.1 Reduce Active Switches 9
1.3.2 Increase Active Switches 10
1.3.3 Reconnection of the Active Bridges 11
1.4 Dual-Active-Bridge Converter with Modified Transformer 12
1.4.1 Multi-winding Transformer 12
1.4.2 Multi-core Transformer 14
1.4.3 Controllable Transformer 14
1.5 Dual-Active-Bridge Converter with Auxiliary HF Networks 15
1.5.1 Non-Resonant Networks 15
1.5.2 Resonant Networks 16
1.6 Dual-Active-Bridge Converter with Modular Structure 18
1.6.1 The Conventional Modular Structure of DAB Converter 18
1.6.2 The Hybrid Modular Structure of DAB Converter 18

1.7 Summary and Future Trends of the DAB Converter 19
1.7.1 Summary of the Modification Methods of the DAB Converter 19
1.7.2 Evolvement History and Future Trends 19
1.8 Conclusion and Future Work 23
1.9 Organization of the Book 23
References 25

Part I Dual-Active-Bridge Converter with Modified IO Types

2 Design Considerations for Plus-Phase-Shift Controlled Current-Fed DAB Converter 31
2.1 Converter Topology and Operation Principle 31
2.1.1 Topology and Modulation Schedule Using PPS Control 32
2.1.2 Working Stages of the Converter 33
2.1.3 Power Expressions of the Converter 35
2.2 Soft-Switching Analysis and Optimized Design 37
2.2.1 ZVS Range Analysis 37
2.2.2 DC and Leakage Inductances Optimization 41
2.3 Comparison of Loss 44
2.3.1 Driving Loss 44
2.3.2 Switches Loss 44
2.3.3 Inductor Loss 45
2.3.4 Transformer Loss 47
2.3.5 Current Parameters Comparison 48
2.3.6 Power Loss Comparison 48
2.3.7 Comparison Analysis of Bidirectional Converter 49
2.4 System Parameters Design 50
2.4.1 DC Inductance L_{dc} Design 50
2.4.2 Turns Ratios of Transformer Design 51
2.4.3 Leakage Inductance L_r Design 51
2.4.4 Clamping Capacitor C_c Selection 52
2.4.5 Swithes Selection 53
2.5 Experiment Verification 53
2.6 Conclusion and Discussion 57
References 60

3 Dead-Time Optimization and Magnetizing Current Design for a Current-Fed Dual Active Bridge DC–DC Converter 61
3.1 Converter Topology and Operation Principle 61
3.1.1 Topology and Modulation Schedule Using PPS Control 62
3.1.2 Operating Modes of the Converter 63

3.2 ZVS Analysis Considering Junction Capacitors and Dead Time ... 64
3.2.1 ZVS Analysis Considering Junction Capacitors ... 64
3.2.2 ZVS Analysis Considering Dead Time ... 67
3.3 Utilizing Magnetizing Inductance to Enlarge ZVS Range ... 69
3.4 Design Considerations ... 73
3.4.1 Turn Ratio of Transformer ... 73
3.4.2 DC Inductance and Leakage Inductance ... 73
3.4.3 The Dead Time and Magnetizing Inductance ... 75
3.4.4 Control Strategy ... 77
3.5 Experimental Results and Discussions ... 78
3.6 Conclusion ... 84
References ... 85

Part II Dual-Active-Bridge Converter with Modified Active Bridge

4 Hybrid-Bridge-Based DAB Converter with Modified Voltage Matching Control and Inductor Current Injections ... 89
4.1 Topology and Conventional VM Control ... 89
4.1.1 Topology and Operation Principle of VM Control ... 90
4.1.2 ZVS Analysis Considering Switch Junction Capacitors ... 92
4.1.3 ZVS Region Under Conventional VM Control with Considering Switch Junction Capacitors ... 96
4.2 Magnetizing Inductor to Eliminate the Effect of Junction Capacitor ... 96
4.3 Modified VM Control to Achieve ZVS Under Backward Power Conditions ... 100
4.3.1 Principle of Modified VM Control ... 100
4.3.2 Implementation of the Modified VM Control ... 103
4.4 Experimental Results ... 104
4.5 Conclusion ... 112
References ... 112

Part III Dual-Active-Bridge Converter with Modified Transformer

5 Three-Winding-Coupled-Inductor-Based Dual Active Bridge DC–DC Converter with Full Load Range ZVS Under Wide Voltage Range ... 117
5.1 Converter Topology and Operation Principle ... 117
5.1.1 Topology and Modulation Schedule Using Phase-Shift Control ... 118
5.1.2 Working Stages of the Converter ... 120

5.2 Power Characteristics and ZVS Analysis 123
5.2.1 Transmission Power with PPS Control 123
5.2.2 ZVS Range Analysis 124
5.3 Design Considerations 125
5.3.1 ZVS Design Related to Leq and Leakage Inductor 125
5.3.2 Inductor RMS Current Design Considerations 125
5.3.3 Inductor RMS Current Design Considerations 128
5.3.4 Inductor RMS Current Design Considerations 131
5.4 System Parameters Design 133
5.4.1 Equivalent Inductance Leq Design 134
5.4.2 Turns Ratios of TWCI Design 134
5.4.3 Leakage Inductance Lrs Design 134
5.4.4 Clamping Capacitor Selection 134
5.4.5 Clamping Capacitor Selection 135
5.5 Experimental Results 135
5.6 Conclusion 136
References 139

6 Modulated Coupled Inductor Based ISOP Dual-Active-Bridge Converter with Hybrid SPS Control 141
6.1 Converter Topology and Operation Principle 141
6.1.1 Topology and Working Principle of the Proposed Converter 141
6.1.2 Model Analysis of the Modulated Coupled Inductor 143
6.1.3 Power Transfer Characteristics of the Proposed Converter 145
6.2 ZVS Conditions Analysis and Control Boundary of the Hybrid SPS Control 146
6.2.1 ZVS Analysis Considering Junction Capacitance and Magnetizing Current Injection 146
6.2.2 RMS Current Derivation 149
6.2.3 Control Boundary Between Mode I and Mode II 149
6.3 Parameters Design Consideration 152
6.3.1 Voltage Ratio G and Coupled Coefficient k_c Design to Achieve Full Load Range ZVS for Primary Side Switches with Small RMS Current 152
6.3.2 Magnetizing Inductance Lm Design to Achieve Full Load Range ZVS for Secondary Side Switches 154
6.4 Comparison 155
6.5 Experimental Verification 155
6.6 Conclusion 159
References 160

7 Magnetizing Current Injection Based Push–Pull Dual Active Bridge Converter with Optimized Control 161
7.1 Topology and Operation of Modified PPDAB Converter Under EPS Contorl 161
7.1.1 Topology and Typical Waveforms for Modified PPDAB Converter with Magnetizing Current Injection 163
7.1.2 Operation Principle Analysis of EPS Control Utilizing Magnetizing Current 163
7.2 Magnetizing Current Injection Analysis for the Modified PPDAB Converter 165
7.2.1 ZVS Region Analysis Considering the Switch Junction Capacitors and the Magnetizing Inductance 165
7.2.2 Disconnection in the ZVS Region Caused by the Switch Junction Capacitors Without Magnetizing Inductance 168
7.2.3 Magnetizing Current Injection for PPDAB Converter to Connect the Separated Two ZVS Regions 170
7.3 Power Control Laws to Obtain Low RMS Current and Easy Engineering Implementation Based on Magnetizing Current Injection 171
7.3.1 Obtain a Power Control Law with Minimum Transformer Winding Current 171
7.3.2 Optimal Power Control Law with Full Load Range ZVS and Easy Engineering Implementation 173
7.3.3 Comparison of Control Laws 176
7.3.4 Optimized Control Strategy Implementation 176
7.4 Design Consideration of the Modified PPDAB Converter 178
7.4.1 Voltage Conversion Ratio k 178
7.4.2 Turns Ratio N 178
7.4.3 Leakage Inductor L 178
7.4.4 Magnetizing Inductance L_m 179
7.5 Experimental Results 179
7.6 Conclusion 188
References 189

Part IV Dual-Active-Bridge Converter with Additional Auxiliary High-Frequency Networks

8 Light-Load Efficiency Enhancement of High-Frequency Dual-Active-Bridge Converter Under SPS Control 193
8.1 Basic Concept of the Proposed Method 193
8.2 ZVS Analysis with Magnetizing Inductance 195

8.2.1 ZVS Condition Analysis Considering Magnetizing Inductance and Switch Junction Capacitors ... 195
8.2.2 ZVS Range Extension with Magnetizing Inductance ... 196
8.3 Design Consideration to Achieve Full Load Range ZVS with Minimum RMS Current ... 198
8.3.1 Optimized Conversion Gain G Design to Achieve ZVS for Primary Side Switches with Minimum Primary RMS Current ... 198
8.3.2 Magnetizing Current Injection Optimization for ZVS of Secondary Side Switches ... 199
8.4 Experimental Results and Comparison ... 200
8.5 Conclusions ... 204
References ... 205

9 Optimized EPS Control to Achieve Full Load Range ZVS with Seamless Transition for Dual Active Bridge Converters ... 207
9.1 Topology and Conventional EPS Control ... 207
9.2 ZVS Analysis Considering Junction Capacitors ... 208
9.2.1 Instant Current Expressions for Leakage Inductance ... 208
9.2.2 ZVS Restrictions Considering Switch Junction Capacitors ... 209
9.2.3 Discontinuous Power Transfer in ZVS Region Without Magnetizing Current ... 212
9.2.4 Magnetizing Current Injection to Achieve Entire Load Range ZVS ... 213
9.2.5 ZVS Seamless Power Control Law to Achieve ZVS with Low Rms Current and Maximum Power Transfer Capability ... 216
9.3 Design Consideration ... 219
9.3.1 Turns Ratio N ... 219
9.3.2 Leakage Inductance L_k ... 219
9.3.3 Magnetizing Inductance L_m ... 220
9.3.4 Implementation of the Control Strategy ... 220
9.4 Comparisons ... 221
9.5 Experimental Results ... 223
9.6 Conclusion and Future Work ... 227
References ... 230

10 An Optimized DPS Control for Dual Active Bridge Converters to Secure Full Load Range ZVS with Low Current Stress ... 231
10.1 ZVS Modeling Considering Junction Capacitors Without Magnetizing Current ... 231
10.1.1 Basic Operation Under DPS Control ... 231
10.1.2 Basic Mathematical Model ... 232

10.1.3 ZVS Analysis Considering Switch Junction Capacitors 234
10.2 ZVS Modeling and Power Control Law Utilizing Magnetizing Current 236
10.2.1 ZVS Modeling Considering Magnetizing Inductance 236
10.2.2 Power Control Law to Obtain Low Current Stress Based on Lagrange Multiplier Method 240
10.3 Comparative Analysis Among Different Control Methods 244
10.4 Parameter Design 246
10.5 Experimental Results 247
10.5.1 Steady State Operation 247
10.5.2 Steady State Operation 249
10.5.3 Loss Breakdown and Efficiency 250
10.5.4 Comparisons 252
10.6 Conclusion and Future Work 253
References 254

Part V Modular Dual-Active-Bridge Converter

11 Single-Stage DAB-LLC Hybrid Bidirectional Converter with Tight Voltage Regulation Under DCX Operation 257
11.1 System Structure and Converter Modeling 257
11.1.1 System Structure 257
11.1.2 Converter Modeling 258
11.1.3 Transfer Function 261
11.1.4 Model Verification 262
11.2 Controller Design 263
11.2.1 Controller Design 264
11.2.2 Controller Verification 265
11.3 Parameter Design Consideration 265
11.3.1 Capacitor 265
11.3.2 DAB Converter Inductor 266
11.3.3 Transformer Turns Ratio 267
11.4 Experiments Results 268
11.5 Conclusion 276

12 An LLC-DAB Bidirectional DCX Converter with Wide Load Range ZVS and Reduced Switch Count 277
12.1 Proposed Converter and Operation Principles 277
12.1.1 Converter Topology 277
12.1.2 Forward Power Transmission Mode 279
12.1.3 Backward Power Transmission Mode 282
12.2 ZVS Operation Analysis 283
12.2.1 ZVS Analysis of the Secondary Side Circuit 283
12.2.2 ZVS Analysis of LLC Primary Circuit 286
12.2.3 ZVS Analysis of Auxiliary DAB Circuit 287

12.3 Parameters Design 287
12.3.1 Power Distribution and Transformer Turns Ratios 287
12.3.2 Leakage Inductor L_k 287
12.3.3 Magnetizing Inductor L_m 289
12.4 Experiment Results 290
12.4.1 Steady State Performance 290
12.4.2 Soft Switching Performance and Comparison 293
12.4.3 Dynamic Performance and Efficiency 295
12.4.4 Efficiency Comparison 297
12.5 Conclusion 299
References 300

13 Stability Analysis and Design of Common Phase Shift Control for Input Series Output Parallel Dual Active Bridge with Consideration of Dead-Time Effect 301
13.1 Existence of Steady Operation Point of ISOP DAB System Under CPSC 301
13.1.1 CPSC Strategy for ISOP DAB System 302
13.1.2 Conditions for the Existence of Steady-State of ISOP DAB System Under CPSC 302
13.1.3 Unstable Operation Simulation Under Parameter Mismatch when Line Damping is Considered 304
13.2 ISOP DAB Stability Analysis Considering Dead-Time Effect 306
13.2.1 Model of DAB with Consideration of Dead-Time Effect 306
13.2.2 Large Signal Model of ISOP DAB System with Consideration of the Dead-Time Effect 308
13.2.3 Small Signal Model of ISOP DAB System with Consideration of the Dead-Time Effect 310
13.3 Analysis of Parameter Influence for Steady State and Stability 312
13.3.1 Analysis of the Combination Modes Corresponding to the Transmission Power 313
13.3.2 Analysis of Parameter Influence for Steady State Tability 315
13.3.3 Stability Verification of ISOP DAB System 316
13.4 Experimental Verification 318
13.5 Conclusion 322
References 323

Chapter 1
Introduction to Topology Deduction of Dual-Active-Bridge Converters

Abstract This chapter provides a comprehensive review and classification of the dual-active-bridge (DAB) converter based on the topological modifications. DAB converters inherit favorite features including zero voltage switching (ZVS) capability, wide voltage range, and natural bidirectional power flow. Facing different applications and high market requirements, conventional DAB topologies with various modulations are gradually inadequate, so topological modifications are emerged in literature to deal with these situations. Firstly, basic cells including input/output types, active bridges, transformers, and auxiliary high-frequency networks are introduced. Next, by combining these cells flexibly, modified DAB converters are derived, and they are classified in terms of structural characteristics and discussed in detail. Finally, the characteristics, merits, and demerits of the modified DAB converters are concluded. Besides, evolvement history and future trends are also analyzed. This chapter aims to present a clear picture of the general law and framework of the modified DAB topologies and give the inspiration to derive new DAB topologies to address specific requirements.

Keywords Basic cells · Dual active bridge (DAB) · Methodology · Topological modifications

1.1 Basic Cells of the Dual-Active-Bridge Converter

From the perspective of topology, a DAB converter can be divided into three stages: a high-frequency (HF) inverter stage, an HF power conversion stage with isolation, and an HF rectifier stage. The modified DAB converter diagram is shown in Fig. 1.1. Specifically, the HF inverter and rectifier stages consist of an I/O type and an active bridge respectively. Only HF transformer as HF power conversion stage in conventional DAB converters. While, in the HF power conversion stage of modified topologies, various auxiliary HF networks can also be introduced to change voltage and current waveforms. Consequently, topological modifications can be carried out

G. Xu and X. Chen, *Topology Deduction and Control for Dual-Active-Bridge Converters*, https://doi.org/10.1007/978-981-97-4856-3_1

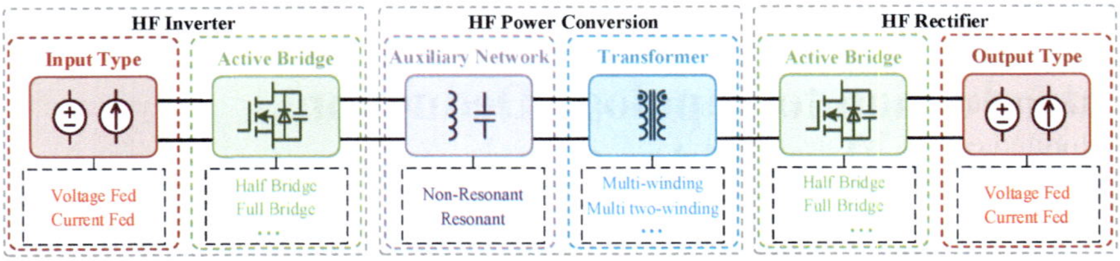

Fig. 1.1 The structure of a DAB converter

Fig. 1.2 I/O types of DAB converters. **a** VF. **b** CF

separately for four parts, including I/O types, active bridges, transformers, and auxiliary HF networks. This section introduces the basic cells of each part, which lays the foundation for the subsequent sections.

1.1.1 The Cells of the Input/Output Type of DAB Converter

Based on the electrical energy source types, the converter can be mainly classified into two types—voltage-fed (VF) DAB converter [1] and current-fed (CF) DAB [2] converters. In VF DAB converters, as the energy storage units, the large capacitors are connected in parallel with both the input and output sides respectively, aiming to maintain voltages within an acceptable range, as depicted in Fig. 1.2a. On the other hand, inductors can also be used as energy storage units. In CF DAB converters, one or more inductors are connected in series with the input side and/or output side to reduce the current ripple at that side depending on the inductance. An example of CF I/O type with one large inductor is shown in Fig. 1.2b.

1.1.2 The Cells of the Active Bridge

Active bridges are generally mainly divided into full bridges, half bridges, and other types. The other types mainly include the other active bridge cells modified with methods such as reconnection among components.

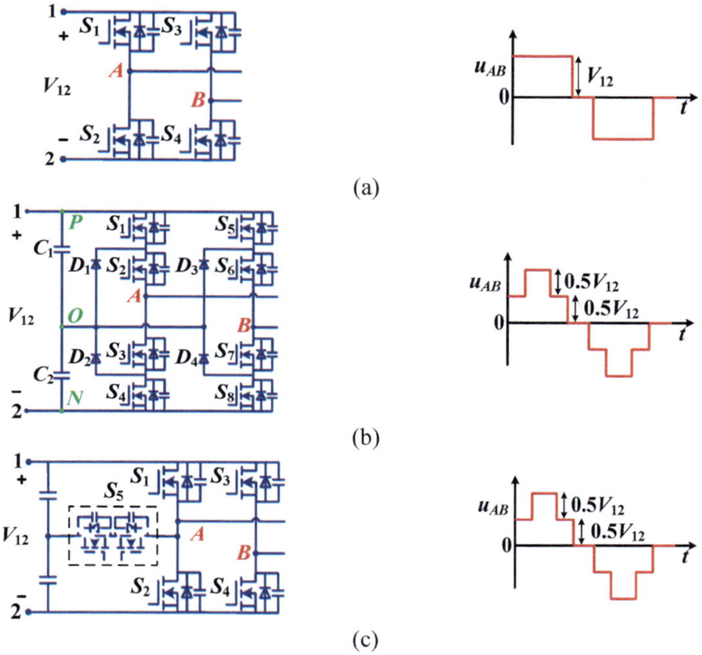

Fig. 1.3 Different cells and voltage waveforms of the full bridge. **a** Conventional structure. **b** NPC structure. **c** T-type structure

(a) Full Bridge Cells

Three relatively common active full bridge cells are listed as examples in Fig. 1.3. Figure 1.3a is the conventional full active bridge [3], and its waveform of the AC output voltage v_{AB} is also shown, with an amplitude of the bus voltage V_{12}. Zero-voltage sequences can also be introduced usually by controlling the phase shift angle between the two bridge legs.

Figure 1.3b presented an example of the neutral point clamped (NPC) full bridge cell [4]. Compared with the conventional full bridge, the NPC full bridge uses four additional switches and clamped diodes to limit the voltage stresses of the switches to half of V_{12}. In this configuration, by controlling the switches, points A and B can be connected to point P /O /N selectively, and then a voltage level of V_{12} /$0.5V_{12}$ /0 can be obtained. Therefore, the NPC full bridge can generate the output voltage v_{AB} with levels of $\pm V_{12}$, $\pm 0.5V_{12}$, and 0.

Another structure capable of generating a three-level voltage is a T-type full bridge [5], depicted in Fig. 1.3c. A four-quadrant switch and two split capacitors form the T-type network. The four-quadrant switch can have different structures, and the one shown in Fig. 1.3c is a back-to-back switch. In the T-type full-bridge cell, the output voltage node (point A) connected to the four-quadrant switch can have three possible voltage levels, which are V_{12}, $0.5V_{12}$, and 0. However, the other output voltage node (point B) cannot generate a voltage level of $0.5V_{12}$. Despite having fewer switches, v_{AB} can be the same as that generated by NPC full bridge, but the voltage stresses on the switches remain V_{12}.

(b) Half Bridge Cells

Figure 1.4 illustrates three commonly used half-bridge cells as examples. Figure 1.4a shows the simplest half-bridge structure [6], which consists of only two switches. This active bridge cell can only generate a unipolar voltage by chopping the source voltage. The unipolar waveform with two voltage levels (V_{12} and 0) is also shown in Fig. 1.4a.

By replacing the switches in one of the bridge legs in Fig. 1.4a with large capacitors, a capacitor-split half bridge cell is constructed [7], which is shown in Fig. 1.4b. The amplitude of the output voltage is half of that in Fig. 1.4a, and the zero-voltage level cannot be introduced.

The active bridge in Fig. 1.4c is a half-bridge version of the NPC bridge shown in Fig. 1.4b. In comparison to the capacitor-split half bridge, the NPC half-bridge can introduce zero-voltage sequences [8]. As a result, the voltage levels of v_{AB} include $\pm 0.5V_{12}$ and zero, and the voltage stresses on the switches are $0.5V_{12}$.

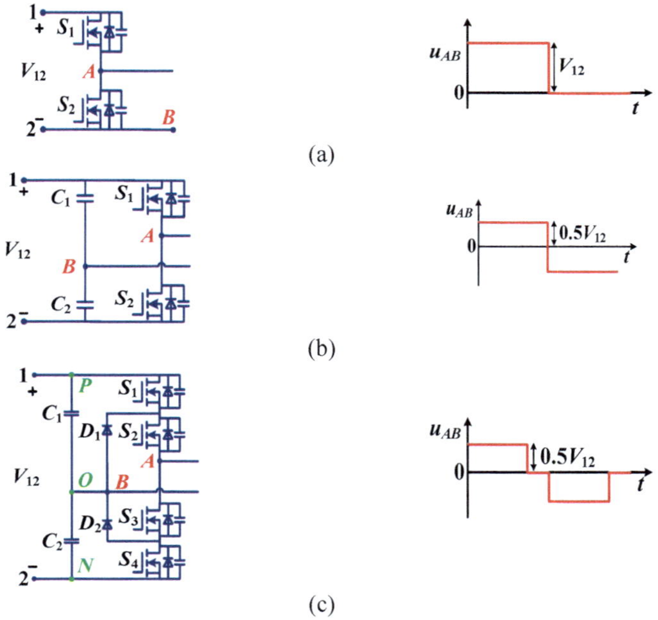

Fig. 1.4 Different cells and voltage waveforms of half bridge. **a** The simplest structure. **b** Capacitor-split structure. **c** NPC structure

Similarly, there are several other half-bridge cells available, but they will not be discussed in this section. Also, the switches are not limited to MOSFETs.

(c) Other Active Bridge Cells

With modified connections, other types of active bridge cells have been proposed. Push–pull structures [9], for example, are an interesting variation of active bridge cells, and stacked active bridges with series-connected bridge legs have also been studied [10].

1.1.3 The Cells of the High-Frequency Transformer

The basic function of the HF transformer in DAB converters is to provide galvanic isolation and adjust the voltage gain. In order to achieve better performance characteristics, researchers have been working on improving the transformer design. The conventional two-winding transformer is shown in Fig. 1.5a [11]. The structure of dual transformers in Fig. 1.5b is an example of the structure of multiple two-winding transformers [12]. Figure 1.5c shows a multi-winding transformer [13].

The HF transformers with controllable parameters are illustrated in Fig. 1.6. Figure 1.6a shows an adjustable-tap transformer [14], in which the turns ratio can be adjusted using relays. On the other hand, a transformer with a controlled variable leakage inductance is depicted in Fig. 1.6b [15].

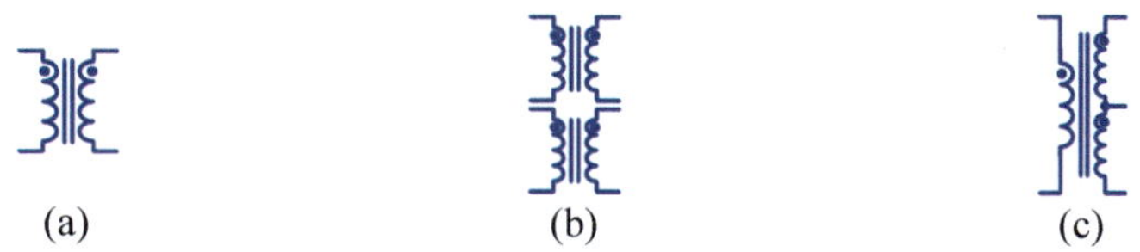

Fig. 1.5 Different structure types of the transformer. **a** Conventional transformer. **b** Dual transformers. **c** Multi-winding transformer

Fig. 1.6 Modified transformers with controllable parameters. **a** With an adjustable tap. **b** With controllable leakage inductance

1.1.4 The Cells of Auxiliary High-Frequency Networks

The auxiliary HF networks can be mainly divided into two categories, which are non-resonant networks and resonant networks. In non-resonant networks, the inductances and capacitances are typically relatively large, and the capacitors essentially function as DC-blocking capacitors rather than resonant capacitors.

Non-resonant networks can include a variety of passive and active components, such as inductors, capacitors, diodes, and switches. Among these components, inductors and capacitors are commonly used. Inductors are often employed to inject reactive current, which allows for more energy to be delivered to charge/discharge the switch junction capacitors, effectively enlarging the ZVS range [16]. Meanwhile, DC-blocking capacitors are frequently placed in series with the leakage inductor to introduce voltage offsets using corresponding modulation strategies [17], which can also extend the ZVS range within a wide voltage range.

By combining switches with inductors or capacitors, switch-controlled components can be constructed, as shown in Fig. 1.7. The switch-controlled inductor, shown in Fig. 1.7a, is composed of a back-to-back switch and an inductor in series [18], and the switch-controlled capacitor [19], shown in Fig. 1.7b, is composed of a back-to-back switch and a capacitor in parallel. These networks allow for real-time adjustment of inductance and capacitance.

The resonant networks primarily comprise inductors and capacitors, which enable the current waveform to approach a sine waveform, leading to reduced peak current and the dc component of the transformer. As a result, switching loss and eddy loss are reduced, and the fundamental harmonic analysis (FHA) becomes more helpful without significant loss of accuracy. Several resonant networks have been studied, such as LC network [20], LCL network [21], CLLC network [22], etc. Three popular resonant networks are illustrated in Fig. 1.8. The resonant networks can also be combined with switches to alter the resonant frequency. Furthermore, there are numerous types of non-resonant and resonant networks that have been proposed, which will not be listed in this section.

Fig. 1.7 Two examples of the controlled variable networks. **a** Switch-controlled inductor. **b** Switch-controlled capacitor

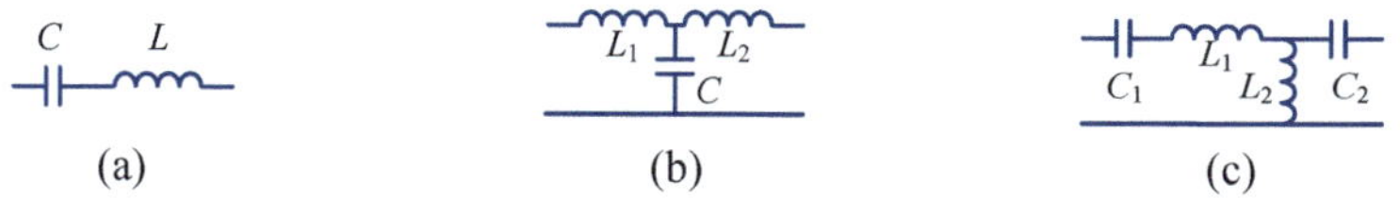

Fig. 1.8 Three popular resonant networks used in DAB/DAB3 converters. **a** LC network. **b** LCL network. **c** CLLC network

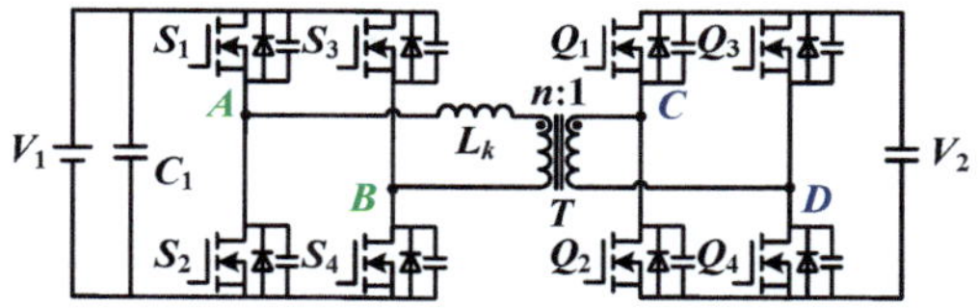

Fig. 1.9 The conventional VF-DAB converter

1.2 Dual-Active-Bridge Converter with Modified I/O Type

1.2.1 Voltage-Fed DAB Converter

The conventional DAB converter are voltage-fed (VF) topologies, which is shown in Fig. 1.9. For the VF structure, it has a major drawback of large current ripples in the input and output side. These large current ripples not only reduce the efficiency of power transfer but also shorten the lifetime of the applications [23]. To address this issue, the I/O type can be modified from VF to CF by changing the energy storage unit from capacitor to inductor. With the additional inductance, the current on the CF side can be directly designed to satisfy the ripple requirement. However, this modification results in a reduction in power density, and ZVS achievement becomes more complicated due to the DC bias on the inductor.

1.2.2 Current-Fed DAB Converter

An L-type CF DAB converter featuring an active clamped circuit is illustrated in Fig. 1.10 [24]. In this configuration, an inductor is placed between the DC bus and the full active bridge. An active clamped circuit is formed by connecting a capacitor in series with a switch, and then in parallel with the bridge legs on the CF side. Current ripple can be controlled by designing the bus inductance, switching frequency, and duty cycle of the switches. The clamped circuit provides an additional commutation path, thereby preventing dramatic changes in the currents of the input and leakage inductors during commutation and avoiding voltage spikes.

To simplify the structure, a modified L-type CF DAB converter without active clamped circuits is shown in Fig. 1.11 [25]. This structure lacks inherent voltage clamp ability. During the switching action, if the currents in the DC inductor and leakage inductor do not match but are directly connected, the switches on the CF side

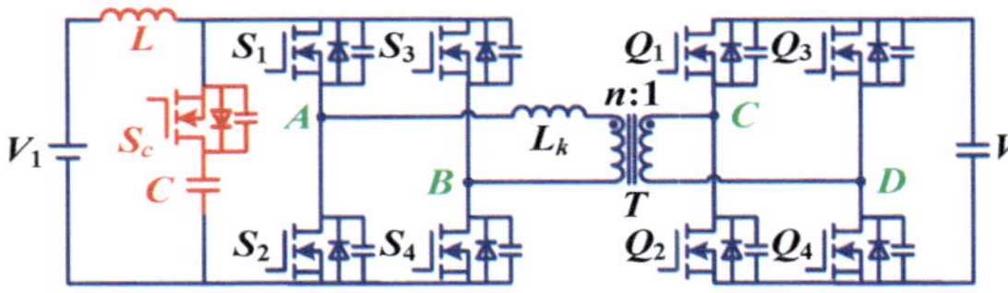

Fig. 1.10 CF-DAB converter with active clamp circuit

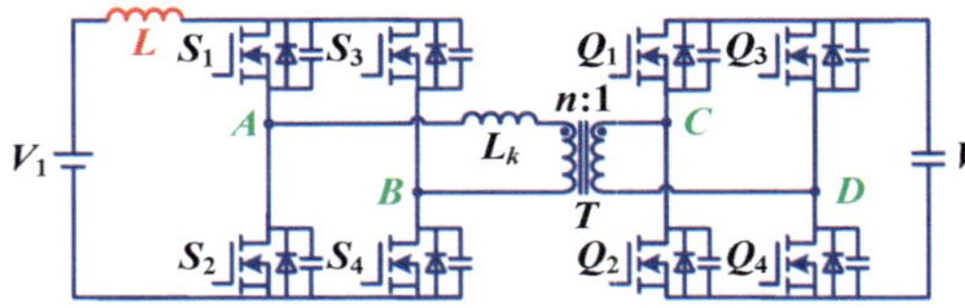

Fig. 1.11 L-type CF-DAB converter

will suffer from high voltage spikes. To address this issue, secondary-side clamping techniques were proposed in [2] using single pulse width modulation (PWM). As a result, the current flowing through the switches on the CF side when they are turned off can be reduced to zero, eliminating the voltage spike without the need for an auxiliary circuit.

In LL-type CF DAB converter, unlike L-type CF DAB converter, there are two inductors and a small clamped capacitor [26] as shown in Fig. 1.12. The inductors are respectively connected to the middle nodes of the two bridge legs. This topology can be considered as a combination of two buck/boost converters and a VF DAB converter with two common bridge legs. The two inductor currents are opposite in phase, resulting in reduced total current ripple. This reduces current ripple on the CF side, which can be made zero with a 50% duty cycle, and the inductors can be made more compact. Besides, the positive current biases in the inductors are beneficial to achieve ZVS of the upper switches, whereas they are disadvantageous for the lower switches. What's more, the LL-type CF DAB converter is operated as a partially isolated multi-port converter for hybrid energy storage systems by connecting a DC source or load to the clamped capacitor C_c in Fig. 1.12.

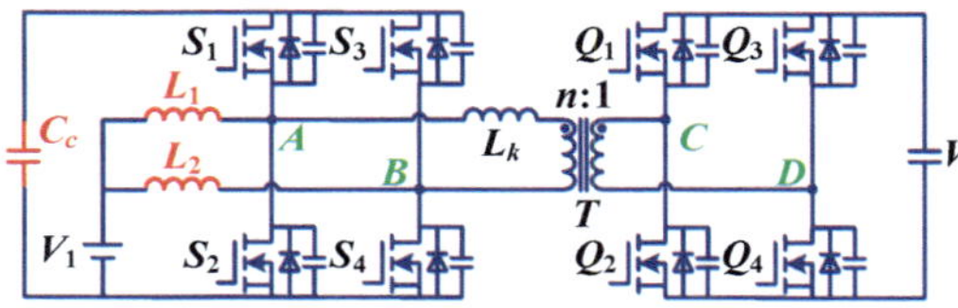

Fig. 1.12 LL-type CF-DAB converter

1.3 Dual-Active-Bridge Converter with Modified Active Bridges

The conventional DAB converter is composed of two full active bridges arranged in an H configuration, with each bridge consisting of four active switches. When dealing with DC sources, the number of active switches can be adjusted by either reducing or increasing them, and the connection can also be reconfigured from the H type. For interfacing with AC sources, the active switches can also be modified. In this section, the active bridges with reduced active switches, increased switches, and reconfigured connections are discussed in detail.

1.3.1 Reduce Active Switches

The conventional active bridge cell in a DAB converter is composed of four active switches. If the number of active switches in a cell is less than four, it can be classified as a bridge cell with reduced active switches. With fewer switches, the driver circuits are reduced, reliability is improved, and control flexibility is also degraded. The typical examples are half bridges and semi bridges.

(a) Half bridges

A half bridge cell is typically created by replacing one active bridge leg with a passive bridge leg made up of two capacitors [27]. The split capacitors prevent any DC voltage component from being applied to the transformer. This structure has the least number of switches and a simple structure. The losses of not only the switches but also their peripherals can be reduced. However, this configuration also has two major drawbacks. One is that twice current stress needs to be withstood with the same rated power and voltage. Another is that the split DC capacitors must carry the full load current, which can lead to significant concerns over capacitor losses and lifetime. Therefore, this configuration is better suited for low-power or high-voltage applications, and have been developed for uninterruptible power supplies, small distributed generators, on-board battery chargers, etc.

CF half bridge is studied in [28]. As depicted in Fig. 1.13, the structure of the CF side can be considered as a combination of a buck/boost converter and a capacitor-split half bridge cell. The same CF half bridges are employed on both sides, resulting in a fully symmetrical topology shown in Fig. 1.13. Similar to the full bridge topology, the magnetic integration technique can also be utilized in CF half bridge DAB converters. For example, in [29], only one coupled inductor is used to replace the input inductor, the transformer, and the leakage inductor, thereby reducing the count of magnetic components.

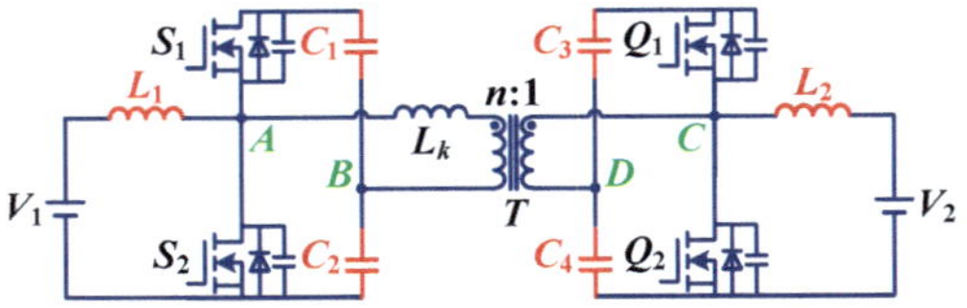

Fig. 1.13 Fully symmetrical CF half bridge DAB converter

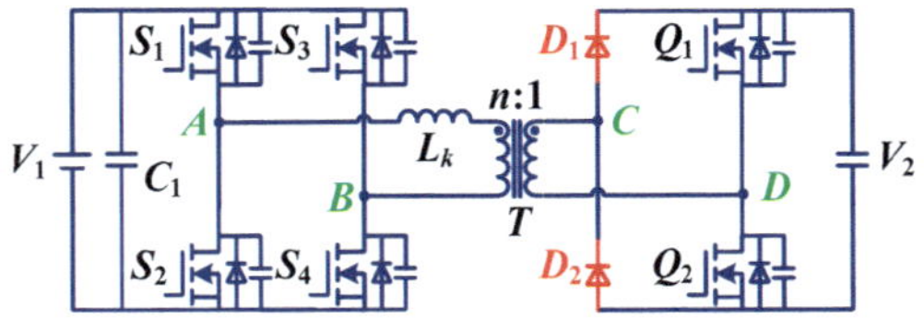

Fig. 1.14 The semi-DAB in with the former construction method

(b) Semi Bridges

By replacing part of the switches on the load side with diodes, a topology called semi-DAB can be constructed. There are two kinds of semi-DAB structures: one involves replacing an active bridge leg with a passive bridge leg, while the other replaces top switches with diodes. Although the reduced number of switches may result in less flexible control, these structures offer many benefits. Not only the driver circuit complexity can be reduced, but also the reverse current on the load side can be eliminated and the required output filter capacitance can be reduced, which makes semi-DAB to be an excellent candidate in some unidirectional power flow applications. They have been applied for battery chargers, distributed dc grid-connected PV generation systems, fuel stacks, and PV array generation systems. The topology of the former was studied in [30], which is shown in Fig. 1.14. While the operating principles are similar to the latter structure, this structure has a more complex driving circuit and may suffer from output short-through.

1.3.2 Increase Active Switches

With DAB topologies utilizing more than four switches, the bridge cells can be considered as active bridge cells with increased switches. As the number of active switches increases, the control degrees of freedom increase accordingly. Various objectives are pursued with increased switches, such as reducing voltage stress and increasing the voltage level of the leakage inductor. In terms of structure characteristics, four types of active bridges with increased switches are introduced, including NPC bridges, T-type bridges, modular multilevel bridges, and others.

(a) NPC bridges

A conventional DAB converter may face difficulties in handling excessive voltage due to limitations of power semiconductor device development. To address this issue, NPC bridge cells were presented. NPC full and half bridge cells with clamped diodes. By doubling the number of active switches, the voltage stresses on the switches can

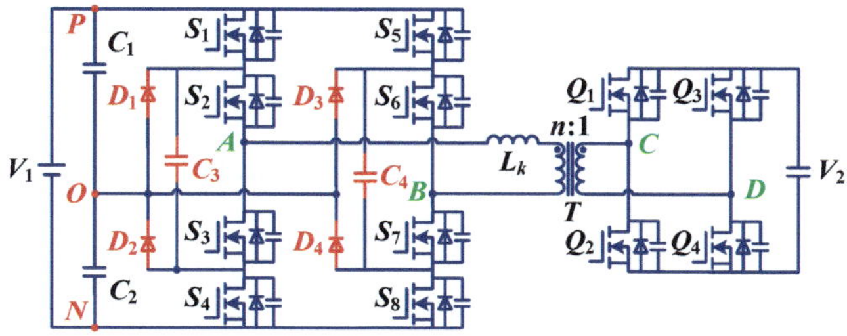

Fig. 1.15 DAB converter with NPC bridge utilizing clamped diodes and flying capacitors

be reduced by half. Thus, the converters can be applied to high voltage applications or lower voltage withstanding switches with better performance characteristics can be used [31]. Additionally, this approach provides increased control flexibility, lower *dv/dt*, and reduced RMS current. However, clamped circuits are necessary, which can include not only clamped diodes but also flying capacitors. Meanwhile, the NPC bridge cell with both clamped diodes and flying capacitors has been studied, as illustrated in Fig. 1.15.

However, the use of additional components results in increased losses and more complex control requirements. Furthermore, because the gate signals, device characteristics, and parameter deviation will not be the same, the voltages through the two DC-link capacitors cannot be balanced ideally. This imbalance can lead to asymmetry in the AC output voltage (u_{AB} in Fig. 1.15) and unexpected power flow. At the same time, the voltage stresses of the switches also become unbalanced, and overvoltage across devices will occur if the imbalance is severe.

(b) T-type Bridges

T-type bridge cells are also capable of generating voltage with multiple voltage levels, which typically comprise two split capacitors, a four-quadrant switch, and an active bridge. While the voltage stresses of the primary switches are doubled compared to NPC structures, the number of switches is reduced, and a voltage-clamped circuit is unnecessary, with capacitor voltage imbalance concerns being negligible. However, full range ZVS of the additional switches still cannot be achieved. The topology of a DAB converter with a T-type full bridge cell was studied in [32], whose four-quadrant switches are back-to-back switches. Two back-to-back switches are connected between the middle node of split capacitors and points A and B respectively, allowing the points to provide the same three voltage levels.

The T-type full bridge cell, as shown in Fig. 1.16.

1.3.3 Reconnection of the Active Bridges

The conventional active bridge can be seen as the devices connected with the typical connection. Therefore, in addition to modifications based on switches, the connection

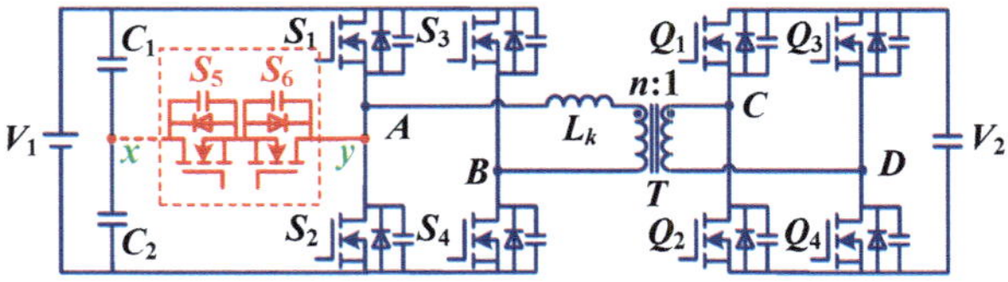

Fig. 1.16 DAB converter with T-type full bridge

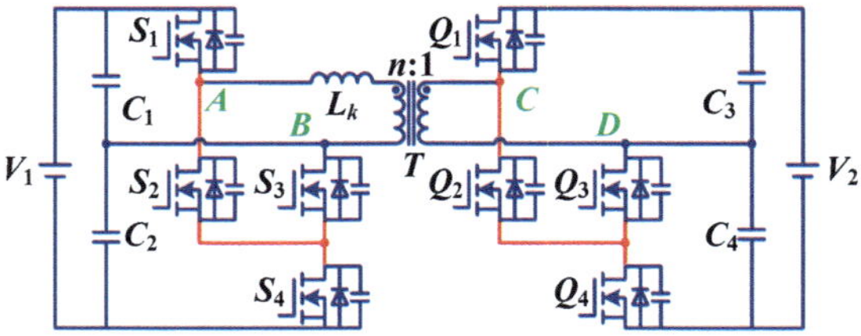

Fig. 1.17 The DAB topology with modified connection

can also be reconfigured. The major target of reconnecting the active bridges is to reduce voltage stress. A split-capacitor DAB converter was proposed in [33], as shown in Fig. 1.17. Compared with a conventional DAB converter, the leakage current in this converter is twice, but the leakage inductance is reduced to a quarter, and the voltage stresses of some switches (S_2–S_4, Q_2–Q_4) are halved.

1.4 Dual-Active-Bridge Converter with Modified Transformer

The conventional DAB converter employs a two-winding transformer. By adding more windings to the magnetic core or using multiple two-winding transformers, additional voltage ports can be provided. Compared with multi-winding transformers, multiple two-winding transformers have special concerns about the power distribution of each transformer, but they can alleviate the magnetic short-circuit issues. Additionally, the leakage inductance and turns ratio, which are essential parameters of the transformer, can be modified to be controllable. In this section, the DAB converters based on the multi-winding transformer, multi-two-winding transformers, and transformer with controllable parameters are analyzed.

1.4.1 Multi-winding Transformer

In a multi-winding transformer, multiple windings are wound around a single magnetic core. This configuration allows for a more compact layout compared to using multiple two-winding transformers, but all the ports of the transformer are coupled. The additional windings can be connected with additional components or bridges. Usually, additional components are utilized in two ports converters, while additional bridges are preferred to be applied in multiport converters.

As known, the half-bridge structure has fewer switches, but the amplitude of the AC output voltage is half of the bus voltage. Utilizing one more winding, push–pull bridges with reduced switches were introduced. The voltage amplitude of the push–pull bridge can be equal to the bus voltage. Usually, clamp circuits are often required to absorb the energy stored in the leakage inductor, avoiding voltage spikes generated when the switches are turned off. Push–pull forward DAB converters with a capacitor-split half bridge and full bridge were studied in [34], as shown in Fig. 1.18. Only two switches are used in the push–pull structure, and the amplitude of the voltage applied to the transformer is equal to bus voltage, so the power transmission capacity is greater than the capacitor-split half bridge. Besides, this structure suppresses the voltage spike just by adding a capacitor.

Triple active bridge (TAB) converters are also derived based on a multi-winding transformer [35]. As shown in Fig. 1.19, the converter has three identical active bridges, which are conventional full bridges. Under single-phase-shift control, the converter is unsuitable for wide voltage variations applications if soft-switching preferably needs to be maintained, just like a conventional DAB converter.

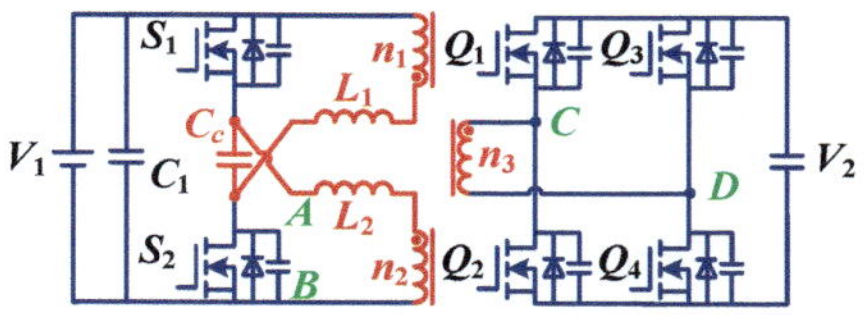

Fig. 1.18 The topology of a push–pull DAB converter

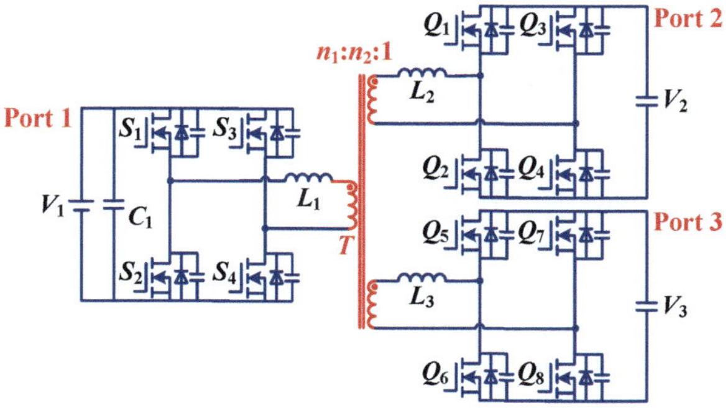

Fig. 1.19 A three-port DAB converter with a multi-winding transformer

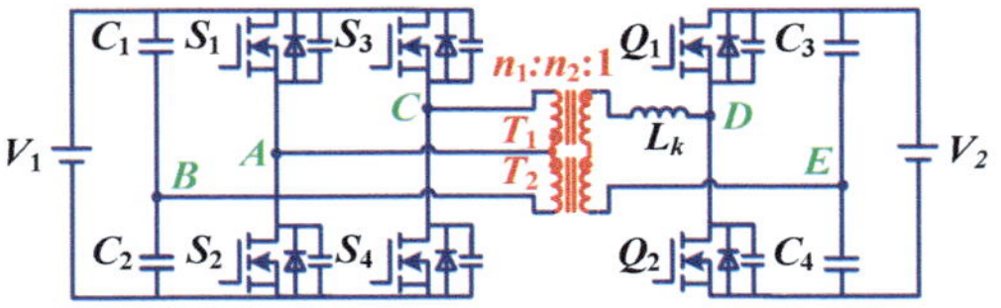

Fig. 1.20 The DAB converter with dual transformers

1.4.2 Multi-core Transformer

Multi-two-winding transformers have more than one magnetic core, each of which has two windings. Utilizing more magnetic cores, the power transfer of each transformer is decoupled, and magnetic short-circuit issues can be alleviated. Similarly, they can be divided into two ports and multi-port DAB converter.

Dual transformers are one kind of multi-core transformer, and are usually used in two ports DAB converters to increase the number of voltage levels across leakage inductor. In dual transformers, the windings on one side are usually connected in series, and those on the other side are in parallel. On the parallel side, the third winding can be connected to an extra bridge leg, such as an active leg, capacitor leg, or diode leg. The topology in [36], illustrated in Fig. 1.20, utilizes a third winding on the primary side connected to split capacitors to generate voltage levels of $(0.5n_1 + n_2)V_1$ and $0.5n_1V_1$ on the secondary side. The amplitudes of voltage levels can be changed by designing the turns ratios, reducing RMS current. Besides, when switch junction capacitors are ignored, ZVS can be achieved for all the switches within the input voltage range from 120 to 240 V when the power flow is forward, while ZVS will be lost for two switches when the power is negative.

1.4.3 Controllable Transformer

The values of turns ratio and leakage inductance are typically fixed in conventional DAB converters. However, in some advanced designs, additional control circuits can be implemented to enable real-time adjustment of these parameters. Therefore, these transformers are classified as transformers with controllable parameters.

By utilizing a transformer with a controllable turns ratio, the voltage gain can be adjusted in real-time to compensate for changes in input or output voltage, ensuring that the voltage gain stays close to unity, thus improving the performance of the converter. In [37], the turns ratio of the HF transformer can be adjusted using relays, which can result in discontinuous changes in the turns ratio and sudden changes in current. However, by using an adjustable tap, the two-port voltages of the transformer can be better matched when the input or output voltage changes. As a result, a GaN-based prototype was built under SPS control, and ZVS is achieved for all switches within input voltage from 10 to 50 V except for light load, and the power transfer capability is increased.

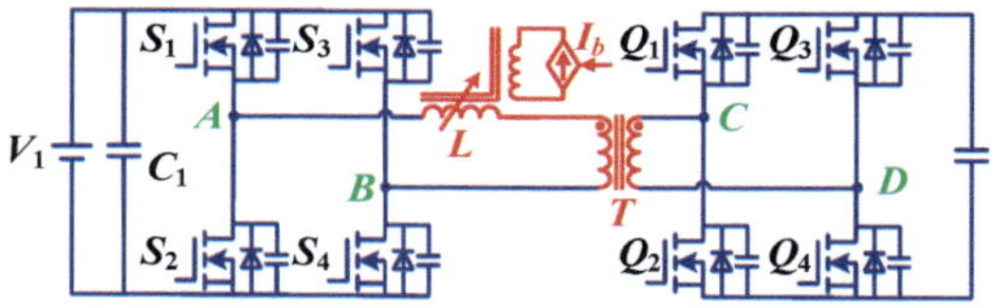

Fig. 1.21 DAB converter with controlled variable leakage inductance

A controlled variable leakage inductor was used in [38] for energy transfer, and the topology is shown in Fig. 1.21. The variable inductor consists of a double E-core, a main winding, and two control windings. Utilizing the output current as a bias current to control the magnetic flux, the inductance can vary continuously. The DAB converter is enabled to operate with optimized leakage inductance and phase shift angle over the full load range, resulting in enhanced ZVS operation, reduced circulating power and improved dynamic performance.

1.5 Dual-Active-Bridge Converter with Auxiliary HF Networks

Auxiliary networks are usually placed between the active bridge and the HF transformer, to change the shapes of the voltage and/or current waveforms in the AC link. Preliminarily, auxiliary networks are categorized into non-resonant networks and resonant networks. Non-resonant networks are usually used to introduce auxiliary currents or voltage bias, while resonant networks are mainly used to reduce the HF components of leakage inductance current. Therefore, the non-resonant networks are further divided into current injection, voltage bias introducing, and others. On the other hand, the resonant networks are divided into LC networks, LCL networks, CLLC networks, and others.

1.5.1 Non-Resonant Networks

In non-resonant networks, the inductances and capacitances are usually designed to be relatively large. Inductive networks are usually used to inject auxiliary current, while capacitive networks are mainly used to introduce voltage bias and block DC bias, and other non-resonant networks are used for different effects.

The reason for the failure of ZVS is due to insufficient commutating current to charge/discharge switch junction capacitors during the dead time. Auxiliary inductance networks can be utilized as current sources to directly increase the commutating current, ensuring that the junction capacitors are fully charged/discharged. However, this results in a lower power density and introduces additional losses due to the added magnetic components. To inject a small circulating current, the auxiliary inductor can be connected in parallel with the voltage sources, such as L_1 and L_2 in Fig. 1.22.

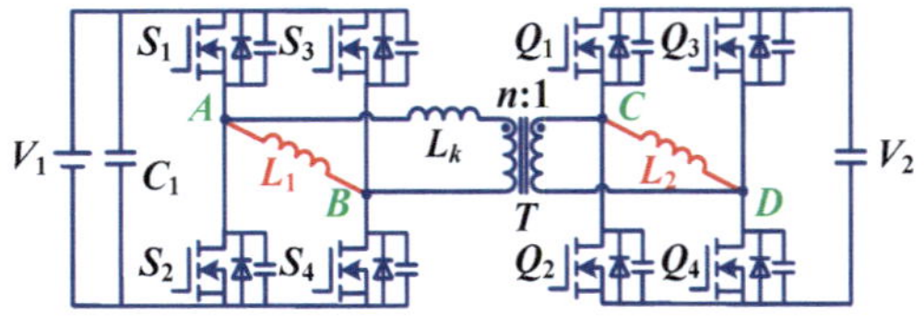

Fig. 1.22 The DAB converter with auxiliary inductors

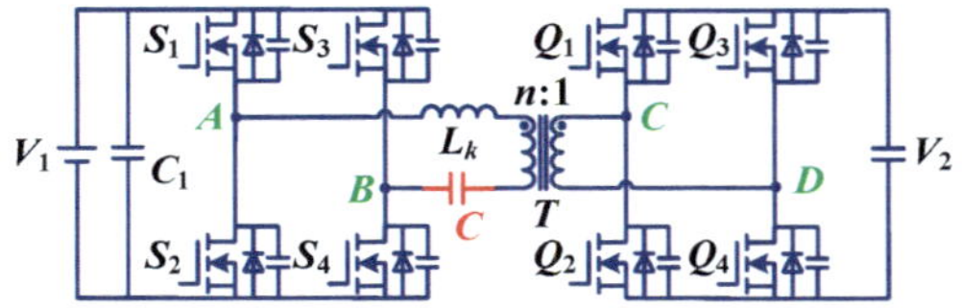

Fig. 1.23 DAB converter with DC blocking capacitors or LC resonant tank

This method has been used in [39] to enlarge the ZVS range under TPS control. And ZVS achievement can be ensured for the prototype linked to the 230Vac half sine source. Besides, utilizing the magnetizing inductance is a simple method to inject a small circulating current, which has the same effect as adding L_2 without any auxiliary magnetic component.

The DC-blocking capacitors are usually connected in series with one or both sides of the transformer, which can block the DC component applied to the transformer, avoiding magnetic saturation. Moreover, with corresponding modulation strategies, voltage offset can be introduced across the blocking capacitor. By utilizing these voltage offsets, the operating mode can be switched from full bridge mode to half bridge mode. And then, the voltages applying to the two ports of the leakage inductor are more matched, helping to reduce the RMS current and extend the ZVS range. DAB converters with DC blocking capacitors have been studied in [40]. The DAB converter with a dc blocking capacitor is illustrated in Fig. 1.23.

1.5.2 Resonant Networks

Large switching turn-off loss is one of the major drawbacks of conventional DAB/DAB3 converters. To address this issue, resonant networks can be introduced. With the resonant networks, HF components of the current can be reduced, the shape of the current waveform is closer to a sine waveform, and then the peak current can be reduced. Also, the analysis can be simpler based on fundamental harmonic.

The most straightforward method to enable the inductor current close to sinusoid in the DAB converter is to add a capacitor in series with the leakage inductor. The conventional DAB converter with an LC network has been studied in [41], as shown in Fig. 1.23. The waveforms of the leakage current and the resonant capacitor voltage are depicted in Fig. 1.24a. This converter still faces the issues with ZVS range and reactive power as the conventional DAB converter. However, lower RMS current can

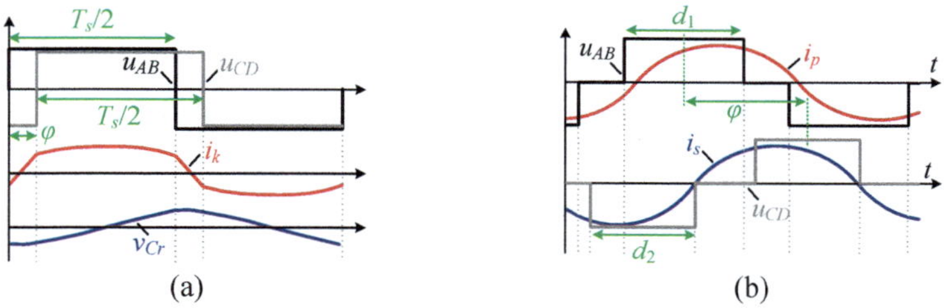

Fig. 1.24 Operation waveforms with resonant networks. **a** With LC network. **b** With LCL or CLC network

be obtained, especially under heavy loads. Also, switch turn-off currents become lower.

The typical DAB converter with an LCL network is illustrated in Fig. 1.25. Unlike the LC DAB converter, the LCL network allows for different shapes of currents in the primary and secondary sides, as illustrated in Fig. 1.24b. One of the main advantages of the LCL DAB converter is the improved power factor in both primary and secondary sides, resulting in a substantial reduction in reactive power [42]. However, the increase in magnetic components brings corresponding loss.

The DAB converter with the CLLC network is illustrated in Fig. 1.26. The operating characteristic is conceptually similar to the DAB converter with the LCL network, and the operation waveforms can also be shown in Fig. 1.24b. Magnetizing inductance is included in CLC and CLLC networks, and another inductor is usually added in the CLLC network to make the topology symmetrical. The DAB converters with CLC or CLLC network were discussed from the aspects of operation principle, modeling and control, ZVS range extension, reactive power reduction, and so on [43].

Fig. 1.25 DAB converter with LCL resonant network

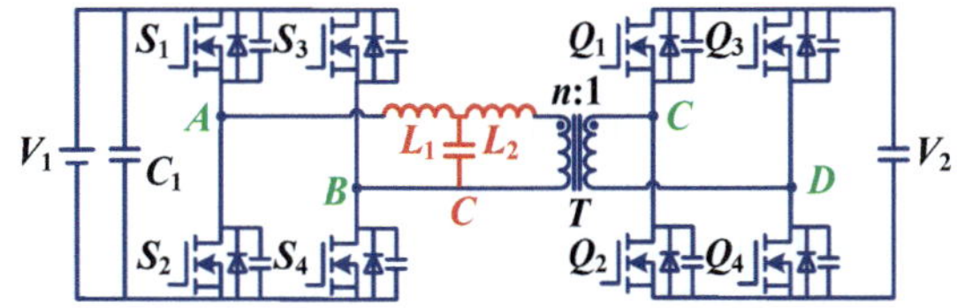

Fig. 1.26 DAB converter with CLC resonant network

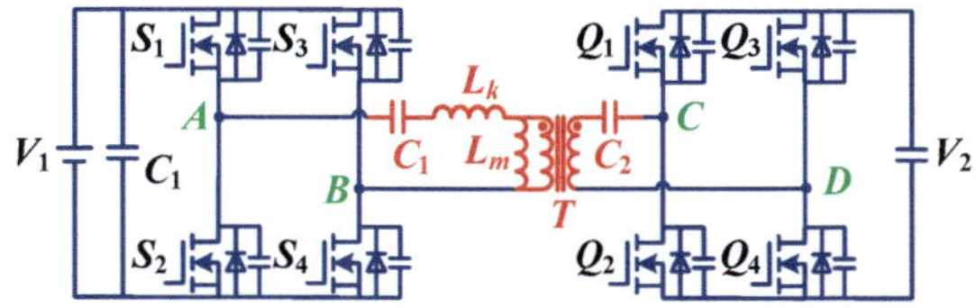

1.6 Dual-Active-Bridge Converter with Modular Structure

1.6.1 The Conventional Modular Structure of DAB Converter

The conventional modular DAB converter is mainly obtained through modular connection between the same type of DAB converter. And there are four kinds of modular structure, including input-serial-output-parallel (ISOP) [44], input-parallel-output-serial (IPOS) [45], input-serial-output-serial (ISOS) [46], and input-parallel-output-parallel (IPOP) [47]. The ISOP and IPOS structure are suitable for power transmission between different voltage levels. The ISOS structure is suitable for applications where the input and output are both high voltages. And the IPOP structure is suitable for applications where the input and output require high current.

However, the modulation and control strategy of each sub-module are independent of each other, and the overall performance of the converter, such as ZVS range, effective current value, turn off current, etc., is still limited by the optimization performance of a single module converter.

1.6.2 The Hybrid Modular Structure of DAB Converter

Due to the overall performance of conventional modular converters being limited by the optimization level of individual converters, some researchers have also combined different types of topologies in modular structure to enhance the overall performance of combination converters by complementing the advantages of different topologies.

Some scholars have combined PWM type converters and resonant converters to obtain Sigma converters [48]. The resonant converter operates in constant frequency mode to achieve high voltage gain and high conversion efficiency, and the overall voltage gain of the converter is controlled by adjusting the duty cycle of the PWM converter. However, this type of converter loses its electrical isolation function due to the use of non-isolated PWM converters. In order to achieve electrical isolation, based on the concept of Sigma converters, non-isolated PWM converters are replaced with low-power isolated DC converters [49]. However, in order to achieve adjustable voltage, low-power isolated DC converters still need to be cascaded with PWM converters to control the overall voltage gain of the converter. This makes it difficult for this type of converter to be used in high-voltage and high-power application, and PWM converters also have the problem of hard switching and cannot operate in both directions. To solve this problem, researchers have combined LLC resonant converters and DAB converter [50]. The LLC resonant converter in this type of combination converter operates in constant frequency mode to achieve high-performance power conversion, and most of the power is transmitted by the LLC resonant converter. The DAB converter is used for power and voltage control due to its flexible control.

1.7 Summary and Future Trends of the DAB Converter

1.7.1 Summary of the Modification Methods of the DAB Converter

DAB converter topologies can be modified using four methods corresponding to the four cells: I/O types, active bridges, HF transformers, and auxiliary HF networks. Table 1.1 summarizes the characteristics, major merits, major demerits, and suitable applications of each method.

In conclusion, the analysis and comparisons show the following.

(a) CF DAB converters exhibit low current ripple in I/O ports, making them suitable for current ripple sensitive applications. However, voltage spikes must be taken into consideration.
(b) The converters with modified active bridges mainly have four types. The first type involves reducing the number of active switches, which has the advantage of a simple structure. The second type is to increase the number of switches, with NPC, and T-type being the most common. All of these structures can reduce current stress, but only the T-type structure cannot reduce voltage stress. The third type is reconnection, which reduces voltage stress and generally results in smaller converter sizes.
(c) The transformer can be modified to a multi-winding transformer, multi-two-winding transformers, or a transformer with controllable parameters. Compared with a multi-winding transformer, the multiport DAB converters with multi-two-winding transformers can alleviate the magnetic short-circuit issues, whereas power distribution would be a concern. While, with adjusted tap or controlled variable leakage inductance, the converters can be applied to wide voltage range applications with extended ZVS range.
(d) Non-resonant networks are commonly added to DAB converters with the requirement of a wide voltage range. Among non-resonant networks, inductors are mainly used to inject current to extend the ZVS range, while capacitors are usually used to introduce voltage bias to match the voltages better, but the power density will reduce. LC, LCL, and CLLC networks are common resonant networks that can be applied to universal applications. These networks produce a current waveform that closely resembles a sine wave, thereby reducing peak current and the DC component of the transformer. Moreover, LCL and CLLC networks can effectively reduce reactive power on the two sides.

1.7.2 Evolvement History and Future Trends

From the first time the conventional DAB converters were proposed to the present, the number of publications with different methods at five-year intervals is shown in

Table 1.1 Conclusion of The Four Topological Modification Methods

Topological Modification Methods		Characteristics	Major Merits	Major Demerits
① Modifying I/O Types	CF DAB	Large inductors are used to provide current sources	Smooth currents on the current fed sides	More magnetic components. Clamped circuits may be needed
	CF DAB3			
② Modifying Active Bridges	Reduced Active Switches	One bridge leg is replaced by split capacitors	Reduced driver circuits and improved reliability	Increased current stress. Capacitors carry the full load current
		One side switches or bridge legs in load side are replaced by diodes	No reverse current in load side	Unidirectional power transfer and reduced control flexibility
	Increased Switches	NPC structure	Reduced voltage stress and current stress	Voltage imbalance problem. Increased losses of the additional switches
		T-type structure	Reduced current stress	Full range ZVS of the additional switches cannot be achieved
		MMC based structure	Modularity and scalability	Voltage imbalance problem among each SM
	Reconnection without Modified Switches	The bridge legs are not in parallel	Reduced voltage stress	Increased current stress
		Two active bridges of two sides are connected in additive polarity in series		Galvanic isolation failure
	Modified Switches for AC Sources	The switches in AC side are changed to four-quadrant switches	Only one stage in AC application	Increased control complexity
③ Modifying Transformer	Multi-winding Transformers	Push–pull structure	Simple structure and few switches	DC current component applied to transformer

(continued)

Table 1.1 (continued)

Topological Modification Methods		Characteristics	Major Merits	Major Demerits
		Y-Δ or Δ-Δ connections in DAB3	Extended ZVS range in Y-Δ Low winding rms current in Δ-Δ	Higher current stress under light load in Y-Δ. Higher winding voltage in Δ-Δ
		The additional windings are linked to additional bridges	Easy to provide multiple ports	High circulating power between the source ports
	Multi Two-winding Transformers	The additional transformers are linked to more bridge legs	Voltage levels can be adjusted flexibly	More magnetic components. Special concerns on power distribution of each transformer
		The additional transformers are linked to additional bridges	Provide multiple ports without magnetic short-circuit	More magnetic components
	Transformer with Controllable Parameters	The transformer has adjustable tap	Extended ZVS range within wide voltage range	Additional control circuit. The value of turns ratio cannot vary continuously
		The leakage inductance can be controlled in real time	Operated with optimized leakage inductance	Additional control circuit
④ Adding Auxiliary HF Networks	Non-resonant Networks	Inductors are added to inject auxiliary currents	Extended ZVS range within wide voltage range	Increase volume and loss with more inductors
		Capacitors are added to inject voltage offsets and create more voltage match points	Extended ZVS range and reduced conduction loss	Capacitors carry the full load current
	Resonant Networks	LC, LCL or CLLC networks are used to enable the inductor current close to sinusoid	Reduced switching turn-off loss. Unity power factor is possible	Additional losses of the added components

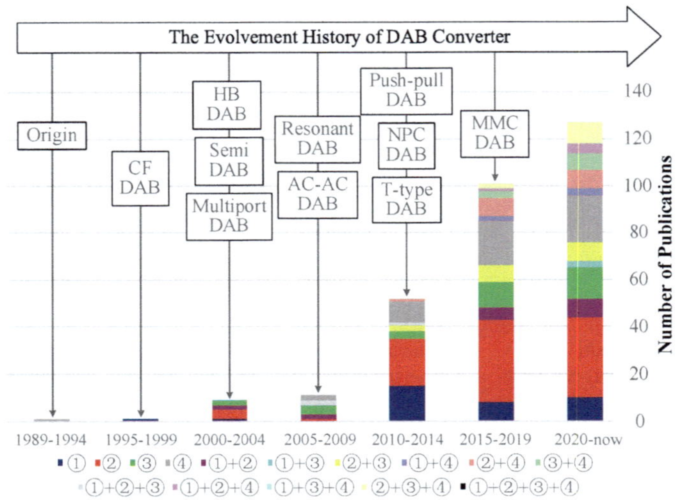

Fig. 1.27 The evolvement history of the DAB converter

Fig. 1.27. It is worth noting that the publications are searched in Web of Science and IEEE Xplore, and conference papers are not fully included. The four methods, represented by ① to ④, are demonstrated on the timeline alongside some typical topologies. For example, the CF DAB converter first appeared between 1995 and 1999, followed by half bridge DAB, semi DAB, and multiport DAB between 2000 and 2004. In the following five-year interval, resonant DAB and AC-AC DAB were proposed.

In future development, high efficiency remains a perpetual requirement, and higher power density is still the research priority. Besides, investigating high power level DAB converters can also be an interesting research topic for future exploration.

High efficiency: High conversion efficiency is a significant indicator for evaluating the performance of a DAB converters. In the last three years (from 2020 to now), over 50% of the publications in Fig. 1.27 include extending the ZVS range as one of the major goals to improve efficiency. As known, ZVS achievement not only boosts efficiency but also enhances system reliability by reducing HF switching noise and EMI emissions. Based on the switch junction capacitance C_{oss}, ZVS models can be divided into three cases. The first approach is to ignore C_{oss}, which is applied in over 60% of the publications in Fig. 1.27. In this case, the polarity of the commutating current is used to determine ZVS regions. However, these ZVS results tend to be overly optimistic. Since energy will dissipate in voltage sources under some operating conditions, the initial inductor energy needs to be large enough to fully charge/discharge the junction capacitor. To calculate the initial inductor current, the second case considers C_{oss} as a fixed value. This assumption simplifies the analysis, because C_{oss} will vary a lot during the resonant procedure in practice. So, in the third case, C_{oss} is nonlinear with drain-source voltage. Using these ZVS models, two major problems with the second approach, namely reactive power increase due to large fixed C_{oss} and ZVS failure with small fixed C_{oss}, can be mitigated. Nevertheless, the calculation procedure is complicated and challenging to implement in a digital controller. Thus, developing a fully soft switching technique under a wide voltage range remains a challenge task.

High power density: High power density is a key research priority in DAB/DAB3 converter development. Over the last three years, about 40% of the prototypes in Fig. 1.27 utilized wide bandgap (WBG) semiconductor devices, 80% of which are SiC-based while the rest are GaN-based. WBG devices allow for higher switching frequencies, which in turn enables significant reduction in the size of magnetic components. Meanwhile, around 50% of the topologies developed in the last three years included at least one auxiliary magnetic component. Magnetic integration techniques are conducive to compacting the overall size of the magnetic components, thereby improving power density even further.

1.8 Conclusion and Future Work

This chapter presents a comprehensive review of the various topological modification methods for DAB converters. These topologies are derived based on four cells, which are I/O types, active bridges, HF transformers, and auxiliary HF net-works. This chapter first lists the typical structures of each cell, and then classifies the four methods corresponding to these cells. Each method is further classified based on the topology characteristics, with a discussion of the motivations, merits, and demerits of each method. Then, the historical evolvement of DAB topologies is discussed, and future trends are analyzed based on the topologies from 2020 to present. While this chapter focuses on single DAB converters, it is worth noting that there have been many modular DAB converters presented in the literature. These modular converters can also be considered as modified DAB converters to a certain extent. Conventional and modified DAB converters, as well as other converters, can serve as submodules of modular DAB converters. Besides, these submodules can be connected with different configurations, such as high series low parallel, input series output series, input parallel output parallel, and more.

1.9 Organization of the Book

From the topology point of view, this book can be divided into six parts. This chapter provides a comprehensive review and classification for dual-active-bridge converters, and the topologies derivation law is also studied. From the perspective of topology modification methods, the remaining content can be divided into 5 parts. The first part is focus on the control and optimization for the converter with the modified IO types which are from Chaps. 2 to 3. The second part presents the novel topology and control strategy with modified active bridges in Chap. 4. The topologies and control methods in third part are studied by modifying transformers, which are from Chaps. 5 to 7. The modification method of adding auxiliary high-frequency networks is presented in fourth part which are from Chaps. 8 to 10. The fifth part is focus on the modular structure of the bidirectional DC-DC converters which are from Chaps. 11 to 13. The content for each chapter is organized as follows.

Chapter 2 is focused on a current-fed dual active bridge dc-dc converter based on pulse width modulation plus phase shift control. Besides, based on maximum power transfer, peak current and root mean square current analysis, optimal DC inductance and leakage inductance design to obtain low inductor RMS current is proposed.

Chapter 3 introduces the dead-time optimization methods and the magnetizing current design for a current-fed dual active bridge converter under the pulse-width plus phase-shift control to achieve full load range zero-voltage-switching in a wide voltage range.

In Chap. 4, a modified voltage matching control and current injections method are proposed for the hybrid-bridge-based DAB converter, which is aimed at achieving full load range zero voltage switching with con-sideration of all the switch junction capacitors.

In Chap. 5, a three-winding-coupled -inductor-based dual active bridge DC-DC converter is proposed. Similar to the current-fed DAB converter, the proposed TWCI DAB has small input current ripple and wide input voltage range.

Chapter 6 focuses on the Input-serial-output-parallel dual-active-bridge converter in the DC power distribution system. To improve the efficiency, an ISOP DAB converter based on a modulated coupled inductor, and a corresponding hybrid SPS control are proposed.

In Chap. 7, a modified push–pull dual active bridge dc-dc converter and its optimized control are proposed to achieve full load range zero voltage switching. Considering the switch junction capacitors, the added current is injected by the magnetizing inductance to achieve full load range ZVS.

Chapter 8 focus on light-load efficiency enhancement of high-frequency dual-active-bridge converter. Aiming at increasing the light load efficiency of high-frequency DAB converter under SPS control, this chapter proposes a concept to achieve full load range zero-voltage-switching with minimum RMS current in certain voltage variation range.

In Chap. 9, to solve the problem of limited zero voltage switching range and achieve seamless transition for extended-phase-shift control, the magnetizing current is utilized and an optimized control law is proposed. With the injection of magnetizing current, the two ZVS regions under EPS control will be creatively intersected with a large region, instead of disconnected or just intersected at a particular point as in previous studies.

In Chap. 10, the magnetizing inductance is designed to create possibility of full load range ZVS achievement for all power switches. Based on ZVS analysis considering switch junction capacitors and power characteristics analysis, a power control law premised on full load range ZVS achievement is proposed to achieve low current stress.

In Chap. 11, to achieve tight voltage regulation and fully utilize the advantages of LLC DCX operation under bidirectional isolated power application, a single-stage DAB-LLC hybrid bidirectional converter based on the sigma converter structure is proposed.

In Chap. 12, an LLC-DAB bidirectional converter with reduced switch count as a DC-DC transformer is proposed. It can achieve voltage regulation, full load range zero voltage switching for eight main switches and wide load range ZVS for four auxiliary switches, with consideration of the switch junction capacitors under different power direction.

Chapter 13 focuses on the input-series-output-parallel connected dual active bridge converters. In this chapter, it is firstly revealed that the stability of the ISOP DAB converter under common phase ratio control can be ensured by consideration of the dead-time. Then based on the derived model, the stability analysis and the parameters design are further performed.

It should be noted that the effectiveness of all the converter topologies and control methods are verified with experimental results.

References

1. Sha, D., Zhang, J., & Xu, Y. (2017). Improved boundary operation for voltage-fed semi-DAB with ZVS achievement and nonactive power reduction. *IEEE Transactions on Industrial Electronics, 64*(8), 6179–6189.
2. Zhang, Y., Wang, Z., Li, Y. W., Hou, N., & Cheng, M. (2020). Decoupled dual-PWM control for naturally commutated current-fed dual-active-bridge DC/DC converter. *IEEE Journal of Emerging and Selected Topics in Power Electronics, 8*(4), 4246–4259.
3. Mou, D., et al. (2023). Modeling and analysis of hybrid dual active bridge converter to optimize efficiency over whole operating range. *IEEE Journal of Emerging and Selected Topics in Power Electronics, 11*(1), 432–441.
4. Chaurasiya, S., & Singh, B. (2023). A bidirectional fast EV charger for wide voltage range using three-level DAB based on current and voltage stress optimization. *IEEE Transactions on Transportation Electrification, 9*(1), 1330–1340.
5. Li, C., Liu, J., Du, S., Dang, H., Deng, Z., & Chen, H. (2023). A T-type DAB-based isolated DC-DC-AC three-port converter with high power efficiency. *IEEE Transactions on Power Electronics, 38*(11), 14178–14194.
6. Jeong, Y., Lee, M.-S., Park, J.-D., Kim, J.-K., & Rorre, R. A. L. (2020). Hold-up time compensation circuit of half-bridge LLC resonant converter for high light-load efficiency. *IEEE Transactions on Power Electronics, 35*(12), 13126–13135.
7. Xu, G., Sha, D., Xu, Y., & Liao, X. (2018). Hybrid-bridge-based dab converter with voltage match control for wide voltage conversion gain application. *IEEE Transactions on Power Electronics, 33*(2), 1378–1388.
8. Feng, H., et al. (2022). Passive capacitor voltage balancing of sic-based three-level dual-active-bridge converter using hybrid NPC-flying capacitor structure. *IEEE Transactions on Power Electronics, 37*(4), 4183–4194.
9. Zhang, Z., Thomsen, O. C., & Andersen, M. A. E. (2012). Optimal design of a push-pull-forward half-bridge (PPFHB) bidirectional DC–DC converter with variable input voltage. *IEEE Transactions on Industrial Electronics, 59*(7), 2761–2771.

10. Abramson, R. A., Gunter, S. J., Otten, D. M., Afridi, K. K., & Perreault, D. J. (2018). Design and evaluation of a reconfigurable stacked active bridge DC–DC converter for efficient wide load range operation. *IEEE Transactions on Power Electronics, 33*(12), 10428–10448.
11. Lu, Z., Xu, G., Xiong, W., Sun, Y., & Su, M. (2023). A Quasi-two-stage isolated bidirectional buck-DAB converter for wide input voltage range. *IEEE Transactions on Power Electronics, 38*(2), 1384–1390.
12. Chen, D., Deng, J., Li, M., Wang, Z., & Wang, S. (2022). An enhanced dual active bridge converter with full domain ZVS by utilizing a simple segment control for wide voltage range applications. *IEEE Transactions on Industrial Electronics, 69*(7), 6817–6827.
13. Sha, G., et al. (2021). Research on multi-port DC–DC converter based on modular multilevel converter and cascaded H bridges for MVDC applications. *IEEE Access, 9*, 95006–95022.
14. Shiva, S. M., Gorla, N. B. Y., Das, P., & Panda, S. K. (2015). Tap changing transformer based dual active bridge bi-directional DC-DC converter. *9th International Conference on Power Electronics and ECCE Asia* (pp. 2025–2030). IEEE.
15. Saeed, S., Garcia, J., & Georgious, R. (2021). Dual-active-bridge isolated DC–DC converter with variable inductor for wide load range operation. *IEEE Transactions on Power Electronics, 36*(7), 8028–8043.
16. Everts, J., Krismer, F., Van den Keybus, J., Driesen, J., & Kolar, J. W. (2014). Optimal ZVS modulation of single-phase single-stage bidirectional DAB AC–DC converters. *IEEE Transactions on Power Electronics, 29*(8), 3954–3970.
17. Liu, P., & Duan, S. (2020). A hybrid modulation strategy providing lower inductor current for the DAB converter with the Aid of DC blocking capacitors. *IEEE Transactions on Power Electronics, 35*(4), 4309–4320.
18. Higa, H., & Itoh, J.-I. (2017). Extension of zero-voltage-switching range in dual active bridge converter by switched auxiliary inductance. *Energy Conversion Congress and Exposition* (pp. 5324–5331). IEEE.
19. Hu, Z., Qiu, Y., Liu, Y.-F., & Sen, P. C. (2014). A Control strategy and design method for interleaved LLC converters operating at variable switching frequency. *IEEE Transactions on Power Electronics, 29*(8), 4426–4437.
20. Wu, J., Li, Y., Sun, X. & Liu, F., A new dual-bridge series resonant DC–DC converter with dual tank. *IEEE Transactions on Power Electronics, 33*(5), 3884–3897.
21. Chan, Y. P., Yaqoob, M., Wong, C. S., & Loo, K. H. (2020). Realization of high-efficiency dual-active-bridge converter with reconfigurable multilevel modulation scheme. *IEEE Journal of Emerging and Selected Topics in Power Electronics, 8*(2), 1178–1192.
22. Malan, W. L., Vilathgamuwa, D. M., & Walker, G. R. (2016). Modeling and control of a resonant dual active bridge with a tuned CLLC network. *IEEE Transactions on Power Electronics, 31*(10), 7297–7310.
23. Wang, Z., Zheng, Z., & Li, C. (2022). A high-step-up low-ripple and high-efficiency DC-DC converter for fuel-cell vehicles. *IEEE Transactions on Power Electronics, 37*(3), 3555–3569.
24. Beldjajev, V., Lehtla, T. & Zakis, J. (2013). Impact of component losses on the efficiency of the LC-filter based dual active bridge for the isolation stage of power electronic transformer. *International Conference-Workshop Compatibility and Power Electronics* (pp. 132–137). IEEE.
25. Xuewei, P., & Rathore, A. K. (2014). Novel bidirectional snubberless naturally commutated soft-switching current-fed full-bridge isolated DC/DC converter for fuel cell vehicles. *IEEE Transactions on Industrial Electronics, 61*(5), 2307–2315.
26. Daneshpajooh, H., Bakhshai, A. & Jain, P. (2012). Modified dual active bridge bidirectional DC–DC converter with optimal efficiency. *Twenty-Seventh Annual IEEE Applied Power Electronics Conference and Exposition* (pp. 1348–1354). IEEE.
27. Aghajani, A. A., Zare Kashani, N., Eldoromi, M. & Moti Birjandi, A. A. (2022). A novel half-full-bridge split-capacitor DC–DC converter based on dual-active-bridge topology. *13th Power Electronics, Drive Systems, and Technologies Conference* (pp. 240–244). IEEE.
28. Ma, G., Qu, W., Yu, G., Liu, Y., Liang, N., & Li, W. (2009). A zero-voltage-switching bidirectional DC–DC converter with state analysis and soft-switching-oriented design consideration. *IEEE Transactions on Industrial Electronics, 56*(6), 2174–2184.

29. Xu, G., Wang, J., Wang, H., Liu, Y., Liu, D., & Su, M. (2022). Coupled-inductor-based dual active bridge converter with soft switching capability and low component count. *IEEE Journal of Emerging and Selected Topics in Power Electronics, 10*(4), 4771–4782.
30. Zhang, J. M., Zhang, F., Xie, X. G., Jiao, D. Z. & Qian, Z. (2002). A novel ZVS DC/DC converter for high power applications. *Seventeenth Annual IEEE Applied Power Electronics Conference and Exposition* (vol.2, pp. 635–640). IEEE.
31. Jin, L., & Chen, C. (2020). Optimal phase-shift control of three-level dual-active bridge DC–DC converter with minimum reflux power for wide voltage conversion ratio range application. *IET Power Electronics, 13*, 1649–1655.
32. Higa, H., Takuma, S., Orikawa, K. & Itoh, J.-I. (2015). Dual active bridge DC-DC converter using both full and half bridge topologies to achieve high efficiency for wide load. *Energy Conversion Congress and Exposition.* (pp. 6344–6351). IEEE.
33. Kim, K., & Cha, H. (2021). Split-capacitor dual-active-bridge converter. *IEEE Transactions on Industrial Electronics, 68*(2), 1445–1453.
34. Ali, A., & Liao, Y. (2021). Optimized control for modified push-pull dual active bridge converter to achieve wide ZVS range and low current stress. *IEEE Access, 9*, 140258–140267.
35. Duarte, J. L., Hendrix, M., & Simoes, M. G. (2007). Three-port bidirectional converter for hybrid fuel cell systems. *IEEE Transactions on Power Electronics, 22*(2), 480–487.
36. Xu, G., Sha, D., Xu, Y., & Liao, X. (2018). Dual-Transformer-based DAB converter with wide ZVS range for wide voltage conversion gain application. *IEEE Transactions on Industrial Electronics, 65*(4), 3306–3316.
37. Jafari, A., Nikoo, M. S., Karakaya, F., & Matioli, E. (2020). Enhanced DAB for efficiency preservation using adjustable-tap high-frequency transformer. *IEEE Transactions on Power Electronics, 35*(7), 6673–6677.
38. Saeed, S., & Garcia, J. (2019). Extended operational range of dual-active-bridge converters by using variable magnetic devices. *Applied Power Electronics Conference and Exposition*, (pp. 1629–1634). IEEE.
39. Muthuraj, S. S., Kanakesh, V. K., Das, P., & Panda, S. K. (2017). Triple phase shift control of an LLL tank based bidirectional dual active bridge converter. *IEEE Transactions on Power Electronics, 32*(10), 8035–8053.
40. Qin, Z., Shen, Y., Loh, P. C., Wang, H., & Blaabjerg, F. (2018). A dual active bridge converter with an extended high-efficiency range by DC blocking capacitor voltage control. *IEEE Transactions on Power Electronics, 33*(7), 5949–5966.
41. Wu, J., Wen, P., Sun, X., & Yan, X. (2020). Reactive power optimization control for bidirectional dual-tank resonant DC–DC converters for fuel cells system. *IEEE Transactions on Power Electronics, 35*(9), 9202–9214.
42. Wang, C., Zsurzsan, T.-G., & Zhang, Z. (2023). Genetic algorithm assisted parametric design of splitting inductance in high frequency GaN-based dual active bridge converter. *IEEE Transactions on Industrial Electronics, 70*(1), 522–531.
43. Twiname, R. P., Thrimawithana, D. J., Madawala, U. K., & Baguley, C. A. (2015). A dual-active bridge topology with a tuned CLC network. *IEEE Transactions on Power Electronics, 30*(12), 6543–6550.
44. Chen, P., Jiang, W., Ruan, X., et al. (2023). Active damping control and parameter design method for ISOP-based load converter in DC distribution system. *IEEE Journal of Emerging and Selected Topics in Power Electronics, 11*(1), 465–477.
45. Li, Y., Shao, S., Chen, H., et al. (2020). High-gain high-efficiency IPOS LLC converter with coupled transformer and current sharing capability. *CPSS Transactions on Power Electronics and Applications, 5*(1), 63–73.
46. Sha, D., Guo, Z., Luo, T., et al. (2014). A general control strategy for input-series–output-series modular DC–DC converters. *IEEE Transactions on Power Electronics, 29*(7), 3766–3775.
47. Xia, Y., Yu, M., Zhang, Y., et al. (2020). Decentralized suppression strategy of circulating currents among IPOP single-phase DC/AC converters. *IEEE Journal of Emerging and Selected Topics in Power Electronics, 8*(2), 1571–1583.

48. Ahmed, M. H., Fei, C., Lee, F. C., et al. (2020). Single-stage high-efficiency 48/1 V sigma converter with integrated magnetics. *IEEE Transactions on Industrial Electronics, 67*(1), 192–202.
49. Wu, Z., Wang, Z., Zhang, Y., et al. (2022). A high efficiency and high power density DC transformer topology with output regulation capability. *IEEE Transactions on Power Electronics, 37*(7), 8232–8247.
50. Sun, C., Zhang, X., Zhang, J., et al. (2022). Hybrid input-series–output-series modular DC–DC converter constituted by resonant and nonresonant dual active bridge modules. *IEEE Transactions on Industrial Electronics, 69*(1), 1062–1069.

Part I
Dual-Active-Bridge Converter with Modified IO Types

Chapter 2
Design Considerations for Plus-Phase-Shift Controlled Current-Fed DAB Converter

Abstract Current-fed dual active bridge dc–dc converters are suited for renewable energy systems such as solar photovoltaic and fuel cells, due to the advantages of small input current ripple and wide input voltage range. In this chapter, a current-fed dual active bridge dc–dc converter based on pulse width modulation plus phase shift (PPS) control is studied. The working principle and zero-voltage switching (ZVS) conditions of the converter are analyzed in depth, and it is proved mathematically for the first time that all switches can naturally achieve zero-voltage-switching (ZVS) independently of DC inductance and leakage inductance under wide input voltage range within full load range. Based on maximum power transfer, peak current and root mean square (RMS) current analysis, optimal DC inductance and leakage inductance design to obtain low inductor RMS current is proposed. In addition, the detailed loss breakdown was implemented by comparing with the non-optimized case. The effectiveness of the theoretical analysis and design method have been verified by experimental results.

Keywords Current-fed dual active bridge (CFDAB) · Parameter optimization design · Pulse width modulation plus phase shift (PPS) control · Soft-switching analysis

2.1 Converter Topology and Operation Principle

For PPS control of CF-DAB converter, besides the PWM modulation, the implementation of soft-switching is another hot research issue because it can significantly improve the efficiency of the converter. The basic working principle and general ZVS operation are analyzed in CF-DAB under PPS control [1]. However, detail mathematical ZVS analysis and parameter design are not studied. For example, the ZVS conditions for all switches are only given by the basic currents limits, and no detailed mathematic expressions are presented [1]. Furthermore, about the soft-switching analysis method, [2–5] give the ZVS condition, and use the graphical plots to get the region that can realize the ZVS. However, the ZVS mathematical proof

G. Xu and X. Chen, *Topology Deduction and Control for Dual-Active-Bridge Converters*, https://doi.org/10.1007/978-981-97-4856-3_2

remains to be studied. For example, the graphic plots are based on particular knowing specifications, which could be regarded as a "case study". Actually, it is not certain what will be the ZVS performance if the application specifications are changed. In addition, when analyzing the wide voltage range or full load range performance, only some voltage points or load points are plotted to determine the region, which could also be regarded as a "case study". But there is no specific mathematical proof of whether ZVS can be realized with the full load range.

In this Chapter, the ZVS performance and optimized inductance design of the CF-DAB converter under PPS control have been studied in detail. It is proved mathematically for the first time that the ZVS of all switches can be achieved naturally under wide input voltage range within full load range.

2.1.1 Topology and Modulation Schedule Using PPS Control

The topology of CF-DAB converter is depicted in Fig. 2.1. For the low voltage side (LVS), V_L is the LVS voltage. L_{dc1} and L_{dc2} are DC boost inductors, and $L_{dc1} = L_{dc2} = L_{dc}$. If L_{dc1} and L_{dc2} have different values, the unbalanced current will occur between i_{Ldc1} and i_{Ldc2}.To achieve better current sharing performance, the current difference of the two inductance can be used to compensate the duty cycle of each bridge arm to balance currents. C_c is the clamping capacitor, and the voltage of the clamping capacitor is V_{Cc}. The duty ratio of the bottom switches Q_1 and Q_2 is d. The inductor L_r denotes the leakage inductance. In the high voltage side (HVS), MOSFETs S_1–S_4 form a full bridge.

Figure 2.2 shows the steady-state waveforms of the converter operating in boost mode when power flows from the LVS to the HVS. The wide range input voltage V_L is boosted to clamping voltage V_{Cc} to match the HVS voltage V_H by the turns of transformer. The duty cycle for the HVS switches operate 50%. And the phase-shift angle exists between the LVS and the HVS.

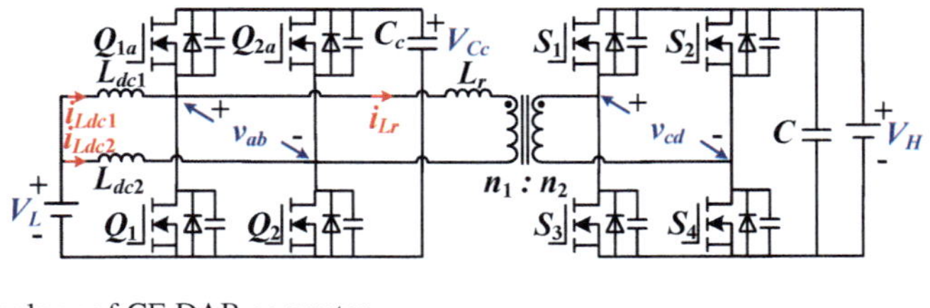

Fig. 2.1 Topology of CF-DAB converter

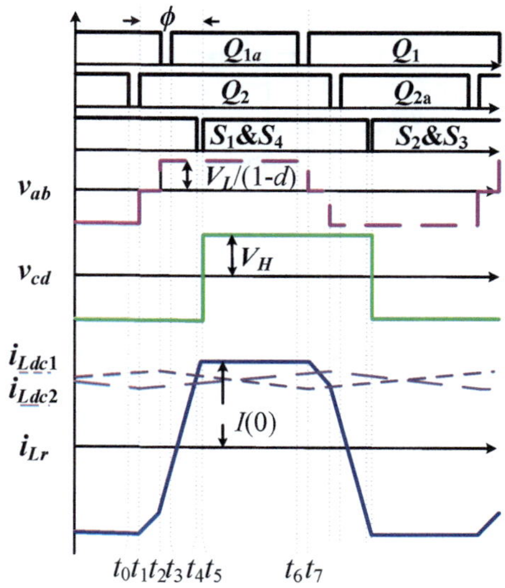

Fig. 2.2 Steady-state waveforms of PPS for CF-DAB in boost mode. (interval A: ϕ ∈ [(2d–1) π, π])

2.1.2 *Working Stages of the Converter*

The circuit mode is shown in Fig. 2.3, and the detailed circuit analysis is as follows. And the junction capacitors of all the switches are not considered.

Stage 1 (Before t_0): During this stage, Q_1, Q_{2a}, S_2 and S_3 are turned on. v_{ab} is equal to $-V_{Cc}$, v_{cd} is equal to $-V_H$. Due to the voltage matching at this time, the leakage inductor voltage is zero, which also means that the slope of the leakage inductor current is zero. Power flows from the LVS to the HVS, and $i_{Lr} = -I\ (0)$.

Stage 2 (t_0–t_1): Q_{2a} is turned off at t_0. At this mode, the sum of i_{Lr} and i_{Ldc2} charges C_{2a} and discharges C_2 until the body diode of Q_2 is conducted. The condition for $Q_2$to achieve ZVS conduction is $|\ i_{Lr}\ (t_0)\ | > i_{Ldc2}\ (t_0)$.

Stage 3 (t_1–t_2): At t_1, Q_2 turns on under ZVS. v_{ab} is zero, the leakage inductor current decreases in the reverse direction until Q_1 turns off. At this stage, leakage inductor voltage and the leakage inductor current can be obtained as

$$v_{Lr} = \frac{n_1 V_H}{n_2}, i_{Lr} = -I(0) + \frac{n_1 V_H}{n_2 L_r}(t - t_1) \tag{2.1}$$

And the duration of this stage is $(2d–1)\ \pi/\omega$, ω is the switching angular frequency.

Stage 4 (t_2–t_3): Q_1 is turned off at t_2. The sum of $-i_{Lr}$ and i_{Ldc1} charges C_1 and discharges C_{1a} until the body diode D_{1a} of Q_{1a} is turned on. The condition for Q_{1a} to achieve ZVS conduction is $i_{Lr}\ (t_2) < i_{Ldc1}\ (t_2)$. During this mode, the voltage and current of the leakage inductor can be expressed as

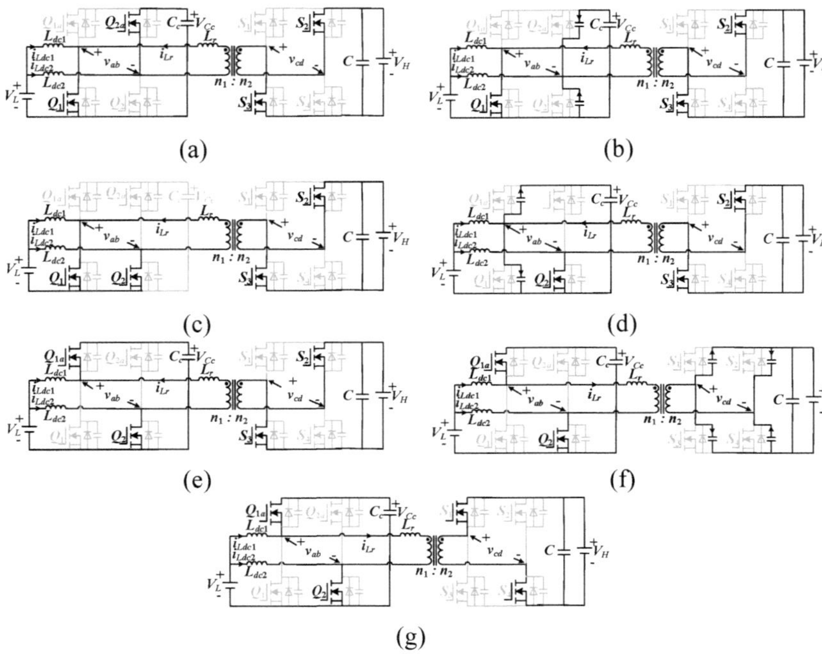

Fig. 2.3 Operation modes of the boost mode. **a** *Before* t_0. **b** t_0–t_1. **c** t_1–t_2. **d** t_2–t_3. **e** t_3–t_4. **f** t_4–t_5. **g** t_5–t_6

$$v_{Lr} = \frac{2n_1 V_H}{n_2}, i_{Lr} = -I(0) + \frac{n_1 V_H}{n_2 L_r}\frac{(2d-1)\pi}{\omega} + \frac{2n_1 V_H}{n_2 L_r}\left(t - t_1 - \frac{(2d-1)\pi}{\omega}\right) \tag{2.2}$$

Stage 5 (t_3–t_4): At t_3, Q_{1a} is turned on under ZVS. The leakage inductor voltage is the same as the previous mode, so the slope of leakage current and the expression are the same. At this stage, the leakage inductor current rises to a positive value.

Stage 6 (t_4–t_5): At t_4, S_2 and S_3 are turned off. The leakage inductor current charges/discharges the junction capacitor. The condition for S_1 and S_4 to achieve ZVS is i_{Lr} $(t_4) > 0$.

Stage 7 (t_5–t_6): At t_5, S_1 and S_4 are turned on under ZVS. v_{cd} is equal to V_H. At this time, the leakage inductor voltage is zero, which also means that the slope of the leakage inductor current is zero. Power flows from the LVS to the HVS, and $i_{Lr} = I$ (0). The working mode of the second half cycle is similar to the first half cycle.

2.1.3 Power Expressions of the Converter

Define that the rising edge of v_{ab} is leading v_{cd} as a positive ϕ. The total volt-seconds applied the L_{dc1} or L_{dc2} of LVS over one switching period are

$$V_L d + (V_L - V_{Cc})(1 - d) = 0 \tag{2.3}$$

Then the duty cycle of Q_1 and Q_2 can be calculated as

$$d = 1 - \frac{V_L}{V_{Cc}} \tag{2.4}$$

PPS control strategy is realized with two individual controllers, which is the decoupling control diagram as shown in Fig. 2.4. Voltage matching is realized strictly by voltage matching control shown in the voltage matching control block. The reference is expressed as $v_H n_1/n_2$ and the relationship between clamp voltage V_{Cc} and HVS voltage V_H in steady state is written as

$$\frac{V_{Cc}}{V_H} = \frac{n_1}{n_2} \tag{2.5}$$

The power can be calculated as.

(1) Phase shift angle $\phi \subset [0, (2d\text{–}1)\,\pi]$.

Based on the same method, when the phase shift angle $\phi \subset [0, (2d\text{–}1)\,\pi]$, the current of leakage inductor in a whole cycle can be written as

$$-I(0) + \frac{n_1 V_H}{n_2 L_r}\frac{\phi}{2\pi}T_s - \frac{n_1 V_H}{n_2 L_r}\left(\frac{(2d-1)}{2} - \frac{\phi}{2\pi}\right)T_s = I(0) \tag{2.6}$$

Through integration of the transformer voltage and current, the power can be calculated as

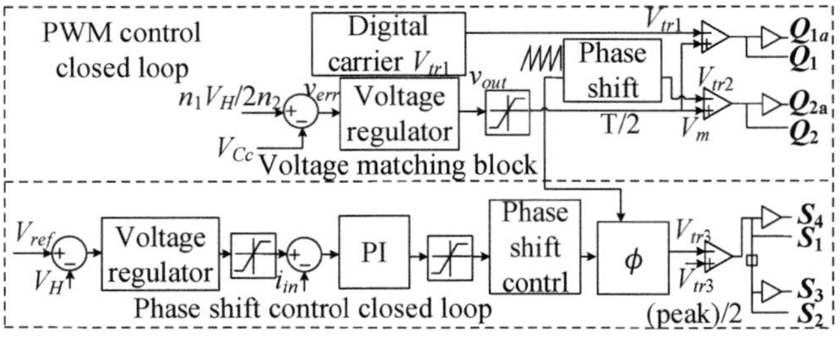

Fig. 2.4 The diagram of the decoupling control

$$P = \frac{1}{T_s}\int_0^{T_s} i_{Lr} v_{ab} dt = \frac{2}{T_s}\left[\int_{\frac{(2d-1)}{2}T_s}^{\frac{T_s}{2}} I(0)\frac{n_1 V_H}{n_2} dt\right] \tag{2.7}$$

Substituting (2.6) into (2.7), the output power is can be written as

$$P = \frac{(n_1 V_H)^2(1-d)[\phi-(d-0.5)\pi]T_s}{(n_2)^2 \pi L_r} \tag{2.8}$$

(2) Phase shift angle $\phi \subset [(2d-1)\pi, \pi]$.

According to the analyses of working modes in section II-A, the current of leakage inductor in a whole cycle can be written as

$$-I(0) + \frac{n_1 V_H}{n_2 L_r}\frac{(2d-1)}{2}T_s + \frac{2n_1 V_H}{n_2 L_r}\left(\frac{\phi}{2\pi} - \frac{(2d-1)}{2}\right)T_s = I(0) \tag{2.9}$$

Through integration of the transformer voltage and current, the power can be calculated as

$$P = \frac{1}{T_s}\int_0^{T_s} i_{Lr} v_{ab} dt = \frac{2}{T_s}\left[\begin{array}{l}\int_{\frac{(2d-1)}{2}T_s}^{\frac{\phi}{2\pi}T_s}\left[\begin{array}{l}-I(0) + \frac{n_1 V_H}{n_2 L_r}\frac{(2d-1)}{2}T_s + \\ \frac{2n_1 V_H}{n_2 L_r}\left(\frac{\phi}{2\pi} - \frac{(2d-1)}{2}\right)T_s\end{array}\right]\frac{n_1 V_H}{n_2} dt \\ + \int_{\frac{\phi}{2\pi}T_s}^{\frac{T_s}{2}} I(0)\frac{n_1 V_H}{n_2} dt\end{array}\right] \tag{2.10}$$

Substituting (2.9) into (2.10), the output power is can be written as

$$P = \frac{(n_1 V_H)^2\left[-\phi^2 + 2d\pi\phi - d(2d-1)\pi^2\right]T_s}{2(n_2)^2\pi^2 L_r} \tag{2.11}$$

Thus, the power can be written as

$$P = \begin{cases} \frac{(n_1 V_H)^2(1-d)[\phi-(d-0.5)\pi]T_s}{(n_2)^2\pi L_r} & \phi \in [0, (2d-1)\pi] \\ \frac{(n_1 V_H)^2\left[-\phi^2+2d\pi\phi-d(2d-1)\pi^2\right]T_s}{2(n_2)^2\pi^2 L_r} & \phi \in [(2d-1)\pi, \pi] \end{cases} \tag{2.12}$$

Figure 2.5 illustrates the power transfer curves under different V_L. When the voltage is matched and ϕ range is $[0, 0.5\pi]$, P increases monotonically as ϕ increases.

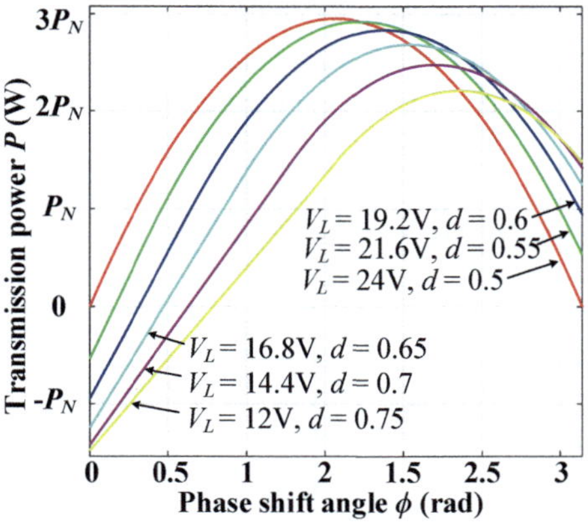

Fig. 2.5 Curves of power and phase shift angle

2.2 Soft-Switching Analysis and Optimized Design

2.2.1 ZVS Range Analysis

In this section, the conditions of ZVS for all switches are derived based on the DC inductor current and leakage inductor current. In addition, it is proved mathematically that all switches can naturally achieve ZVS at wide input voltage range within full load range. The study proves that, when the voltages are matched, the implementation of the soft-switching is independent of any parameters, including DC inductance, leakage inductance, load power and other component parameters. The analysis process is shown as follows.

In order to achieve ZVS of Q_1 and Q_2, the current limit of ZVS $I_{\_ZVS\_Q1\&Q2} = i_{Ldc2}(t_0) + i_{Lr}(t_0) < 0$ needs to be satisfied. Based on mode analysis and calculation of instant current value, the current at switch turn-on for charging (discharging) the junction capacitor C_{2a} (C_2) to achieve ZVS of Q_1 and Q_2 is written as

$$I_{ZVS\_Q_1\&Q_2} = i_{Ldc2}(t_0) + i_{Lr}(t_0)$$
$$= \begin{cases} \frac{n_1 V_H d(d-1)T_s}{2n_2 L} & P \le P_{div} \\ \frac{P}{2V_L} + \frac{n_2 V_L^2 T_s}{2n_1 V_H L_{dc}} - \frac{V_L T_s}{2L_{dc}} - \frac{n_1 V_H T_s}{4n_2 L_r} + \frac{1}{2L_r}\sqrt{\left(\frac{n_1 V_H}{n_2} T_s\right)^2 d(1-d) - 2L_r P T_s} & P \ge P_{div} \end{cases} \tag{2.13}$$

P_{div} is determined by (2.12) with $\phi = (2d-1)\pi$. And it can be written as

$$P_{div} = -\frac{n_1^2 V_H^2 T_s(2d^2 - 3d + 1)}{2n_2^2 L_r} \tag{2.14}$$

The current limit of ZVS $I_{\_ZVS\_Q1a\&Q2a} = i_{Lr}(t_2) - i_{Ldc1}(t_2) < 0$ needs to be satisfied to achieve ZVS of Q_{1a} and Q_{2a}. The same method can be used to obtain the current at switch turn-on for charging (discharging) the top switching junction capacitor C_{1a} (C_1) to achieve ZVS of Q_{1a} and Q_{2a} is written as

$$I_{ZVS\_Q_{1a}\&Q_{2a}} = i_{Lr}(t_2) - i_{Ldc1}(t_2) = \begin{cases} \frac{n_1 V_H d(d-1)T_s}{2n_2 L} & P \leq P_{div} \\ -\frac{P}{2V_L} + \frac{n_2 V_L^2 T_s}{2n_1 V_H L_{dc}} - \frac{V_L T_s}{2L_{dc}} - \frac{n_1(3-4d)V_H T_s}{4n_2 L_r} + \frac{1}{2L_r}\sqrt{\left(\frac{n_1 V_H}{n_2} T_s\right)^2 d(1-d) - 2L_r P T_s} & P \geq P_{div} \end{cases} \tag{2.15}$$

And the current limit of ZVS $I_{\_ZVS\_S1\&S2\&S3\&S4} = i_{Lr}(t_4) > 0$ needs to be satisfied to achieve ZVS of S_1–S_4. The ZVS current at switch turn-on for charging (discharging) the junction capacitor of the high-voltage side full bridge switch is written as

$$I_{ZVS\_S_1\&S_2\&S_3\&S_4} = i_{Lr}(t_4) = \begin{cases} \frac{n_1 V_H (2d-1)T_s}{4n_2 L_r} & P \leq P_{div} \\ \frac{\frac{n_1 V_H T_s}{n_2} - 2\sqrt{\frac{n_1 V_L V_H}{n_2} T_s^2 - 2L_r P T_s - T_s^2 V_L^2}}{4L_r} & P \geq P_{div} \end{cases} \tag{2.16}$$

To obtain the minimal value of (2.13), rewrite the above current at switch turn-on $I_{\_ZVS\_Q1\&Q2}$ into a univariate function with load power P as independent variable, and its derivative can be expressed as

$$\frac{\partial I_{ZVS\_Q_1\&Q_2}(P)}{\partial P} = \begin{cases} 0 & P \leq P_{div} \\ \frac{1}{2V_L} - \frac{T_s}{2\sqrt{\left(\frac{n_1 V_H}{n_2} T_s\right)^2 d(1-d) - 2L_r P T_s}} & P \geq P_{div} \end{cases} \tag{2.17}$$

As can be seen, $I'_{\_ZVS\_Q1\&Q2}$ is equal to 0 when P is belonging to [0, P_{div}]. When P is in the region [0, P_{div}], the value of $I_{\_ZVS\_Q1\&Q2}$ is constant. When P is in the region [P_{div}, P_N], according to (2.4) and (2.5), $I'_{\_ZVS\_Q1\&Q2}$ can be written as

$$I'_{ZVS\_Q_1\&Q_2} = \frac{1}{2V_L}\left(1 - \frac{1}{\sqrt{1 - \frac{1}{1-d} - \frac{2L_r P}{V_L^2 T_s}}}\right) \tag{2.18}$$

Since (1/(1–d)) and ($2L_r P/TV_L$^2) of above formula are both greater than zero in (2.18), the condition of ZVS can be further written as

$$0 < 1 - \frac{1}{1-d} - \frac{2L_r P}{V_L^2 T_s} \leq 1 \tag{2.19}$$

According to (2.18) and (2.19), $I'_{\_ZVS\_Q1\&Q2}$ is less than zero when P is in the region $[P_{div}, P_N]$. Therefore, $I_{\_ZVS\_Q1\&Q2}$ is decreasing monotonically over this interval, and since $I_{\_ZVS\_Q1\&Q2}$ is continuous over the entire interval $[0, P_N]$, this means that as long as $I_{\_ZVS\_Q1\&Q2}$ is less than zero when $P = 0$, $I_{\_ZVS\_Q1\&Q2}$ will always be smaller than zero, which means Q_1 and Q_2 can achieve ZVS in full load range. As can be seen that in (2.13), $(d-1)$ is less zero when $P \leq P_{div}$. Therefore, switches Q_1 and Q_2 can naturally achieve ZVS at full load range.

The same analysis is applied to the switches Q_{1a} and Q_{2a}, the derivative of current at switch turn-on $I_{\_ZVS\_Q1a\&Q2a}$ with load power P as independent variable can be obtained as

$$\frac{\partial I_{ZVS\_Q_{1a}\&Q_{2a}}(P)}{\partial P} = \begin{cases} 0 & P \leq P_{div} \\ -\frac{1}{2V_L} - \frac{T_s}{2\sqrt{\left(\frac{n_1 V_H}{n_2} T_s\right)^2 d(1-d) - 2L_r P T_s}} & P \geq P_{div} \end{cases} \tag{2.20}$$

$I_{\_ZVS\_S1\&S2\&S3\&S4}$ is monotonically increasing and continuous over the entire interval $[0, P_N]$, so as long as $I_{\_ZVS\_S1\&S2\&S3\&S4}$ is greater than zero or equal to zero when $P = 0$, switches S_1, S_2, S_3 and S_4 can realize ZVS at full load range. As can be seen that in ($I_{ZVS\_S_1\&S_2\&S_3\&S_4} = i_{Lr}(t_4) = \begin{cases} \frac{n_1 V_H (2d-1) T_s}{4 n_2 L_r} & P \leq P_{div} \\ \frac{\frac{n_1 V_H T_s}{n_2} - 2\sqrt{\frac{n_1 V_L T_H}{n_2} T_s^2 - 2L_r P T_s - T_s^2 V_L^2}}{4L_r} & P \geq P_{div} \end{cases}$ (2.16), $(2d-1)$ is bigger than zero when $P \leq P_{div}$. Therefore, switches S_1, S_2, S_3 and S_4 can naturally achieve ZVS at full load range.

The same analysis is applied to the switches S_1, S_2, S_3 and S_4, the derivative of current at switch turn-on $I_{\_ZVS\_S1\&S2\&S3\&S4}$ with load power P as independent variable can be obtained as

$$\frac{\partial I_{ZVS\_S_1\&S_2\&S_3\&S_4}(P)}{\partial P} = \begin{cases} 0 & P \leq P_{div} \\ \frac{T_s}{2\sqrt{\left(\frac{n_1 V_H}{n_2} T_s\right)^2 d(1-d) - 2L_r P T_s}} & P \geq P_{div} \end{cases} \tag{2.21}$$

$I_{\_ZVS\_S1\&S2\&S3\&S4}$ is monotonically increasing and continuous over the entire interval $[0, P_N]$, so as long as $I_{\_ZVS\_S1\&S2\&S3\&S4}$ is greater than zero or equal to zero when $P = 0$, switches S_1, S_2, S_3 and S_4 can realize ZVS at full load range. As can be seen that in (2.16), $(2d-1)$ is bigger than zero when $P \leq P_{div}$. Therefore, switches S_1, S_2, S_3 and S_4 can naturally achieve ZVS at full load range.

To verify the theoretical proof, take the following parameters as an example: V_L = 12~24 V, $V_H = 240$ V, $f = 100$ kHz, $P_N = 650$ W. Figure 2.6 shows the current at switch turn-on. When $V_L = 12$ V, P is belonging to $[0, P_{div}]$, ZVS of Q_1 and Q_2 is shown in Fig. 2.6a. The current at switch turn-on $I_{\_ZVS\_Q1\&Q2} = -0.9\text{A} < 0$ with very large DC inductance $L_{dc} = 50$ μH under no load, and the current at switch turn-on $I_{\_ZVS\_Q1\&Q2} = -15\text{A} < 0$ with very small DC inductance $L_{dc} = 3$ μH under no load, this means that switches Q_1 and Q_2 can achieve ZVS when $V_L = 12$ V under different DC inductances with no load. Meanwhile, the current at switch turn-on $I_{\_ZVS\_Q1\&Q2}$ is also less than zero at full load, which means that switches Q_1 and Q_2 can achieve

ZVS when $V_L = 12$ V under different DC inductances within full load range. When $V_L = 24$ V, P belongs to $[P_{div}, P_N]$, ZVS of Q_1 and Q_2 is shown in Fig. 2.6b. The current at switch turn-on $I_{\_ZVS\_Q1\&Q2} = -1.2\text{A} < 0$ with very large DC inductance $L_{dc} = 50$ μH and $L_r = 1.5$ μH under no load, and the current at switch turn-on $I_{\_ZVS\_Q1\&Q2} = -20\text{A} < 0$ with very small DC inductance $L_{dc} = 3$ μH and $L_r = 1.5$ μH under no load, this means that switches Q_1 and Q_2 can achieve ZVS when $V_L = 24$ V under different DC inductances with no load. Meanwhile, the current at switch turn-on $I_{\_ZVS\_Q1\&Q2} = -2.595\text{A} < 0$ with very large DC inductance $L_{dc} = 50$ μH and $L_r = 1.5$ μH under full load, and the current at switch turn-on $I_{\_ZVS\_Q1\&Q2} = -21.39\text{A} < 0$ with very small DC inductance $L_{dc} = 3$ μH and $L_r = 1.5$ μH under full load, this means that switches Q_1 and Q_2 can achieve ZVS when $V_L = 24$ V under different DC inductances within full load range. The current at switch turn-on $I_{\_ZVS\_Q1\&Q2}$ is less than zero not only when leakage inductance $L_r = 1.5$ μH, but also when $L_r = 0.5$ μH and $L_r = 10$ μH. To sum up, Fig. 2.6a, b verify switches Q_1 and Q_2 can achieve ZVS under different DC and leakage inductances at full input voltage range within full load range.

For Q_{1a} and Q_{2a}, the ZVS of Q_{1a} and Q_{2a} is shown in Fig. 2.6c, d. As can be seen, the same theory could verify that switches Q_{1a} and Q_{2a} can achieve ZVS under different DC and leakage inductances at full input voltage range within full load range. For S_1, S_2, S_3 and S_4, ZVS of S_1, S_2, S_3 and S_4 is shown in Fig. 2.6e, f. As can be seen, since the junction capacitor is not considered, ZVS is not achieved at just

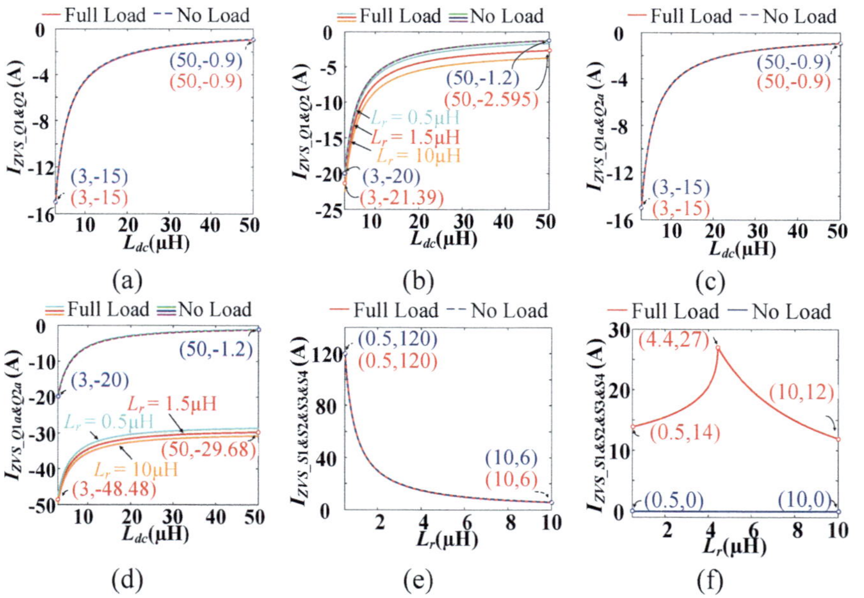

Fig. 2.6 Current at switch turn-on. **a** $Q_1\&Q_2$ $V_L = 12$ V. **b** $Q_1\&Q_2$ when $V_L = 24$ V. **c** $Q_{1a}\&Q_{2a}$ $V_L = 12$ V. **d** $Q_{1a}\&Q_{2a}$ when $V_L = 24$ V. **e** $S_1\&S_2\&S_3\&S_4$ when $V_L = 12$ V. **f** $S_1\&S_2\&S_3\&S_4$ when $V_L = 24$ V. ($V_L = 12$~24 V, $V_H = 240$ V, $P_N = 650$W)

one point for HVS switches, that is $V_L = 24$ V, $P = 0$. In addition, the same theory could verify that switches S_1, S_2, S_3 and S_4 can achieve ZVS under different DC and leakage inductances at other input voltages and load powers.

It has been proved that any inductance parameter can make the converter realize ZVS at full input voltage range within full load range, so how to determine the value of DC and leakage inductance becomes a key problem. Therefore, in order to determine the DC and leakage inductance, the maximum power transmission capability and inductor current stress need to be analyzed.

2.2.2 DC and Leakage Inductances Optimization

In this section, the parameter considerations for DC and leakage inductances are discussed with the goals of achieving both low inductor current stress and maximum power transfer. The related parameters are: $V_L = 12 \sim 24$ V, $V_H = 240$ V, $P_N = 650$ W. According to (2.4), the duty cycle range is 0.5~0.75. In this chapter, the design of converter is under the condition of rated load power. And the power loss is maximum at the rated power. As long as the converter can run safely and reliably when the maximum loss is met, the converter can run safely under different loss conditions in the full load range.

From the relationship between the transmission power and the phase shift angle ϕ in Fig. 2.5, when the phase shift angle $\phi = 0.5\pi$, the relationship between the transmission power versus the leakage inductance and the duty cycle is written as

$$P(L_r, d) = \frac{(n_1 V_H)^2 T_s\left[-8d^2 + 8d - 1\right]}{8n_2^2 L_r} \tag{2.22}$$

Figure 2.7 shows that the leakage inductance range for different d when the maximal power is equal to P_N. Consequently, based on Fig. 2.7, $L_r \leq 2.215$ μH should be satisfied to achieve maximum power transfer at the extreme case.

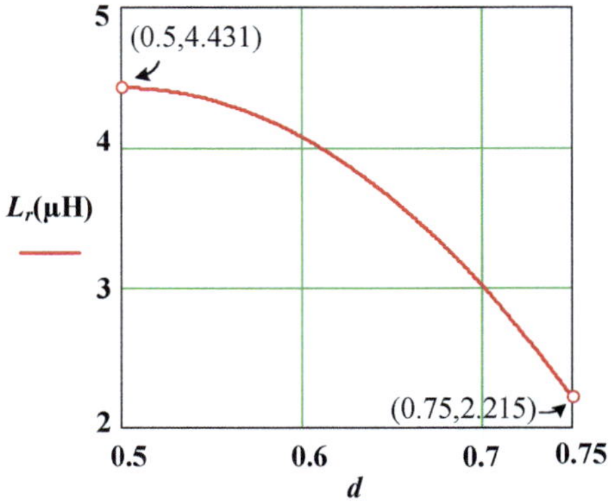

Fig. 2.7 Relation between transmission power to leakage inductance and duty cycle. ($\phi = 0.5\pi$, $P_N = 650$ W)

Under rated power P_N, the RMS current and peak current of the leakage inductor can be calculated as (2.23) and (2.24), respectively.

$$I_{Lr\_RMS} = \begin{cases} \frac{n_1 V_H T_s}{2n_2 \pi L_r}\sqrt{2\phi(1-d)[\phi-(2d-1)\pi]-\frac{\pi^2}{12}(2d-1)^2(4d-5)} & \left\{ \begin{matrix} V_{L\_min} \le V_L \le V_{L\_div} \\ \phi \in [0, (2d-1)\pi] \end{matrix} \right\} \\ \frac{n_1 V_H T_s}{n_2 \pi L_r}\sqrt{\frac{\phi\left[\phi(\pi d-\phi)+3\pi^2 d(1-2d)\right]}{6\pi}+\frac{\pi^2(2d-1)^2(4d-1)}{16}} & \left\{ \begin{matrix} V_{L\_div} \le V_1 \le V_{L\_max} \\ \phi \in [(2d-1)\pi, \pi] \end{matrix} \right\} \end{cases} \tag{2.23}$$

$$I_{Lr\_peak} = \begin{cases} \frac{P}{2V_L} & \phi \in [0, (2d-1)\pi] \\ \frac{n_1 V_H T_s}{4n_2 L_r} - \frac{\sqrt{\frac{n_1 V_H V_L}{n_2} T_s^2 - V_L^2 T_s^2 - 2L_r P T_s}}{2L_r} & \phi \in [(2d-1)\pi, \pi] \end{cases} \tag{2.24}$$

Under PPS modulation, the converter operates in two modes. Thus, the mode boundary of input voltage should be calculated. In order to determine the interval boundary of input voltage, taking $\phi = (2\ d–1)\ \pi$ into ($P = \begin{cases} \frac{(n_1 V_H^2)(1-d)[\phi-(d-0.5)\pi]T_s}{(n_2)^2 \pi L_r} & \phi \in [0, (2d-1)\pi] \\ \frac{(n_1 V_H^2)[-\phi^2+2d\pi\phi-(2d-1)\pi^2]T_s}{2(n_2)^2 \pi L_r} & \phi \in [(2d-1)\pi, \pi] \end{cases}$ (2.12), the output power with duty cycle as independent variable can be expressed as

$$P(d) = -\frac{(n_1 V_H)^2 (2d^2 - 3d + 1) T_s}{(n_2)^2 L_r} \tag{2.25}$$

Under rated power P_N, the interval boundary of duty cycle d can be calculated as

$$d_{div} = 0.75 - \frac{n_2}{4n_1 V_H T_s}\sqrt{\frac{T_s^2 n_1^2 V_H^2}{n_2^2} - 16 L_r P_N T_s} \tag{2.26}$$

Then, according to ($d_{div} = 0.75 - \frac{n_2}{4n_1 V_H T_s}\sqrt{\frac{T_s^2 n_1^2 V_H^2}{n_2^2}} - 16 L_r P_N T_s$ (2.26) and ($d = 1 - \frac{V_L}{V_{Cc}}$ (2.4) in the manuscript, the interval boundary of input voltage can be written as $V_{L\_div} = V_{Cc}\ (1–d_{div})$.

In addition, The RMS current of the DC inductor can be calculated as

$$I_{L\_RMS} = \sqrt{\frac{P_N^2}{4V_L^2} + \frac{V_L^2 T_s^2}{12L^2}\left(\frac{n_2 V_L}{n_1 V_H} - 1\right)^2} \tag{2.27}$$

Figure 2.8 illustrates the curves for the inductor current. Figure 2.8a, b illustrates the relationship between the RMS current, peak current and the input voltage for different leakage inductance, respectively. It should be noted that, unlike other control strategies, RMS current of the leakage inductor is related to the input voltage V_L. For example, when V_L is equal to 12 V, smaller leakage inductance leads to larger RMS current. When V_L is equal to 24 V, smaller leakage inductance leads to smaller

RMS current. Besides, lower leakage inductance will lead to smaller peak current at the full input voltage range.

As can be seen, when the leakage inductance L_r is equal to 0.5 μH, the RMS current increases significantly if the input voltage is low; when L_r is larger than 2 μH, the leakage inductor RMS currents change slightly. Considering that more windings will be needed to increase the inductance, the leakage inductance is designed to be 1.5 μH to make a tradeoff between RMS current and inductor losses.

According to the general engineering design, the DC current ripple is set in a range such as 15%–100%, with substitution of all the parameters, the DC inductance region is about 3–11 μH. Figure 2.9 illustrates the RMS current curves of the DC inductor. Similar to leakage inductance, considering the magnetic loss of the DC inductor and the RMS current, $L_{dc} = 5$ μH is finally selected for the DC inductance.

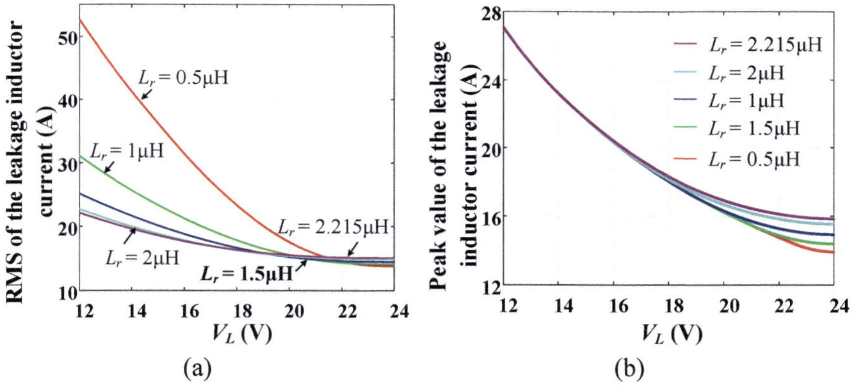

Fig. 2.8 Curves for current of the leakage inductor. **a** RMS currents under different leakage inductor values. **b** peak currents under different leakage inductor values

Fig. 2.9 RMS currents under different DC inductor values

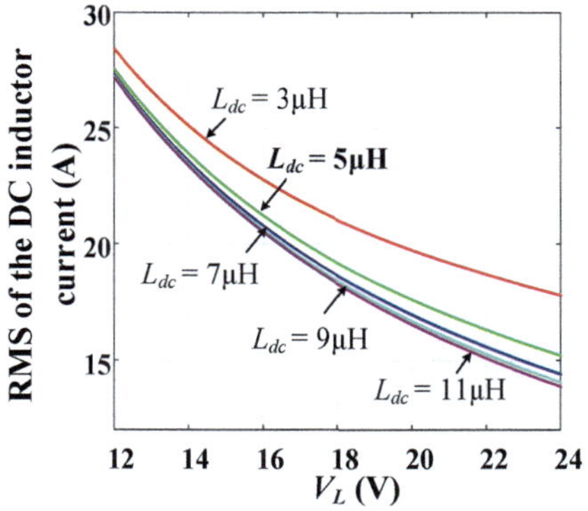

2.3 Comparison of Loss

In this section, the loss breakdown comparisons between the non-optimized case and the optimized design have been analyzed at both light load and full load. The total losses consist of four parts, namely driving loss, switches loss, inductor loss, and transformer loss. The loss breakdown of each part is analyzed as follows.

2.3.1 *Driving Loss*

Driving loss can be calculated as

$$\begin{aligned} P_{dri\_loss} =& P_{dri\_LVS} + P_{dri\_HVS} = 4\left(V_{gs\_on} - V_{gs\_off}\right)Q_{g\_LVS}f_s \\ &+ 4\left(V_{gs\_on} - V_{gs\_off}\right)Q_{g\_HVS}f_s \end{aligned} \tag{2.28}$$

where the positive voltage of driving signal V_{gs_on} is 15 V, the negative voltage of driving signal V_{gs_off} is −5 V. Device type IPP023N10N5 from Infineon company is selected for LVS switches, and the junction capacitor of LVS switch is 1810 pF. Meanwhile, device type UJ3C065030K3S from United-SiC company is selected for HVS switches, and the junction capacitor of LVS switch is 320pF. The gate charge of each LVS switch Q_{g_LVS} is 210nC, the gate charge each HVS switch Q_{g_HVS} is 50nC.

2.3.2 *Switches Loss*

(1) *Switches conduction loss*

Conduction loss can be calculated as

$$\begin{aligned} P_{con\_loss} =& P_{con\_LVS} + P_{con\_HVS} = 2\left(\frac{I_{D\_rms\_Q1a}}{\sqrt{2}}\right)^2 R_{DS\_LVS} \\ &+ 2\left(\frac{I_{D\_rms\_Q2}}{\sqrt{2}}\right)^2 R_{DS\_LVS} + 4\left(\frac{I_{D\_rms\_S1}}{\sqrt{2}}\right)^2 R_{DS\_HVS} \end{aligned} \tag{2.29}$$

The drain-source on-resistance is designed according to room temperature. $R_{DS\_LVS} = 2.3\ \text{m}\Omega$ is selected for on-resistance of LVS switch. $R_{DS\_HVS} = 27\ \text{m}\Omega$ is selected for on-resistance of HVS switch. The RMS current of drain current value of switches is given as the following.

$$
\begin{cases}
I_{D\_rms\_Q1a} = \sqrt{\frac{1}{(1-d)T_s}\left[\int_{t_3}^{t_6} (i_{Ldc1}(t) - i_{Lr}(t))^2 dt\right]} \\
I_{D\_rms\_Q2} = \sqrt{\frac{1}{dT_s}\left[\int_{t_1}^{t_8} (i_{Ldc2}(t) + i_{Lr}(t))^2 dt\right]} \\
I_{D\_rms\_S1} = \sqrt{\frac{2}{T_s}\left[\int_{t_5}^{t_{10}} i_{Lr}(t)^2 dt\right]}
\end{cases}
\tag{2.30}
$$

(2) *Switches turn-on loss*

According to the ZVS range analysis of all switches in III-A, the converter can achieve full load range ZVS with full input voltage range for all switches. Therefore, the switches turn-on loss can be considered as

$$
P_{turn-on\_loss} = 0 \tag{2.31}
$$

(3) *Switches turn-off loss*

The switching turn-off loss can be expressed as

$$
P_{turn-off\_loss} = f_s \begin{bmatrix}
2\int_0^{t_c}\left(I_{d\_Q1a} - \frac{I_{d\_Q1a}}{t_c}t\right)\left(\frac{V_{ds\_Q1a}}{t_c}t\right)dt + \\
2\int_0^{t_c}\left(I_{d\_Q2} - \frac{I_{d\_Q2}}{t_c}t\right)\left(\frac{V_{ds\_Q2}}{t_c}t\right)dt + \\
4\int_0^{t_c}\left(I_{d\_S1} - \frac{I_{d\_S1}}{t_c}t\right)\left(\frac{V_{ds\_S1}}{t_c}t\right)dt
\end{bmatrix} \tag{2.32}
$$

where I_{d_Q1a} and V_{ds_Q1a} are the current and drain-source voltage of switch Q_{1a} at turned off time, respectively. And t_c is the fall time of the switches.

2.3.3 Inductor Loss

(1) *Inductor core loss*

The empirical Steinmetz equation is applied to calculate the core loss of the inductor, which is given by

$$
P_{L\_core\_loss} = k_L f_s^{\alpha_L} B_{L\_p}^{\beta_L} V_{e\_L} \tag{2.33}
$$

where $k_L = 31.32$, $\alpha_L = 1.6$, and $\beta_L = 1.37$ are the Steinmetz coefficients provided by the manufacturer, V_{e_L} is the effective volume of the magnetic core, B_{L_p} is the peak flux density of the inductor, which is expressed as

$$B_{L\_p}=\frac{Li_{L\_\max}}{n_L A_{e\_L}} \tag{2.34}$$

where n_L is the turns of the inductor, A_{e_L} is the effective area of the magnetic core. Core material and size of inductors are shown as Table 2.1

(2) *Inductor copper loss*

The Litz wire is used as the copper material for the two DC inductors and leakage inductor in the converter. The single-turn high-frequency Litz wire is composed of 150 strands of enameled wire, in which the diameter of the single-strand enameled wire is 0.1 mm. And the inductor copper loss can be calculated as

$$P_{L\_copper\_loss} = I^2_{L\_dcrms\_p} R_{L\_dc} + I^2_{L\_acrms\_p} R_{L\_ac} \tag{2.35}$$

where R_{ac_L} is the copper AC resistance of the inductor considering the skin effect, which is expressed as

$$R_{L\_ac} = R_{L\_dc}\Delta\left[\frac{\sinh(2\Delta)+\sin(2\Delta)}{\cosh(2\Delta)-\cos(2\Delta)} + \frac{2\left(N_n N_L^2 - 1\right)}{3}\frac{\sinh(2\Delta)-\sin(2\Delta)}{\cosh(2\Delta)+\cos(2\Delta)}\right] \tag{2.36}$$

The first term in (2.36) describes the skin effect, the second term represents the proximity effect factor [6, 7]. N_n is the number of strands in a bundle for the Litz wire, N_L is the number of bundle layers, Δ is expressed as

$$\Delta = \left(\frac{\pi}{4}\right)^{0.75} \cdot \frac{d_w}{\delta} \cdot \sqrt{0.8} \tag{2.37}$$

d_w is the diameter of the strand for the Litz wire, δ is the skin depth which is calculated as

$$\delta = \sqrt{\frac{2}{2\pi f \mu_0 \sigma_0}} \tag{2.38}$$

R_{L_dc} is the copper DC resistance of the inductor, which is calculated as

Table 2.1 Core material and size of inductors

Inductors	Core Material	Path length L_e (mm)	Cross section A_e (mm^2)	Volume V_e (mm^3)
DC inductors	MPP	81.4 × 2	65.6 × 2	5340 × 2
Leakage inductor	DMR95	57	146	8322

$$R_{L\_dc} = \frac{4n_L l_L}{N_n \pi d_w^2 \sigma_0} \tag{2.39}$$

μ_0 is free-air magnetic permeability, l_L is the mean length per turn for the inductor, σ_0 is the electrical conductivity of the copper.

2.3.4 *Transformer Loss*

(1) *Transformer core loss*

The standard PQ cores and the Litz wire windings constitute the transformer structure. And the sandwich structure is applied to reduce the leakage inductor of transformer. Core material and size of transformer are shown as Table 2.2. The calculation method of the transformer core loss is the same to the inductor core loss, according to the empirical Steinmetz equation. The transformer core loss is calculated as

$$P_{T\_core\_loss} = \frac{k_T f_s^{\alpha_T} B_{T\_p}^{\beta_T} (x - yT + zT^2) V_{e\_T}}{1000} = P_{V\_T} V_e \tag{2.40}$$

where k_T, α_T, and β_T are the Steinmetz coefficients, and the P_{V_T} is power loss density. Both Steinmetz coefficients and power loss density are provided by the manufacturer. However, sometimes the manufacturer only provides the power loss density in the datasheet of transformer for convenience. V_{e_T} is the effective volume of the magnetic core, B_{T_p} is the peak flux density of the transformer, which is expressed as

$$B_{T\_p} = \frac{V_{Cc}(1 - D_{\max})}{n_T A_{e\_T} f_s} \tag{2.41}$$

where n_T is the primary side turns of the transformer, A_{e_T} is the effective area of the magnetic core.

(2) *Transformer copper loss*

The calculation method for the transformer copper loss is the same to the inductor copper loss. The transformer AC resistance is expressed as

Table 2.2 Core Material and Size of Transformer

Core material	Mn-Zn power ferrite	l_{Width} (mm)	28
l_{Length} (mm)	40	l_{Height}(mm)	40
Effective area (mm^2)	210	Effective volume (mm^3)	19,698

Table 2.3 Current Parameters Comparison

Item	Optimized design		Non-optimized case	
Power load	Full load (0.65 kW)	Light load (65 W)	Full load (0.65 kW)	Light load (65 W)
DC inductor RMS current i_{Ldc1} or i_{Ldc2} (A)	27.7	6.1	27.7	6.1
Leakage inductor RMS current i_{Lr} (A)	25.23	16.49	31.24	24.57

$$R_{T\_ac} = R_{T\_dc}\Delta\left[\frac{\sinh(2\Delta)+\sin(2\Delta)}{\cosh(2\Delta)-\cos(2\Delta)} + \frac{2\left(N_nN_L^2-1\right)}{3}\frac{\sinh(\Delta)-\sin(\Delta)}{\cosh(\Delta)+\cos(\Delta)}\right] \tag{2.42}$$

Then the transformer copper loss is expressed as

$$P_{T\_copper} = I_{LVS\_rms}^2 R_{ac\_T\_LVS} + I_{HVS\_rms}^2 R_{ac\_T\_HVS} \tag{2.43}$$

where $R_{ac\_T\_LVS}$ and $R_{ac\_T\_HVS}$ are the transformer AC resistance of the primary side winding and secondary side winding, respectively.

2.3.5 *Current Parameters Comparison*

The current parameters comparison between optimized design and non-optimized case under input voltage $V_L = 12$ V are as shown in Table 2.3.

2.3.6 *Power Loss Comparison*

To verify the effectiveness of the optimization design, parameters with $L_r = 1\ \mu$H is used to be the non-optimized case to make the comparison. The loss breakdown comparisions of the converter between the optimized design and non-optimized case at different loads are shown in Fig. 2.10. As Fig. 2.10a shows, when the converter works at light load (120W), the total loss of the non-optimized case is highter than that of the optimized design about 3.89 W. As Fig. 2.10b shows, when the converter works at full load condition (650 W), the power loss with the non-optimized case is highter than that of the non-optimized case about 6.94 W. Due to the peak current is high when the switch is turned off and the frequency of switches is high, the turn-off loss is high according to (2.32). The turn-off loss consists of turn-off of

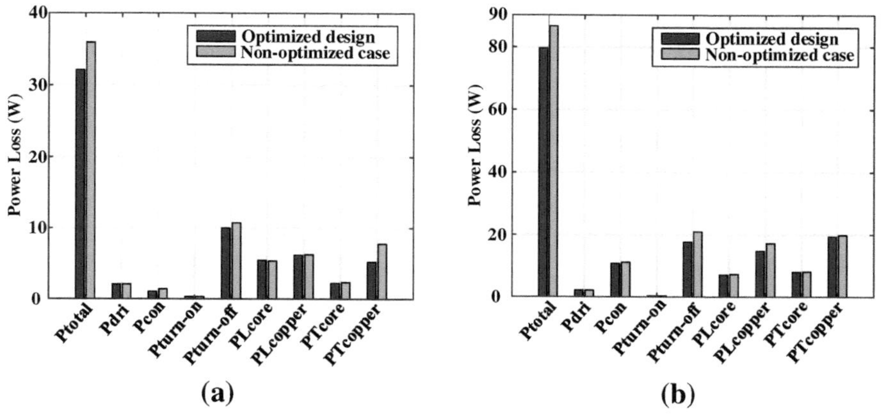

Fig. 2.10 Loss breakdown when $V_L = 12$ V. **a** at light load ($P = 120$ W). **b** at full load ($P = 650$ W)

LVS switches and turn-off of HVS switches. And the turn-off currents of the non-optimized case for primary and secondary sides are larger than the optimized design at the switches turn-off time. Thus, the turn-off loss of non-optimized case is large than the optimized design from Fig. 2.10. According to Fig. 2.8a, RMS currents are different under different values of leakage inductor although ZVS of all switches are achieved independent of leakage inductor. As can be seen, even though ZVS of all switches can be realized, RMS current still decreases with the value of leakage inductance increases near $V_L = 12$ to 22 V. As shown, RMS current with leakage inductance $L_r = 1\ \mu$H is larger than $L_r = 1.5\ \mu$H near $V_L = 12$ to 22 V and peak current of the leakage is also smaller near $V_L = 22$ to 24 V. Switch conduction loss, leakage inductor loss and transformer loss are closely related to RMS current and peak current of leakage inductor. Thus, the total loss of optimized design is smaller than non-optimized case when $V_L = 12$ V and $V_L = 18$ V. The conclusions coincide with the comparison of efficiency curves between optimized design and non-optimized case of the converter from Fig. 2.21.

2.3.7 Comparison Analysis of Bidirectional Converter

The characteristic comparison of related literatures on bidirectional isolated converters is shown in Table 2.4. n_{MOSFETs} is the number of power switches, n_{inductor} is the number of inductors, $n_{\text{transformer}}$ is the number of transformer, and $n_{\text{capacitor}}$ is the number of capacitors. The converter in [8] has a large input current ripple and high current stress of LVS switches although it has the small voltage surges on switches by the function of leakage inductance energy recovery. The converter in [5] uses voltage multiplier cells to achieve a high conversion ratio, but the analysis of ZVS is complex. The converter in [9] has the disadvantage of large circulation current due to

Table 2.4 Comparison of Related Literatures on Bidirectional Converters

Topology	CF-DAB	Reference [8]	Reference [5]	Reference [9]	Reference [10]	Reference [11]
$n_{MOSFETs}$	8	6	10	4	8	8
$n_{inductor}$	3	2	5	2	4	4
$n_{transformer}$	1	1	1	1	1	1
$n_{capacitor}$	2	3	5	5	2	2
$G_{step\text{-}up}$ (V_H/ V_L)	$n_2/n_1(1-d)$	$n_1/n_2(1-d)$^2	$n_1+n_2/n_2(1-d)$	$n_2/n_1(1-d)$	$n_2/n_1(1-d)$	$n_1\sin(D_L\pi)/D_L\sin(D_H\pi)$
Current ripple of V_L	Small	Medium	Small	Medium	Small	Small
ZVS analysis complexity	Low	Low	High	Low	Medium	Medium
Soft switching range	Wide	Wide	Wide	Narrow	Wide	Wide
Control complexity	Low	Low	Medium	Low	High	Medium
Parameter design complexity	Low	Low	Medium	Low	Medium	High

mismatching control of half bridge on primary, secondary sides. In addition, the ZVS range of switch is narrow. ZVS can be achieved for all power switches throughout full range of load even at no-load condition in [10]. However, the working modes are multifarious and control of converter is complex. The converter in [11] can achieved high efficiency. However, the design of resonant tank is difficult.

2.4 System Parameters Design

An experimental prototype has been built in order to verify the effectiveness of the analyses, and it is shown in Fig. 2.11. The system specifications are illustrated in Table 2.5. All the design parameters are designed taking into consideration the safe operation of converter for highest loss, i.e., 12 V LVS voltage and 650 W output power.

2.4.1 DC Inductance L_{dc} Design

Firstly, considering the ZVS performance of the LVS switches. According to Fig. 2.6a, d, the appropriate value range [3 μH, 50 μH] of L_{dc} can be preliminarily selected. Secondly, the RMS current should be discussed based on the allowed current

Fig. 2.11 The experimental prototype

Table 2.5 System Specifications

P	650 W	L_r	1.5 μH
f	100 kHz	L_{dc}	5 μH
V_L	12 ~ 24 V	n	5
V_H	240 V	C_c	60 μF

ripple to obtain the optimized inductance L_{dc} from Fig. 2.9. Taking the ZVS design of the LVS switches, current ripple and RMS current into consideration, L_{eq} is finally designed to be 5 μH.

2.4.2 *Turns Ratios of Transformer Design*

The turn ratio of transformer should be designed to ensure the voltage matching control, that is the voltage of clamping capacitor V_{Cc} matches the output voltage V_H. The clamping voltage $V_{Cc} = 48$ V and the output voltage $V_H = 240$ V. According to (2.5), the turn ratio of transformer is $n = n_2/n_1 = V_H/V_{Cc} = 5$.

2.4.3 *Leakage Inductance L_r Design*

Firstly, the ZVS performance of the HVS switches is affected by the value of the leakage inductance L_r. According to Fig. 2.6e, f, the appropriate value range [0.5 μH, 10 μH] of L_r can be preliminarily selected. Secondly, the leakage inductor is closely related to the output power. Thus, $Lr \leq 2.215$ μH should be satisfied to achieve maximum power transfer form Fig. 2.7. Finally, based on the analysis of RMS current and peak current from Fig. 2.8, the optimized leakage inductor is

Fig. 2.12 The diagram of leakage inductor

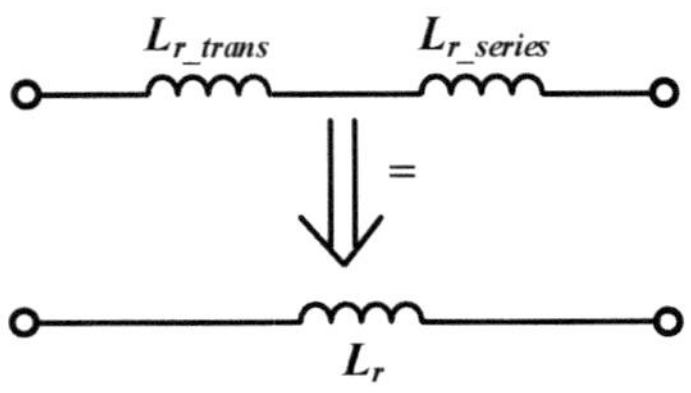

designed by the tradeoff between RMS current and copper loss. Based on considering the ZVS range, maximum power transfer capability, RMS current and peak current, the leakage inductance $L_r = 1.5\ \mu$H is finally selected.

In the experimental prototype of this chapter, the leakage inductor is placed in HVS of converter, that is $L_{rs} = (n_2/n_1)$ ^2 L_r. Through measurement by impedance analyzer, the leakage inductance parameter of primary side and secondary side for the sandwich structural transformer are $L_{r_trans} = 0.38\ \mu$H and $L_{rs_trans} = 8.3\ \mu$H, respectively. Then, the required leakage inductance value is achieved by series connection of an external inductor, that is the leakage inductance L_{rs_trans} of the transformer is compensated by an external series inductance L_{rs_series} to realize the required leakage inductance L_{rs}, as shown in Fig. 2.12.

2.4.4 Clamping Capacitor C_c Selection

The selection of the clamping capacitor C_c should ensure that the voltage ripple across C_c is confined within an allowed range. Due to the complementary of the LVS modulation, only half a switching cycle $[t_0, t_6)$ is analyzed. The clamping voltage $v_{Cc}(t)$ can be calculated by

$$v_{Cc}(t) = \frac{1}{C_c} \int_{t_3}^{t} (i_{Ldc1}(t) - i_{Lr}(t))dt + v_{Cc}(t_3) \tag{2.44}$$

The current i_{Ldc1}–i_{Lr} will charge/discharge Cc. During $[t_3, t_4)$, i_{Ldc1}–i_{Lr} is larger than zero, Cc will be charged and v_{Cc} will rise; while during (t_4, t_6), i_{Ldc1}–i_{Lr} is smaller than zero, Cc is discharged and v_{Cc} will fall. The maximum of v_{Cc} is at t_4, and minimum one is either at t_3 or t_6. So the voltage ripple ΔV_{Cc} across Cc can be illustrated as

$$\Delta V_{Cc} = v_{Cc}(t_4) - \min\{v_{Cc}(t_3), v_{Cc}(t_6)\} \tag{2.45}$$

ΔV_{Cc} is designed to satisfy the following criteria:

$$\Delta V_{Cc}/V_{Cc} \le 1\% \tag{2.46}$$

Calculation for the case of 12 V input and 0.65 kW load, the critical value for Cc can be obtained

$$C_c \geq 32.7\ \mu F \tag{2.47}$$

A near 80% margin is designed. Thus, the clamping capacitor is selected to be 60 μF.

2.4.5 Swithes Selection

The maximum voltage across the LVS switches $V_{Q1\_Q1a\_Q2\_Q2a}$ is the clamping voltage V_{Cc}, that is $V_{Q1\_Q1a\_Q2\_Q2a} = 48$ V. And the maximum voltage across the HVS switches $V_{S1\_S2\_S3\_S4}$ is the output voltage V_H, that is $V_{S1\_S2\_S3\_S4} = 240$ V. The current stresses of LVS switches $I_{Q1} = i_{Ldc1}(t_1)–i_{Lr}(t_1) = 63.4$ A. And the current stresses of HVS switches $I_{S1} = i_{Lr}(t_5)/n_1 = 8$ A. According to the voltage, current stresses and margin should be considered, device type IPP023N10N5 from Infineon company is selected for the LVS switches, and device type UJ3C065030K3S from United-SiC company is selected for the HVS switches.

2.5 Experiment Verification

Figures 2.13, 2.14 and 2.15 describe the steady-state waveforms under different loads with different LVS voltages in boost mode. As seen in Fig. 2.13, the converter works at light load under different voltage conditions, in spite of LVS voltage variation, voltage matching can be achieved, that is, the slope of leakage inductor current i_{Lr} is zero in the power transmission stage. The half load and full load steady-state waveforms are shown in Figs. 2.14 and 2.15 respectively. The phase shift angle exists between v_{ab} and v_{cd}, and it varies with load and LVS voltage V_L. The HVS voltage V_H is fixed in spite of the different loads and V_L.

Figures 2.16, 2.17 and 2.18 show the ZVS waveforms of the switches when LVS voltage $V_L = 12$ V under different loads. As can be seen, ZVS waveforms of the Q_{1a} and Q_{2a} under light load, half load and full load are shown in Fig. 2.16a–c respectively, where the drain-source voltage v_{dsQ1a}, date-source voltage v_{gsQ1a} and current i_{Lr}, i_{Ldc1} are given. As seen, ZVS of the LVS upper switches Q_{1a} and Q_{2a} can be achieved at full load range. ZVS waveforms of LVS bottom switches and HVS switches are shown in Figs. 2.17 and 2.18 respectively. Q_1 and Q_2, S_1–S_4 can achieve ZVS, respectively. Thus all switches can achieve ZVS at full load range when LVS voltage $V_L = 12$ V.

Figures 2.19, 2.20 and 2.21 show the ZVS waveforms of the switches when LVS voltage $V_L = 24$ V under different load. As can be seen, all switches can achieve

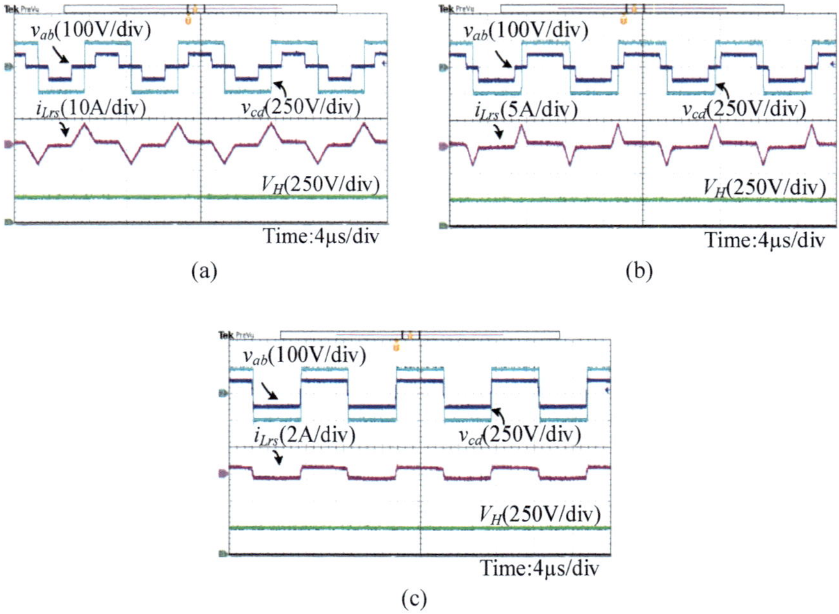

Fig. 2.13 Steady-state waveforms under light load. **a** 12 V. **b** 18 V. **c** 24 V

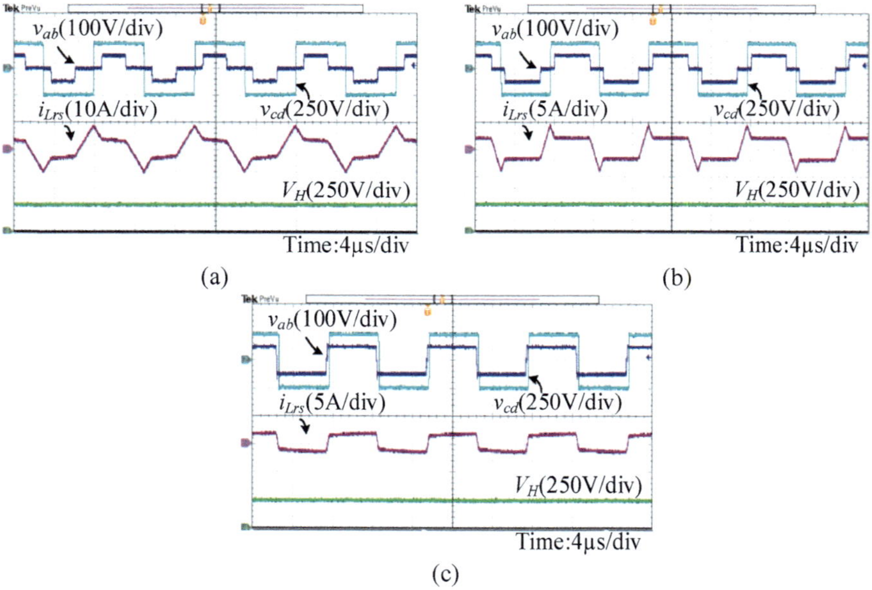

Fig. 2.14 Steady-state waveforms under half load. **a**12 V. **b** 18 V. **c** 24 V

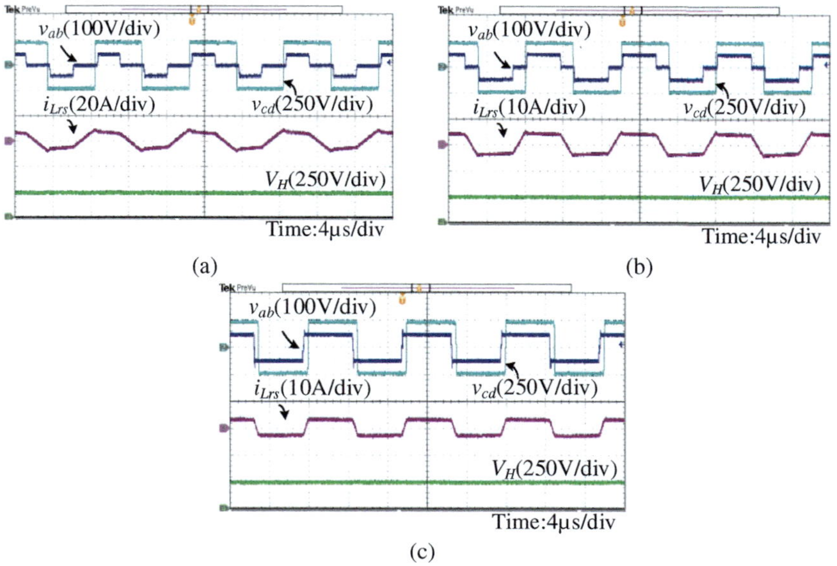

Fig. 2.15 Steady-state waveforms under full load. **a** 12 V. **b** 18 V. **c** 24 V

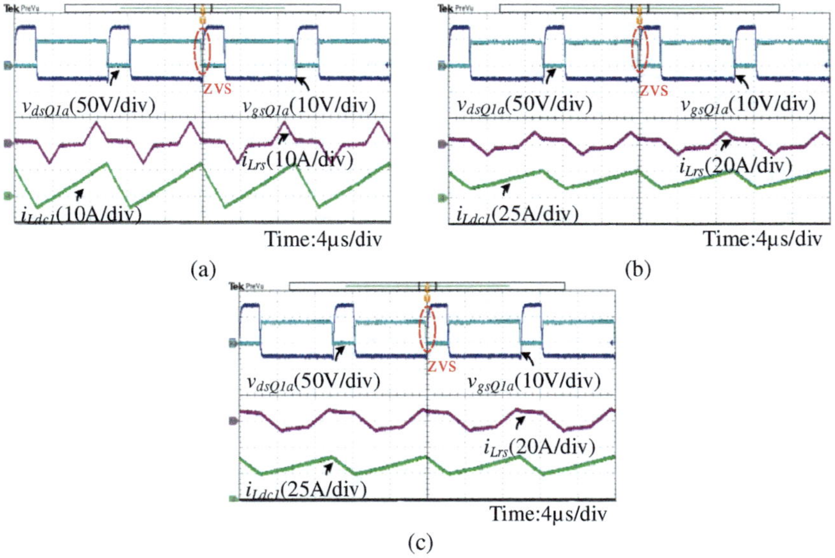

Fig. 2.16 Soft-switching waveforms for Q_{1a}&Q_{2a} when $V_L = 12$ V. **a** light load. **b** half load. **c** full load

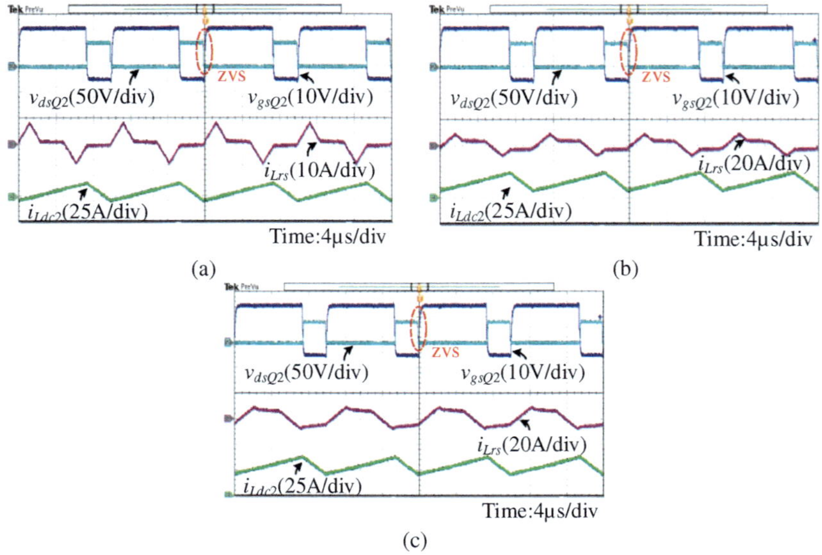

Fig. 2.17 Soft-switching waveforms for Q_1&Q_2 when $V_L = 12$ V. **a** light load. **b** half load. **c** full load

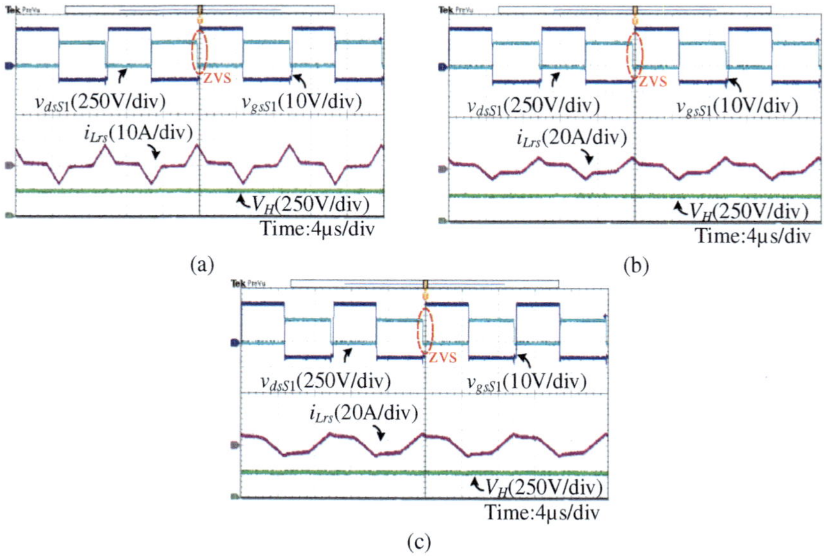

Fig. 2.18 Soft-switching waveforms for S_1&S_2&S_3&S_4 when $V_L = 12$ V. **a** light load. **b** half load. **c** full load

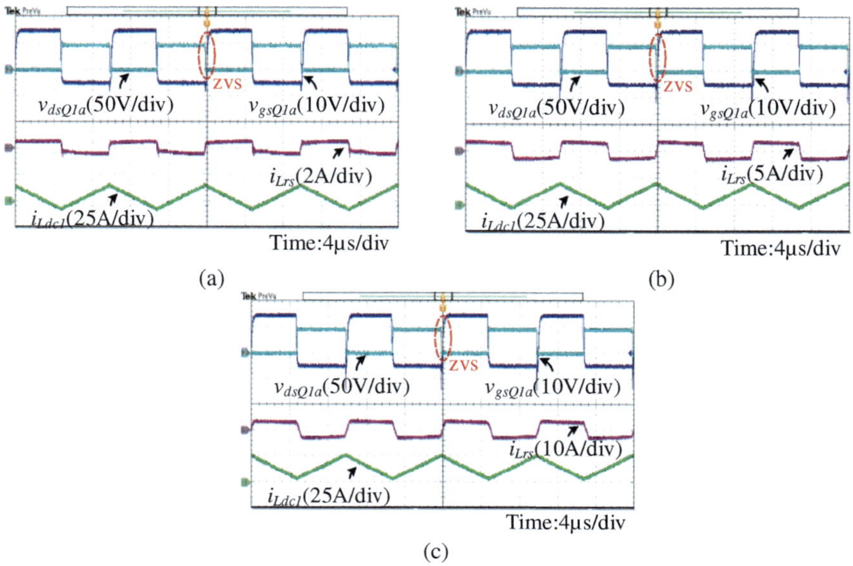

Fig. 2.19 Soft-switching waveforms for Q_{1a}&Q_{2a} when $V_L = 24$ V. **a** light load. **b** half load. **c** full load

ZVS at full load range when LVS voltage $V_L = 24$ V. Thus all switches can achieve ZVS at full load rang with full input voltage range.

The efficiency curves of the prototype using the optimized design and non-optimized case with different LVS voltage and different loads are plotted in Fig. 2.22. The highest efficiency of the optimized design reaches 96% under the 650 W load when $V_L = 24$ V. As can be seen, at $V_L = 12$ V, when working at light load and full load, the optimized design has better efficiency than the non-optimized case with the different LVS input voltage, which is consistent with the loss breakdown in section IV-F. When $V_L = 24$ V, due to the smaller RMS current of leakage inductor, the efficiency of the non-optimized case is slightly higher than the optimized design, but the difference is almost negligible. When the input voltage is 12 V or 18 V, it can be seen that the efficiency of the optimized design is higher than that of the non-optimized case.

2.6 Conclusion and Discussion

In this chapter, the working principle of CF-DAB converter under PPS control is analyzed in detail. The voltage matching condition can always be achieved when input voltage varies, which can reduce circulating conduction loss. The ZVS conditions of all switches of the converter are given, and it is proved mathematically that all switches can naturally achieve ZVS independently of DC inductance and leakage

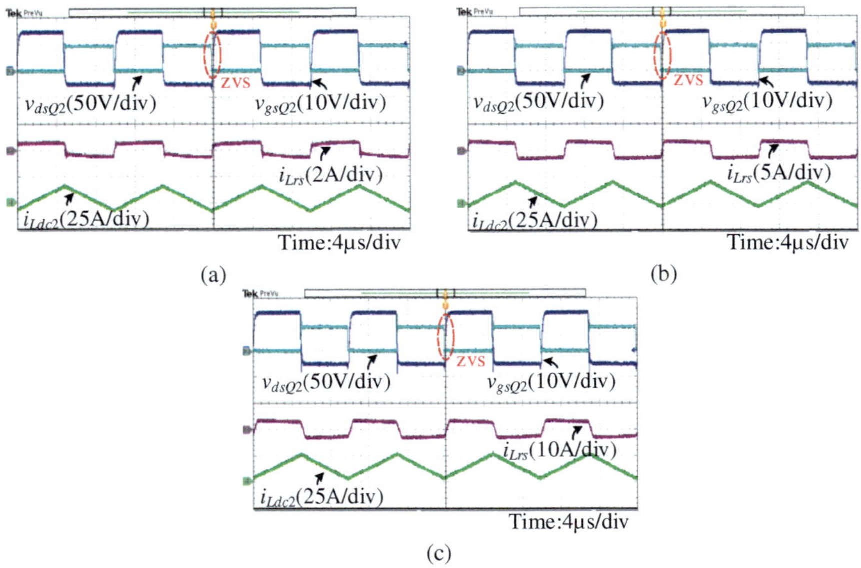

Fig. 2.20 Soft-switching waveforms for $Q_1\&Q_2$ when $V_L = 24$ V. **a** light load. **b** half load. **c** full load

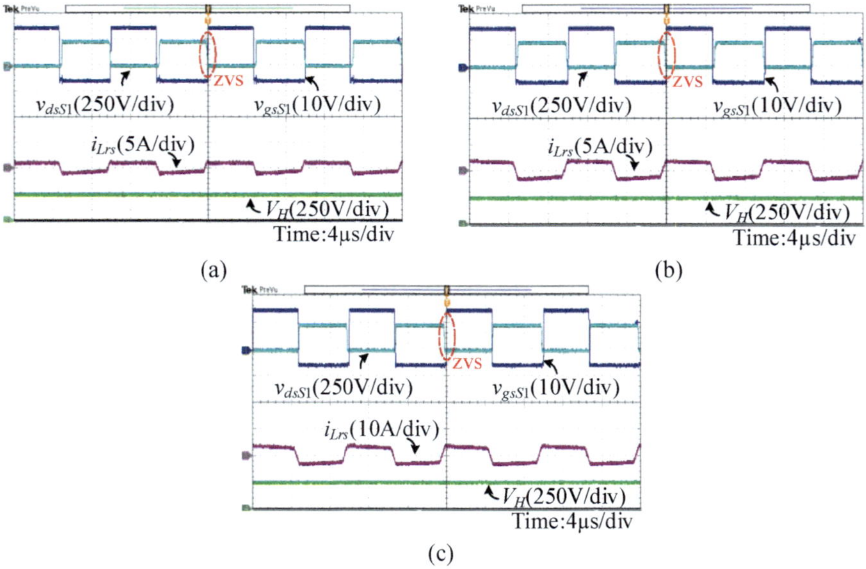

Fig. 2.21 Soft-switching waveforms for $S_1\&S_2\&S_3\&S_4$ when $V_L = 24$ V. **a** light load. **b** half load. **c** full load

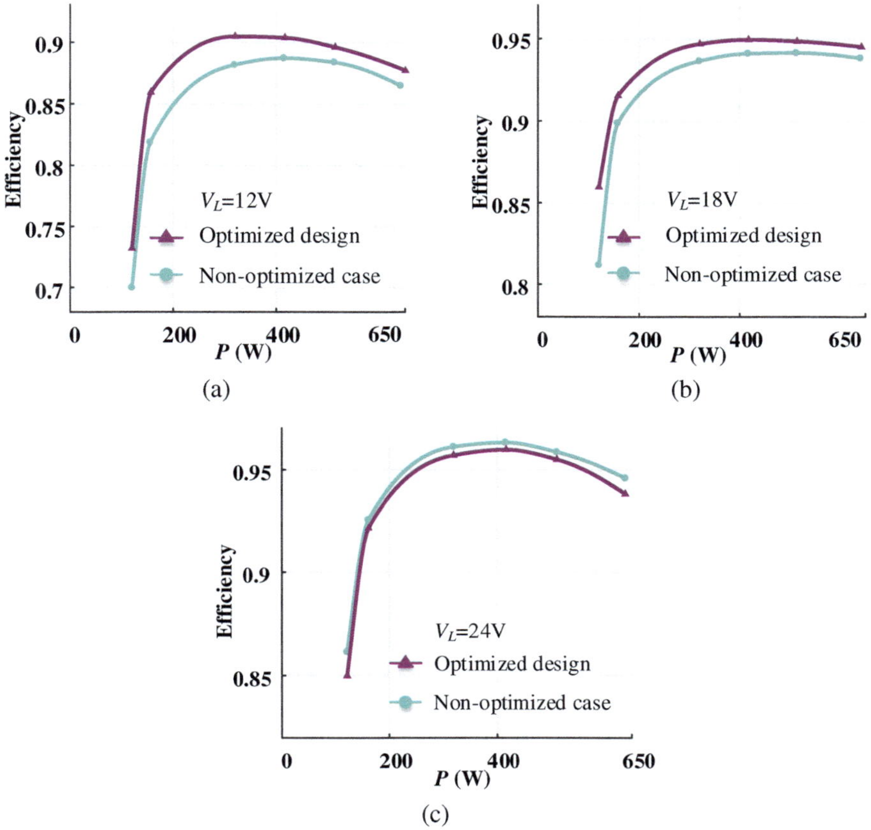

Fig. 2.22 Comparison of efficiency curves between optimized design and non-optimized case of the converter. **a** when $V_L = 12$ V. **b** when $V_L = 18$ V. **c** when $V_L = 24$ V

inductance under full input voltage range with full load range. In addition, the design of leakage and DC inductances are discussed to achieve relatively low RMS current with the precondition of achieving maximum power transfer. The power loss of the converter of the optimized design is analyzed and total power loss is reduced compared with non-optimized case. The experiment results of the prototype verified the validity of the theoretical analysis. In the future work, the junction capacitor can be added to further analyze the ZVS of all switches.

References

1. Xiao, H., & Xie, S. (2008). A ZVS bidirectional DC–DC converter with phase-shift plus PWM control scheme. *IEEE Transactions on Power Electronics, 23*(2), 813–823.
2. Sha, D., Chen, D. & Zhang, J. (2018). A bidirectional three-level DC–DC converter with reduced circulating loss and fully ZVS achievement for battery charging discharging. *IEEE Journal Emerging Selected Topics Power Electronics, 6*(2), 993–1003.
3. Guo, J., Han, H., Xu, G., Cai, Z., Wang, H., Sun, Y., & Su, M. (2020). Design considerations for PPS controlled current-fed DAB converter to achieve full load range ZVS and low inductor current stress. *Energy Conversion Congress and Exposition* (pp. 5971–5975). IEEE.
4. Chen, L., Tarisciotti, L., Costabeber, A., Guan, Q., Wheeler, P., & Zanchetta, P. (2019). Phase-shift modulation for a current-fed isolated DC–DC converter in more electric aircrafts. *IEEE Transactions on Power Electronics, 34*(9), 8528–8543.
5. Chaudhury, T., & Kastha, D. (2020). A high gain multiport DC–DC converter for integrating energy storage devices to dc microgrid. *IEEE Transactions on Power Electronics, 35*(10), 10501–10514.
6. Hurley, W. G., Gath, E., & Breslin, J. G. (2000). Optimizing the AC resistance of multilayer transformer windings with arbitrary current waveforms. *IEEE Transactions on Power Electronics, 15*(2), 369–376.
7. Ouyang, Z., Thomsen, O. C., & Andersen, M. A. E. (2012). Optimal design and tradeoff analysis of planar transformer in high-power DC–DC converters. *IEEE Transactions on Industrial Electronics, 59*(7), 5529–5538.
8. Wu, Y., & Ke, Y. (2021). A novel bidirectional isolated DC–DC converter with high voltage gain and wide input voltage. *IEEE Transactions on Power Electronics, 36*(7), 7973–7985.
9. Hui, L., Fang Zheng, P., & Lawler, J. S. (2003). A natural ZVS medium-power bidirectional DC–DC converter with minimum number of devices. *IEEE Transactions on Industry Applications, 39*(2), 525–535.
10. Sha, D., Wang, X., & Chen, D. (2018). High-Efficiency current-fed dual active bridge DC–DC converter With ZVS achievement throughout full range of load using optimized switching patterns. *IEEE Transactions on Power Electronics, 33*(2), 1347–1357.
11. Gu, L., Zhang, X., & Li, P. (2021). Hybrid-PWM-controlled current-fed bidirectional series resonant converter with low current ripple and wide voltage gain. *IEEE Transactions on Industrial Electronics, 68*(8), 7125–7136.

Chapter 3
Dead-Time Optimization and Magnetizing Current Design for a Current-Fed Dual Active Bridge DC–DC Converter

Abstract This chapter introduces the dead-time optimization methods and the magnetizing current design for a current-fed dual active bridge (CF-DAB) converter under the pulse-width plus phase-shift (PPS) control to achieve full load range zero-voltage-switching (ZVS) in a wide voltage range. In the ZVS analysis, the influence of junction capacitance and dead time is considered, including the dead-time effect on leakage inductance current, which can effectively explain that the CF-DAB converter under PPS control is difficult to achieve ZVS at light load in practice. Based on the analysis, it is first revealed that the CF-DAB converter under PPS control is not sufficient to achieve ZVS only by adjusting the dead time. Based on the new ZVS model and analysis of the dead-time effect, the magnetizing inductance design and dead-time optimization without adding extra control degrees of freedom and auxiliary components are proposed to achieve the full load range ZVS and improve the overall efficiency of the converter in a wide voltage range. Besides, to obtain the low root-mean-square (RMS) current, the design of DC inductance and leakage inductance are also investigated. Finally, a 1 kW prototype is built to verify the effectiveness of the proposed solution.

Keywords Current-Fed dual active bridge (CF · DAB) converter · Zero voltage switching (ZVS) · Dead time effect · Pulse width plus phase shift (PPS) control

3.1 Converter Topology and Operation Principle

Current-fed dual active bridge (CF-DAB) converters are widely used because of the small input current ripple and high-power density. In [1], a current-fed half-bridge bidirectional DC–DC converter with single phase shift (SPS) control to reduce ripple current is proposed. However, there are large voltage spikes when the switch is turned off. Therefore, A CF-DAB converter with an active clamp network [2] is proposed to solve this problem. And under PPS control, the converter can reduce the circulating current loss. In addition, the research of ZVS technology plays an important role in CF-DAB converters, especially for applications with a high switching

G. Xu and X. Chen, *Topology Deduction and Control for Dual-Active-Bridge Converters*, https://doi.org/10.1007/978-981-97-4856-3_3

frequency, as the development of fast switching wide-bandgap devices like GaN [3]. However, in ZVS analysis, the junction capacitors and dead time are usually ignored, which impacts the practical performance of the converter [4]. The ZVS of CF-DAB converters for high-voltage side switches under light load usually is difficult to realize and requires additional control degrees of freedom to achieve [5].

In this Chapter, to achieve ZVS of the CF-DAB converter under full load range, a dead-time optimization and magnetizing current design are proposed. The mechanism of the dead time effect on the ZVS region is investigated for CF-DAB converters under PPS control for the first time, including considering the variation of leakage inductance current during the dead time.

3.1.1 Topology and Modulation Schedule Using PPS Control

Figure 3.1 illustrates the topology of the CF-DAB converter. L_1 and L_2 are used as the DC boost inductors on the current-fed side. C_c is the clamping capacitor. d is defined as the duty cycle of the main switches Q_1 and Q_2. The active clamping circuit is composed of Q_{1a} and Q_{2a}. L_r represents the leakage inductor of the primary side of the transformer. In the high-voltage side (HVS), S_3, S_4, C_a, and C_b constitute the half-bridge converter. It should be noted that the magnetizing inductance is also drawn in the figure since the magnetizing current will be used to enlarge the ZVS region in Sect. 3.3.

The phase shift angle φ is defined to be positive when v_{ab} is leading to v_{cd} in phase. Adjusting the phase shift angle φ can control the magnitude and direction of transmission power. The converter works in the boost mode when energy flows from the low-voltage side (LVS) to the HVS, and when energy flows reversely is defined as buck mode. Figure 3.2 shows the main waveforms in one switching cycle under two operation modes. Owing to the symmetrical topology, the working principles in buck mode are similar to those in boost mode.

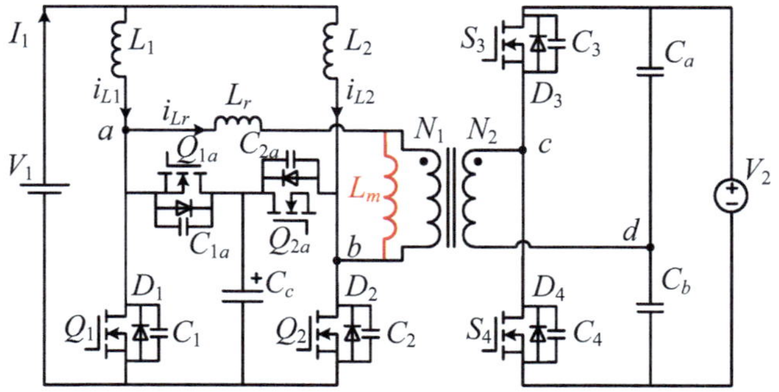

Fig. 3.1 The topology of the CF-DAB converter

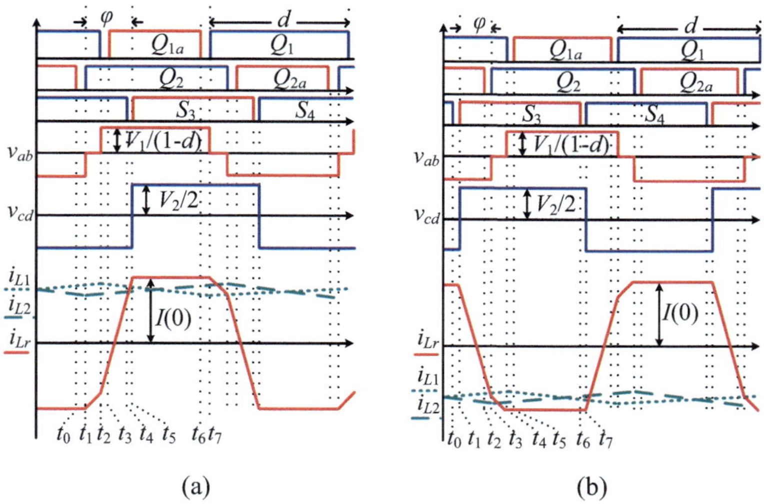

Fig. 3.2 Steady-state waveforms of the converter. **a** Key waveforms in boost mode. **b** Key waveforms in buck mode

3.1.2 Operating Modes of the Converter

Taking the boost mode shown in Fig. 3.2a as an example, due to the symmetry of the topology and working process, only half of the switching cycle is analyzed next, and the circuit operating modes at different times are shown in Fig. 3.3.

Stage 1 (Before t_0) [Fig. 3.3a]: Q_1, Q_{2a} and S_4 conduct. The converter is in the power transfer stage (power flows from the LVS to the HVS) and $i_{Lr} = -I_{Lr}(0)$.

Stage 2 (t_0, t_1) [Fig. 3.3b]: Q_{2a} is turned off at t_0. L_r, C_2 and C_{2a} begin to resonate.

Stage 3 (t_1, t_2) [Fig. 3.3c]: D_2 conduct at t_1. The voltage of C_{2a} is clamped at V_{Cc}. Q_2 can be turned on under zero voltage. During this stage, the current of the leakage inductance can be expressed as

$$i_{Lr} = -I(0) + \frac{N_1 V_2}{2N_2 L_r} \cdot t \tag{3.1}$$

Stage 4 (t_2, t_3) [Fig. 3.3d]: Q_1 is turned off at t_2. L_r, C_1 and C_{1a} begin to resonate.

Stage 5 (t_3, t_4) [Fig. 3.3e]: At t_3, Q_{1a} is turned on under zero voltage. During this stage, the leakage inductance current rises to a positive value.

$$i_{Lr} = -I(0) + \frac{N_1 V_2 (2d-1)\pi}{2N_2 \omega L_r} + \frac{N_1 V_2 [2t - (2d-1)T_s]}{2N_2 L_r} \tag{3.2}$$

Stage 6 (t_4, t_5) [Fig. 3.3f]: S_3 conduct at t_4, and S_3 can achieve ZVS. Similar to Stage 1, the converter is in the power transfer stage, and $i_{Lr} = I_{Lr}(0)$.

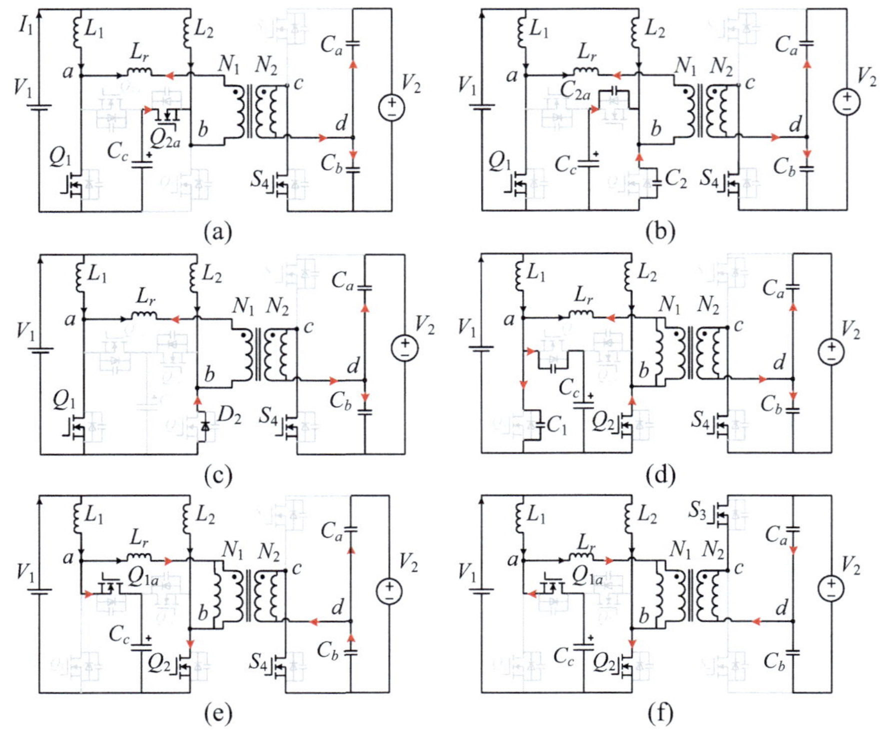

Fig. 3.3 Operation stages in the boost mode I. **a** Stage 1. **b** Stage 2. **c** Stage 3. **d** Stage 4. **e** Stage 5. **f** Stage 6

3.2 ZVS Analysis Considering Junction Capacitors and Dead Time

Compared with the traditional ZVS analysis, this section will analyze the influence of junction capacitors and dead time. For the convenience of analysis, the nonlinear junction capacitance is regarded as a charge-equivalent linear value.

3.2.1 ZVS Analysis Considering Junction Capacitors

In the converter, soft switching is realized by the resonance of the inductor and the junction capacitor. There needs to be a minimum current in the leakage inductor to discharge the voltage on the junction capacitor to be zero. The charging and discharging process of the parasitic capacitor is analyzed in detail for boost mode one ($\varphi \in [(2d-1)\pi, \pi]$). Figure 3.4 illustrates equivalent circuits of the converter during resonance.

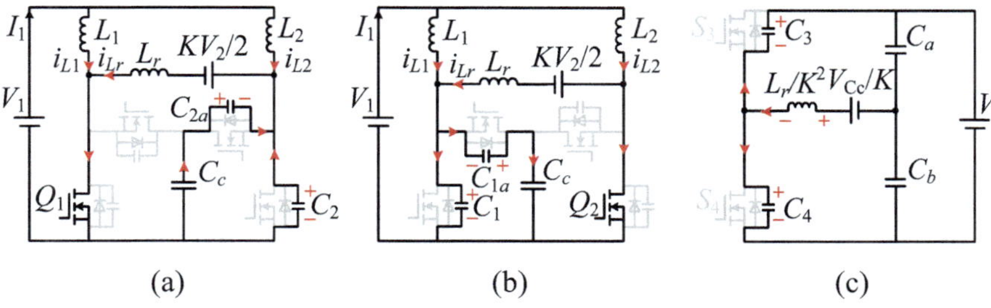

Fig. 3.4 Equivalent circuits for boost mode one. **a** t_0–t_1. **b** t_2–t_3. **c** t_4–t_5

Taking Fig. 3.4a as an instance, Q_{2a} is turned off at t_0. During this time, the sum of i_{Lr} and i_{L2} charges C_{2a} and discharges C_2. The dynamic equation of the equivalent circuit can be described as

$$\begin{cases} L_r\frac{di_{Lr}(t)}{dt} - \frac{KV_2}{2} = -u_{C2}(t) \\ i_{Lr}(t) = (C_2 + C_{1a})\frac{du_{C2}(t)}{dt} - i_{L2} \\ u_{C2}(0) = V_{Cc} \\ i_{Lr}(0) = I_{Lr\_t0} \end{cases} \tag{3.3}$$

By simplifying (3.3), the voltage expression of junction capacitor C_2 can be given by

$$u_{C2}(t) = \sqrt{\frac{L_r}{C_2 + C_{1a}}}(I_{Lr\_t0} + i_{L2})\sin(\sqrt{\frac{1}{L_r(C_2 + C_{1a})}} \cdot t) + \frac{KV_2}{2} \tag{3.4}$$

where I_{Lr_t0} is the value of the leakage inductance current at t_0.

To realize the zero-voltage conduction of the switch Q_2, the voltage of the junction capacitor C_2 should be zero before Q_2 turns on. And the current to achieve ZVS of Q_2 should satisfy as

$$I_{ZVS\_Q_1\&Q_2} = I_{Lr\_t0} + i_{L2} + \frac{KV_2}{2}\sqrt{\frac{C_2 + C_{1a}}{L_r}} \leq 0 \tag{3.5}$$

Based on the operation mode analysis, the expressions of the leakage inductance current I_{Lr_t0} and the DC inductance current i_{L2} when Q_{2a} is turned off can be written as

$$\begin{cases} I_{Lr\_t0} = i_{Lr}(t_0) = \frac{KV_2(-\pi - 2\varphi + 2\pi d)}{4\omega L_r} \\ i_{L2} = \frac{P}{2V_1} - \frac{V_1 dT_s}{2L_2} \end{cases} \tag{3.6}$$

To obtain the maximum value of the ZVS limit current, expressing the phase shift angle by the transmission power P and regarding P as a variable can obtain the derivative as shown below

$$\frac{dI_{ZVS\_Q_1\&Q_2}(P)}{dP}=\frac{\pi}{KV_2}\cdot\frac{\pi(2d-1)-\varphi}{(1-d)\pi\sqrt{\pi^2(d-d^2)-\frac{4\pi\omega L_r P}{K^2V_2^2}}} \quad (3.7)$$

Because the ZVS analysis is discussed in boost mode one, $\pi(2d-1)-\varphi<0$ and the derivative is less than zero, $I_{ZVS\_Q1\&Q2}$ decreases monotonically in this case. When $P=0$, $I_{ZVS\_Q1\&Q2}<0$. Therefore, from the principle of the above ZVS analysis, it is finally found that the switch Q_2 can easily achieve ZVS. Likewise, the ZVS limit considering junction capacitors for Fig. 3.4b, c can also be obtained, as listed in Table 3.1. The same method can be used to find that the other switches can achieve ZVS no matter how heavy the load is.

To test the rationality of the above analysis, take the following parameters as an example: $L_1=L_2=10\ \mu H$, $L_r=3\ \mu H$, $f=100$ kHz, $V_2=270$ V, $P_N=1000$ W, $C_2=1500$ pF, $C_{1a}=378$ pF, $C_3=294$ pF. Figure 3.5 shows the ZVS limit condition of switches at different d according to Table 3.1. For example, although the duty cycle changes, the ZVS limit condition $I_{ZVS\_Q1a\&Q2a}$ is always less than zero as shown in Fig. 3.5b. Therefore, the conclusion that Q_{1a} can realize zero voltage conduction is verified. In conclusion, the converter can achieve the full load range ZVS even with considering junction capacitors. However, from experimental results, it can be found that ZVS will be lost under some particular circumstances, which will be illustrated in Sect. 3.2.2.

Table 3.1 ZVS limit for other switches

Switch	Constraints of ZVS	Value
$Q_{1a}\&Q_{2a}$	$I_{Lr\_t2}-i_{L1}+KV_2\sqrt{\frac{3(C_{1a}+C_2)}{4L_r}}\le 0$	$\begin{cases} I_{Lr\_t2}=\frac{KV_2(-3\pi-2\varphi+6\pi d)}{4\omega L_r} \\ i_{L1}=\frac{P}{2V_1}+\frac{V_1 dT_s}{2L_2}\end{cases}$
$S_3\&S_4$	$I_{Lr\_t4}\ge 0$	$I_{Lr\_t4}=\frac{KV_2(\pi+2\varphi-2\pi d)}{4\omega L_r}$

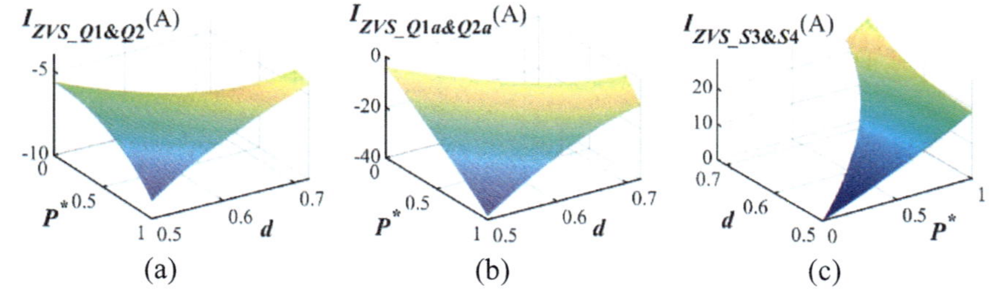

Fig. 3.5 ZVS limit currents of different switches for boost mode one ($L_1=L_2=10\ \mu H$, $L_r=3\ \mu H$, $f=100$ kHz, $V_2=270$ V, $P_N=1000$ W, $C_2=1500$ pF, $C_{1a}=378$ pF, $C_3=294$ pF). **a** $Q_1\&Q_2$. **b** $Q_{1a}\&Q_{2a}$. **c** $S_3\&S_4$

3.2.2 ZVS Analysis Considering Dead Time

To achieve the full load range under soft switching, the switches must satisfy the conditions described in Table 3.1, but only meeting the basic energy conditions does not ensure the realization of ZVS for all switches. The influence of dead time on the leakage inductance current is ignored in traditional analysis. And the leakage inductance current can be affected by the dead time, especially when the duty ratio is close to 0.5 and the phase shift angle is small. Therefore, combined with the dead time and junction capacitors, the changes of i_{Lr} and the corresponding ZVS range are re-analyzed under the light load. In this part, T_{d_p} is the dead time of the primary side switches, T_{d_s} is the dead time of the secondary side switches. i_0, i_1, i_2 represent the leakage inductance current at t_0, t_1, t_2 respectively. u_{C3} is the voltage of C_3.

Figure 3.6 illustrates the waveforms after zooming in on the leakage inductance current change, which can be divided into two cases. It should be noted that the switch currents i_{Q2} ($i_{Q2} = i_{Lr} + i_{L2}$) and i_{Q1a} ($i_{Q1a} = i_{Lr}$-i_{L1}) of the primary side switches are plotted in Fig. 3.6 to facilitate observing the ZVS implementation conditions. Since i_{Q2} and i_{Q1a} are far less than 0 during the resonance process, ZVS is easy to achieve for the primary side switches. In addition, the switch current of the secondary side switch is proportional to the primary side leakage inductance current, that is $i_{S3} = K{\cdot}i_{Lr}$, so the ZVS of the secondary side switch can be determined just according to the leakage inductance current i_{Lr}.

Taking case 1 as an example, the leakage inductance current increases in three stages because of the influence of d, φ, and the dead time. During the first stage, Q_{2a} is turned off at t_0. Then the body diode D_2 of switch Q_2 is turned on, v_{ab} rises to zero. Because S_4 is conducting, the secondary side voltage v_{cd} of the transformer is equal to $-V_2/2$. The leakage inductance current can be represented as

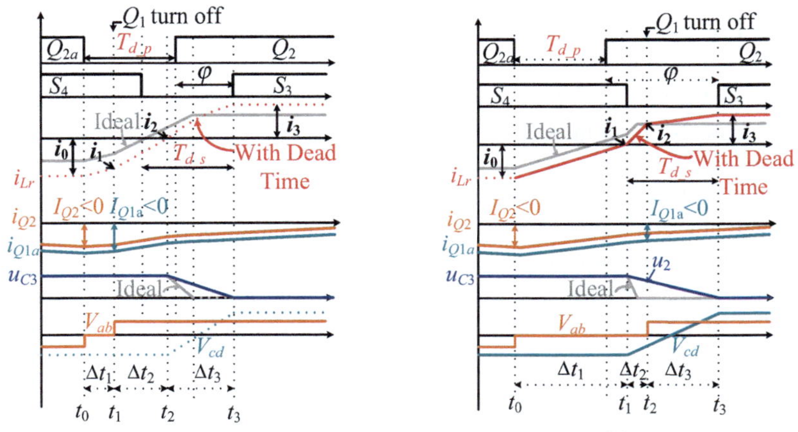

(a) (b)

Fig. 3.6 Waveforms considering dead time. **a** Case 1. **b** Case 2

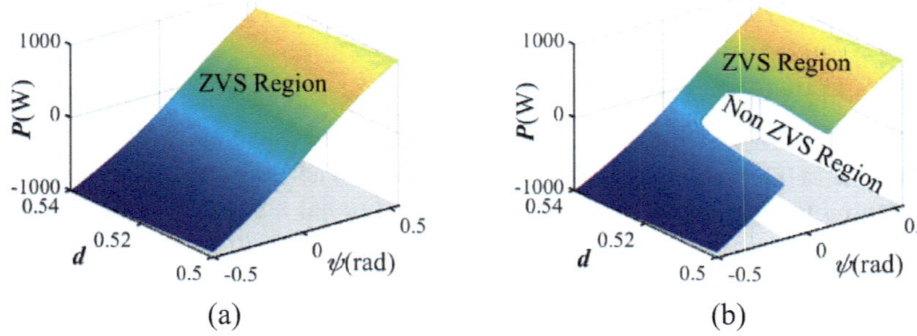

Fig. 3.7 Power transfer curves and corresponding ZVS range ($T_{d\_s} = 150$ ns, $T_{d\_p} = 120$ ns, $P_N = 1000$ W). **a** Traditional analysis. **b** ZVS analysis considering dead time

$$i_{Lr\_1}(t) = i_0 + \frac{KV_2}{2L_r} \cdot t,\ \Delta t_1 = \frac{(2d-1)T_s}{2} \tag{3.8}$$

In the next stage, Q_1 is turned off at t_1. Because the turn-on signal of Q_2 is not coming, the body diode D_2 of Q_2 is still conducting. v_{ab} is equal to V_{Cc}. The leakage inductance current rises linearly and the slope is doubled, and

$$i_{Lr\_2}(t) = i_1 + \frac{KV_2}{L_r} \cdot t,\ \Delta t_2 = -\frac{i_1 L_r}{KV_2} \tag{3.9}$$

In the third stage, i_{Lr} rises to zero at t_2. Therefore, the junction capacitor C_3 of S_3 can be discharged from t_2. Based on the same method as Eq. (3.4), the voltage u_{C3} of C_3 and the leakage inductance current i_{Lr} can be given by

$$\begin{cases} u_{C3}(t) = V_2 \cos \omega_3 t \\ i_{Lr\_3}(t) = \sqrt{\frac{2C_3}{L_r}} V_2 \sin \omega_3 t \end{cases},\ \Delta t_3 = T_{d\_p} + \varphi - \Delta t_1 - \Delta t_2 \tag{3.10}$$

where $\omega_3 = K/(2L_r C_3)^{1/2}$. Similarly, the expressions of i_{Lr} and u_{C3} of each stage for Fig. 3.6b can also be obtained.

According to $i_{Lr}(t_3) = -i_0$, the leakage inductance current values at each moment can be obtained. To find out the corresponding ZVS limit in the different duty cycles, make $u_{C3}(t)$ exactly equal to zero when the turn-on signal of S_3 comes. Therefore, the constraint condition $\Delta t_3 = \pi/2\omega_3$ can be derived.

Based on Fig. 3.6a and the expressions of i_{Lr} and u_{C3}, the critical transferred power P_{cri} to achieve soft switching can be calculated as follows

$$P_{cri} = V_{cc} \cdot \frac{2}{T_s} \left(\int_0^{\Delta t_2} i_{Lr\_2}(t) dt + \int_0^{\Delta t_3} i_{Lr\_3}(t) dt + \int_0^{\frac{T_s}{2} - \Delta t_1 - \Delta t_2 - \Delta t_3} i_3 dt \right) \tag{3.11}$$

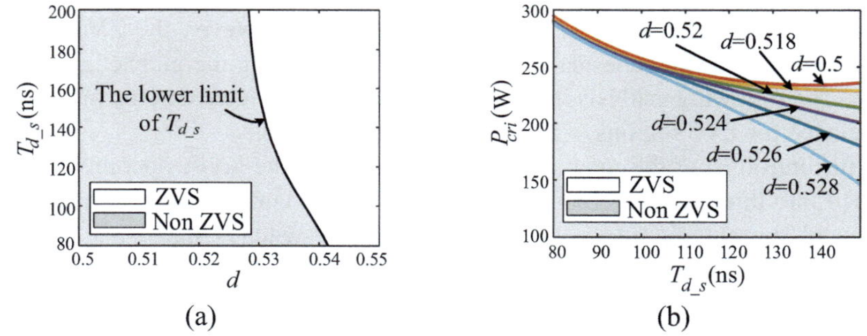

Fig. 3.8 The ZVS range considering dead time. **a** Under different secondary side dead times and duty cycles. **b** The critical transmission power to achieve ZVS ($P_N = 1000$ W)

Figure 3.7b illustrates the ZVS region considering the dead time when $T_{d\_s} = 150$ ns and $T_{d\_p} = 120$ ns. Obviously, the curved surface lacks a part that about 26.4% of the positive power cannot achieve ZVS (The maximum power is 1000 W). The converter cannot achieve soft switching when the duty cycle is close to 0.5 and the phase shift angle is small.

The DC inductance current on the primary side is large, and the resonance process of the primary switches is very short so that the primary side dead time has little impact on ZVS. Therefore, Fig. 3.8 mainly illustrates the impact of the secondary side dead time T_{d_s} on the ZVS range, where the rated power is set to 1000 W. It can be seen from Fig. 3.8a that as the duty cycle d and the dead time T_{d_s} decrease, the converter will not be able to realize ZVS. Then, Fig. 3.8b shows the plot for the relationship between the secondary side dead time T_{d_s} and the critical transferred power P_{cri} to achieve soft switching. The different curves marked in Fig. 3.8b represent the transmission power required to achieve ZVS under different duty cycles. As the size of the dead time T_{d_s} increases, the critical transmission power P_{cri} decreases accordingly. Therefore, by choosing an appropriate size of the dead time, the ZVS range can be extended. However, the dead time cannot increase indefinitely, and ZVS cannot be achieved regardless of the design of the converter parameters. **Therefore, the magnetizing current is proposed to be utilized to expand the ZVS range**, and the specific analysis will be described in the next section.

3.3 Utilizing Magnetizing Inductance to Enlarge ZVS Range

To achieve the full range of ZVS after considering the dead-time effect, the magnetizing inductance is proposed to be redesigned along with designing proper dead time. There are large DC inductance currents on the primary side and the introduction of the magnetizing inductance has little effect on the switch currents of the primary side switches when the primary side switches are turned off. Thus, the primary side switches can easily realize the full load range ZVS. However, the ZVS range of the secondary side switches can be extended after utilizing the magnetizing inductance. The following analysis is mainly to describe the influence of the magnetizing inductance for ZVS regions of the secondary side switches.

After introducing the magnetizing inductance L_m, the boost operation mode is divided into three cases as illustrated in Fig. 3.9. The curves in the dashed box are the switch currents i_{Q2} and i_{Q1a} of the primary side switches, the switch current of the secondary side switch after the magnetizing inductance improvement, and the secondary-side switch current under the conventional method, respectively. Introducing the magnetizing inductance does not affect the ZVS achievement of the primary-side switch. However, it can increase the secondary-side switch current and accelerate the discharge process of the secondary-side junction capacitance. Because of the similarity of the three situations in Fig. 3.9, only the waveform shown in Fig. 3.9a is analyzed in detail. Because the ZVS range is affected by the dead time under light load, this part focuses on the enlarged half-cycle waveform. The leakage inductance current changes in three stages, including $v_{ab} = -V_{Cc}$, $v_{ab} = 0$, $v_{ab} = V_{Cc}$.

Figure 3.10a is the equivalent circuit of the secondary side in the first stage. The relationship between the secondary side current $i_s(t)$ and the primary side current can be described as $i_s(t) = K[i_{Lr}(t)–i_m(t)]$. The voltage of L_m is a square wave converted from the transformer secondary side to the primary side, and the generated magnetizing current is a triangle wave. Take the half period $T_s/2$ of the voltage as an example, the slope of current descending is $di_m(t)/dt = -KV_2/2L_m$, and the value of the magnetizing current at any time can be obtained.

During the first stage, Q_1 and Q_{2a} conduct, and $v_{ab} = -V_{Cc}$. The secondary current can increase to a positive value before S_4 is turned off because of the effect of the magnetizing inductance, making the junction capacitor C_3 of switch S_3 start to discharge earlier, which is more conducive to the realization of ZVS for the switch

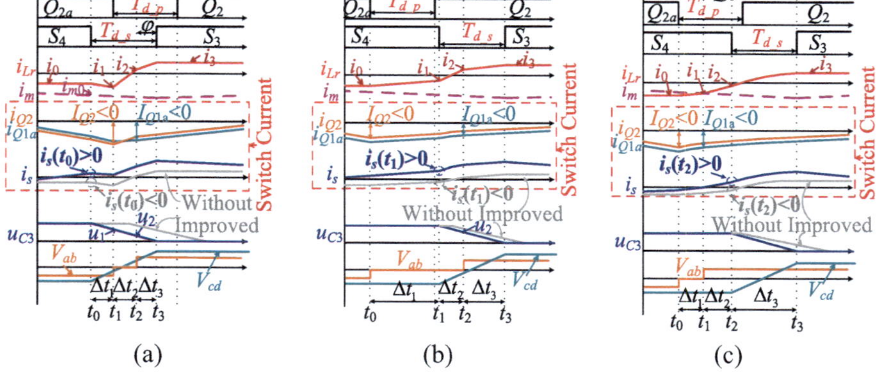

Fig. 3.9 Waveforms considering magnetizing inductance. **a** Case 1. **b** Case 2. **c** Case 3

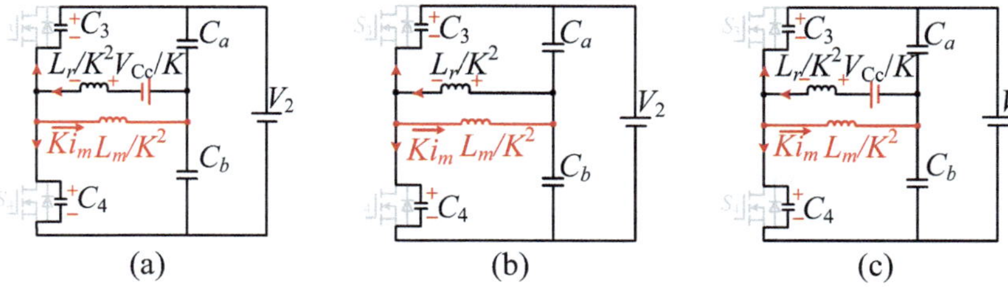

Fig. 3.10 Equivalent circuits for case one. **a** t_0–t_1. **b** t_1–t_2. **c** t_2–t_3

S_3. Therefore, the secondary side current i_s charges C_4 and discharges C_3 at t_0. The dynamic equations of the equivalent circuit shown in Fig. 3.10a can be represented as

$$\begin{cases} \dfrac{L_r}{K^2}\dfrac{dKi_{Lr}(t)}{dt} + \dfrac{V_{Cc}}{K} = u_{C3}(t) - \dfrac{V_2}{2} \\ K[i_{Lr}(t) - (i_{m0} - \dfrac{KV_2}{2L_m}t)] = -2C_3\dfrac{du_{C3}(t)}{dt} \\ u_{C3}(0) = V_2 \\ i_{Lr}(0) = i_0 \end{cases} \tag{3.12}$$

By simplifying (3.12), the voltage of the junction capacitor C_3 and the leakage inductance current i_{Lr} can be given by

$$\begin{cases} u_{C3}(t) = \frac{L_r k_1}{K}\cos\omega_3 t + \sqrt{\frac{L_r}{2C_3}}(-i_0 + i_{m0})\sin\omega_3 t + V_2 - \frac{L_r k_1}{K} \\ i_{Lr\_1}(t) = (i_0 - i_{m0})\cos\omega_3 t + \frac{k_1\sqrt{2L_r C_3}}{K}\sin\omega_3 t + i_{m0} - k_1 t \end{cases} \tag{3.13}$$

where $k_1 = KV_2/2L_m$, $\omega_3 = K/(2L_rC_3)^{1/2}$, $i_0 = i_{Lr}(t_0)$, and i_{m0} is the value of the magnetizing current at t_0.

As seen in Fig. 3.10b, c, resonance equivalent circuits of the other two stages differ only in the converted voltage from the transformer primary side to the secondary side. Likewise, expressions for u_{C3} and i_{Lr} can be obtained. And the constraint condition for the ZVS range can be derived by

$$\begin{cases} u_{C3}(t_3) = 0 \\ \Delta t_2 = \dfrac{(2d-1)T_s}{2} \\ \Delta t_1 + \Delta t_2 + \Delta t_3 = t_s \end{cases} \tag{3.14}$$

Based on waveforms in Fig. 3.9a and the expressions of $i_{Lr}(t)$ and $u_{C3}(t)$, the critical transferred power P_{cri} to realize soft switching can be obtained. Figure 3.11b illustrates the relationship between transmission power P and ZVS region under different d and φ by taking the secondary side dead time of 150 ns and the magnetizing inductance of 57 μH as an example. Compared with the ZVS range of the conventional converter shown in Fig. 3.11a, the ZVS region can be significantly expanded to a full load range after introducing a magnetizing inductance.

Figure 3.12 shows the ZVS range that varies with the secondary side dead time and duty cycle under different magnetizing inductances. As shown in Fig. 3.12, the smaller L_m is, the larger the ZVS range is. When the magnetizing inductance and dead time match well, the converter can achieve full range ZVS, as seen in Fig. 3.12c. Therefore, it is necessary to optimize the design of the parameters of the dead time and magnetizing inductance.

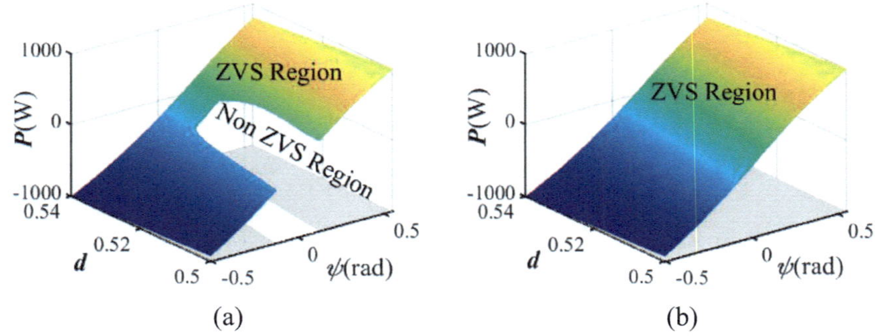

Fig. 3.11 Power transfer curves and corresponding ZVS range ($T_{d\_s} = 150$ ns, $T_{d\_p} = 120$ ns, $P_N =$ 1000 W). **a** Conventional converter without magnetizing inductance. **b** Converter with magnetizing inductance ($L_m = 57$ μH)

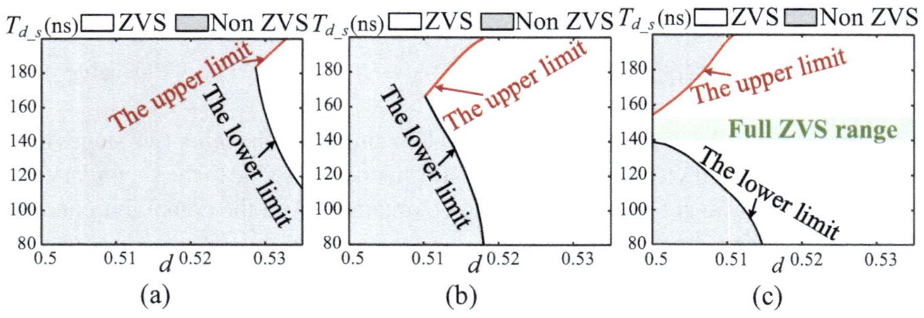

Fig. 3.12 ZVS regions with different magnetizing inductances. **a** $L_m = 100$ μH. **b** $L_m = 80$ μH. **c** $L_m = 58$ μH

3.4 Design Considerations

In this section, A design example of the proposed converter with $V_1 = 18–32$ V, $V_2 = 270$ V and $P_N = 1000$ W is introduced. A 100 kHz switching frequency is adopted.

3.4.1 Turn Ratio of Transformer

When the transformer achieves the voltage amplitude matching, the CF-DAB converter can reduce the circulating loss, so the transformer ratio K should satisfy

$$K = \frac{2V_1}{(1-d)V_2} \tag{3.15}$$

Constrained by the duty cycle $d \geq 0.5$, the turn ratio K is finally designed to be 1:2.1.

3.4.2 DC Inductance and Leakage Inductance

Leakage inductance and DC inductance affect the design of the magnetizing inductance and dead time to achieve ZVS, so it is necessary to determine the reasonable design values of leakage inductance and DC inductance. Figure 3.13 illustrates the ZVS region of the converter without the magnetizing inductance under different leakage inductances by taking the secondary side dead time of 150 ns and the primary side dead time of 120 ns as an example. With the change of the leakage inductance, the transmission power and ZVS range of the converter will also change, which affects the subsequent design of magnetizing inductance and dead time.

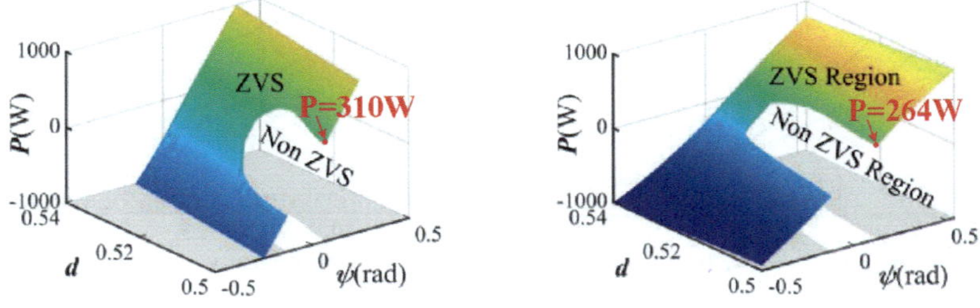

Fig. 3.13 Power transfer curves and corresponding ZVS range under different leakage inductances ($T_{d\_s} = 150$ ns, $T_{d\_p} = 120$ ns, $P_N = 1000$ W, $K = 1/2.1$, $V_2 = 270$ V). **a** $L_r = 1.5$ μH. **b** $L_r = 3$ μH

The leakage inductance determines the maximum transmission power, which is a pivotal part of the converter to achieve bidirectional energy transmission. The value of L_r should be designed to ensure that the maximum transmission power is larger than the rated power, and the expression of the transmission power can be obtained by

$$P = \frac{\pi K^2 V_2^2(-8d^2 + 8d - 1)}{16\omega L_r} \tag{3.16}$$

where $\omega = 2\pi f_s$, $\varphi = 0.5\pi$. According to (3.16), the maximum leakage inductance cannot be larger than 3.166 μH.

In addition, the RMS value of the leakage inductance current under boost mode one can be calculated as follows

$$I_{Lr\_RMS} = \frac{KV_2}{12\omega L_r}\sqrt{[-8\varphi(\varphi^2 - 3\pi\varphi d + 6\pi^2 d^2 - 3\pi^2 d) + 3\pi^3(2d-1)^2(4d-1)]\frac{3}{\pi}} \tag{3.17}$$

The RMS currents of the leakage inductor under the rated power are shown in Fig. 3.14a. As the voltage V_1 in the LVS decreases, the current increases. When the leakage inductance is 1.5 μH, the RMS current raises significantly as voltage V_1 decreases. Since the large leakage inductance will affect the magnetic loss and the wire loss, the leakage inductance of 3 μH is selected.

The DC inductance current is a triangular wave. And the RMS current of the DC inductor can be given by

$$I_{L\_RMS} = \sqrt{i_{L\min} \cdot \frac{\pi KV_2(-d^2 + d)}{\omega L} + \frac{\pi^2 K^2 V_2^2(1-d)^2 d^2}{3\omega^2 L^2} + i_{L\min}^2} \tag{3.18}$$

where $i_{Lmin} = P/2V_1 - V_1 dT_s/2L$.

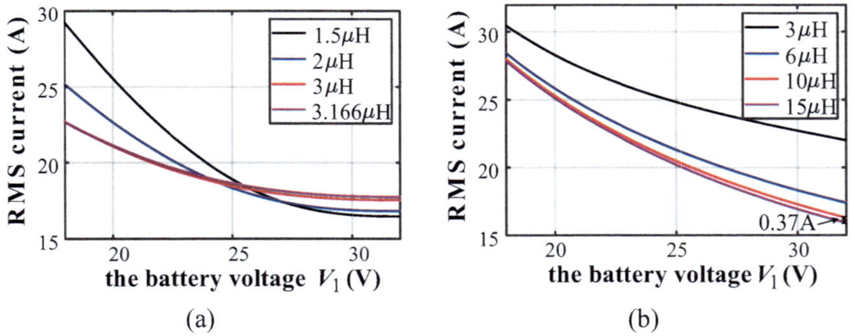

Fig. 3.14 The RMS currents of different inductance values. **a** Leakage inductances. **b** DC inductances

As illustrated in Fig. 3.14b, the smaller the DC inductance, the higher the RMS value. However, with the increase of the DC inductance, the RMS current advantage brought by it gradually decreases. For example, the difference between the RMS current of 10 μH DC inductance and 15 μH DC inductance is at most 0.37 A, while the 15 μH DC inductance may require more windings. In addition, the design range of the primary side dead time of a 15 μH DC inductance narrows down. Therefore, the DC inductance value is selected as 10 μH.

3.4.3 The Dead Time and Magnetizing Inductance

According to the above ZVS analysis, the design of the dead time also has a vital role in the realization of ZVS. Take the dead time design of the switch Q_2 as an example. There is a lower limit for the primary side dead time so that the current can discharge the voltage of parasitic capacitor to zero before the switch Q_2 conducts.

Based on the expression of the voltage $u_{C2}(t)$, make the voltage $u_{C2}(t)$ of junction capacitor C_2 equal to zero and simplify the equation, the shortest time $t_{r\_Q1\&Q2}$ required for the resonance process under boost mode one can be obtained by

$$t_{r\_Q1\&Q2} = \sqrt{L_r(C_2 + C_{1a})} \arcsin \frac{-V_{Cc}}{\sqrt{\frac{L_r}{2C_2}}(I_{Lr\_t0} + i_{L2})} \tag{3.19}$$

where I_{Lr_t0} is the leakage inductance current in the LVS at the initial time, V_{Cc} is the voltage of clamping capacitor.

The actual lower limit of the primary side dead time is composed of the turn-off delay time t_{d_off} of the switch Q_2 and the resonance time. Similarly, the lower limit of the dead time of switch Q_{1a} can be given by

$$\begin{aligned} t_{p\min\_Q1a\&Q2a} =& \sqrt{L_r(C_{1a} + C_2)}(-\arccos \frac{2V_{Cc}}{\sqrt{V_{Cc}^2 + \frac{L_r}{C_{1a}+C_2}(I_{Lr\_t2} - i_{L1})^2}} \\ &+ \arccos \frac{V_{Cc}}{\sqrt{V_{Cc}^2 + \frac{L_r}{C_{1a}+C_2}(I_{Lr\_t2} - i_{L1})^2}}) + t_{d\_off1a} \end{aligned} \tag{3.20}$$

where I_{Lr_t2} is the value of the leakage inductance current at t_2, t_{d_off1a} is the turn-off delay time of switch Q_{1a}.

Likewise, the lower limit of the secondary side dead time can be obtained, based on the ZVS analysis of the converter after adding the magnetizing inductance and the expression of the voltage $u_{C3}(t)$ in Sect. 3.3.

Besides, there is an upper limit for the deadtime because too long deadtime will lead to the reverse charging of the junction capacitor. The switches must be turned on before the currents change to a reverse direction. Table 3.2 shows the conditions that the converter does not produce the anti-resonance phenomenon. The upper limit of

Table 3.2 Upper limits for all switches

Switches	Upper limits of deadtime
$Q_1\&Q_2$	$i_{Lr}(t) + i_{L2}(t) \leq 0$
$Q_{1a}\&Q_{2a}$	$i_{Lr}(t) - i_{L1}(t) \leq 0$
$S_3\&S_4$	$K[i_{Lr}(t) - i_m(t)] \geq 0$

the deadtime can be obtained by substituting the leakage inductance current value and the magnetizing current value described in Sect. 3.3 into the corresponding formulas.

According to the above analysis of the deadtime, Fig. 3.15 shows a reasonable range of the primary side dead time T_{d_p}. In Fig. 3.15, the upper curved surface is the upper limit of the deadtime T_{d_p}, and the lower curved surface is the lower limit of T_{d_p}. The part enclosed by the two curved surfaces can realize the zero-voltage conduction of the primary side switches. Keep a certain margin, take the intermediate value in this range, and the deadtime of the current-fed side switches is designed to be 180 ns.

Figure 3.16 illustrates the relationship of magnetizing inductance L_m, deadtime of secondary side switches T_{d_s}, and the ZVS range. The reduction of magnetizing inductance can help to achieve zero-voltage conduction for the secondary side switches. On the other hand, it will bring a certain circulating current to the converter. Therefore, within the range of meeting ZVS, it is better to design a larger magnetizing inductance. According to Fig. 3.16, it is found that point B is the boundary point of ZVS, and a magnetizing inductance of 60 μH can just achieve ZVS. In Fig. 3.16, point A is a feasible point that can achieve the full range ZVS. Therefore, the magnetizing inductance is designed to 57 μH, and the dead time of switches S_3/ S_4 is designed to 210 ns. It is worth pointing out that the proportion of dead time in the switching period of the DAB converter is usually small, so the dead time has little effect on the maximum frequency of the converter.

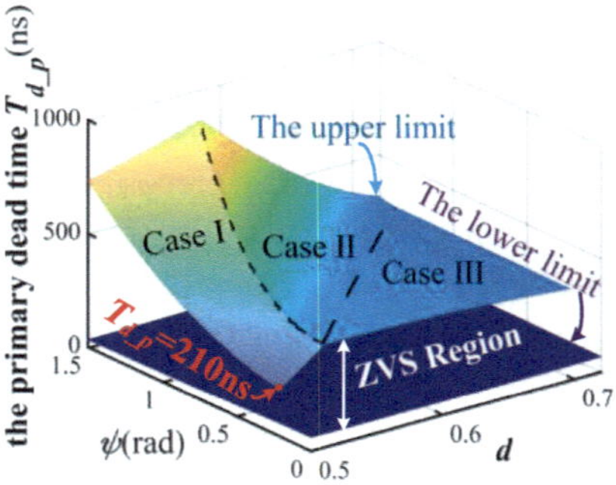

Fig. 3.15 The range of the primary side deadtime to realize ZVS

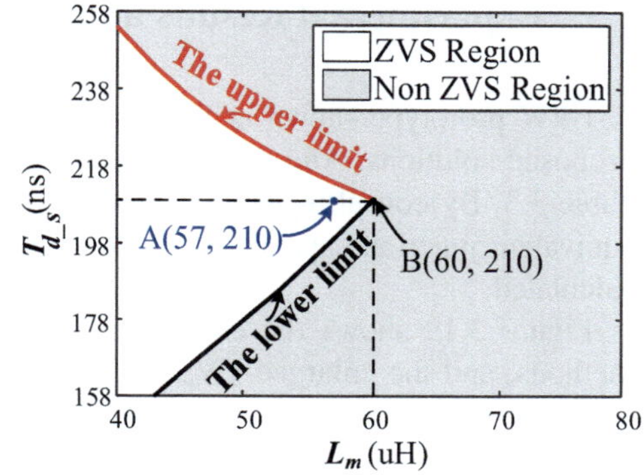

Fig. 3.16 ZVS range with different T_{d_s} and L_m

3.4.4 *Control Strategy*

Figure 3.17 illustrates the control block diagram. This article uses the independent phase shift angle closed-loop control and the duty cycle closed-loop control to achieve decoupling control of phase shift angle φ and duty cycle d. Therefore, the complexity of the control strategy is simplified. Controlling the duty cycle can adjust the voltage of the low-voltage side of the transformer to achieve the voltage amplitude matching of the transformer, which effectively reduces the circulating current and improves the efficiency and power density of the converter. The phase shift angle closed-loop control adjusts the phase shift angle between the two sides of the transformer. Therefore, it can control the flow direction of the energy and realize the seamless switching of the power transmission direction in the bidirectional converter.

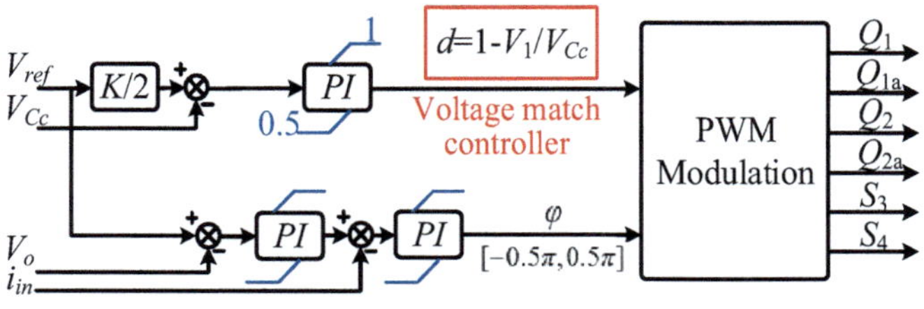

Fig. 3.17 Control block diagram for the proposed control strategy

3.5 Experimental Results and Discussions

A 1 kW prototype shown in Fig. 3.18 was built to verify the effectiveness of the proposed solutions. And the specifications of the CF-DAB converter are listed in Table 3.3. By consulting the C_{oss} curve in datasheets of the switches, the charge-equivalent junction capacitances of Q_2, Q_{2a}, and S_3 under rated operation can be calculated.

Figure 3.19 shows the experimental waveforms of switch S_3 under different methods, and the enlarged ZVS waveforms are placed next to each subfigure for easy observation. It can be seen from Fig. 3.19a that when the converter using the conventional method has a transmission power of about 200 W, the secondary side switches cannot realize the ZVS. Compared with the conventional method, the zero-voltage conduction of the high-voltage side switches is easier after introducing the suitable magnetizing inductance in Fig. 3.19b.

Furthermore, Fig. 3.20 shows the thermal image of the secondary side switches of the prototype in thermal equilibrium under non-ZVS/ZVS conditions when the

Fig. 3.18 Prototype of 1 kW converter

Table 3.3 Specifications of prototype

Parameters	Value
Voltage in LVS (V_1)	18–32 V
Voltage in HVS (V_2)	270 V
Nominal power (P_N)	1000 W
Switching frequency (f_s)	100 kHz
Turn ratio of transformer (K)	1:2.1
Leakage inductance (L_r)	3 μH
DC inductance (L_1/L_2)	10 μH
Switch (Q_1/Q_2)	IPP051N15N5
Switch (Q_{1a}/Q_{2a})	IPP111N15N3
Switch (S_3/S_4)	SCTW90N65G2V

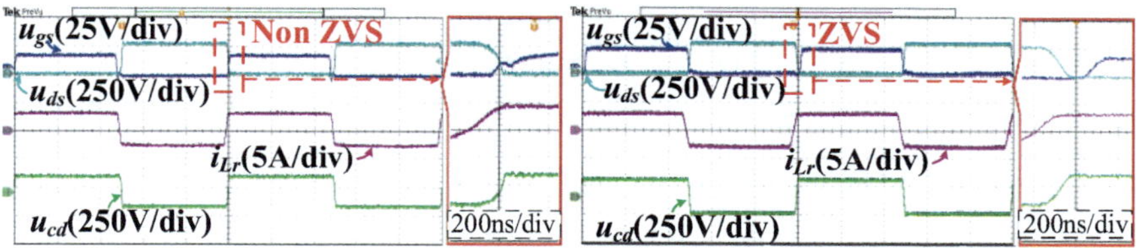

(a) (b)

Fig. 3.19 Experimental waveforms of switch S_3 under different methods when $P = 200$ W and $d = 0.5$. **a** the conventional method. **b** the proposed method

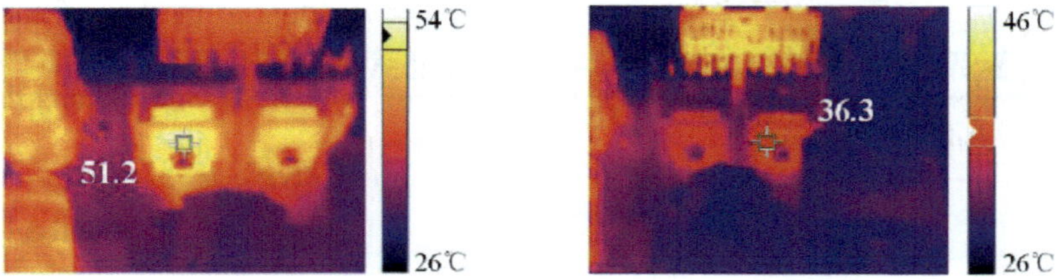

(a) (b)

Fig. 3.20 Thermal image of switches S_3&S_4 ($P = 200$ W). **a** Non-ZVS case. **b** ZVS case

output power is 200 W. The temperature rise of switches is higher under the non-ZVS working condition.

Figure 3.21 is drawn based on Fig. 3.16. In Fig. 3.21, five points were selected for testing on the boundary of the secondary side deadtime to verify the reliability of the theoretical analysis. Figure 3.22 shows the results of the ZVS experiment at these points. As shown in Fig. 3.22, when the dead time and magnetizing inductance are selected in the ZVS area, the secondary side switches can achieve soft switching. For instance, Fig. 3.22a, c illustrate respectively that point A and point D can achieve soft switching. However, when the test point is located in the Non-ZVS area in Fig. 3.21, the switch S_3 cannot fully achieve zero-voltage conduction under light load. The vertical dashed lines in Fig. 3.22b, d mark the turn-on moment of switch S_3. As shown in Fig. 3.22b, the drain-source voltage u_{ds} of the switch S_3 will first drop and then rise a little, and it is greater than zero before S_3 turns on, so S_3 cannot achieve ZVS. Similarly, the waveform at point E in Fig. 3.21 is shown in Fig. 3.22d, where u_{ds} drops slowly and cannot drop to zero before the switch S_3 turn-on signal comes, resulting in the loss of ZVS conditions. In addition, the gate-source voltage u_{gs} in Fig. 3.22b, d both have a relatively obvious Miller platform marked with lines. Therefore, Fig. 3.22 verifies the rationality of the ZVS analysis about the dead time and magnetizing inductance.

In the 1 kW prototype developed in this chapter, to ensure that all the converter switches can achieve soft switching under different loads, consider optimizing the magnetizing inductance to 57 μH and the secondary dead time to 210 ns as shown

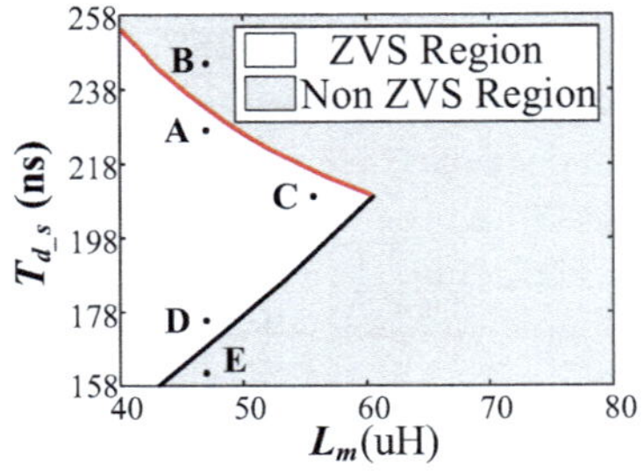

Fig. 3.21 The relationship between T_{d_s} and L_m

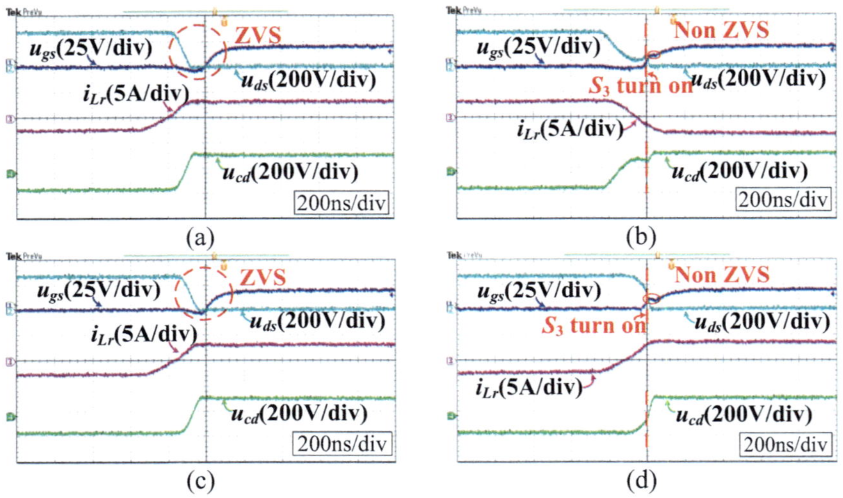

Fig. 3.22 ZVS experimental waveforms of switches S_3&S_4. **a** A ($L_m = 48$ μH, $T_{d\_s} = 230$ ns). **b** B ($L_m = 48$ μH, $T_{d\_s} = 250$ ns). **c** D ($L_m = 48$ μH, $T_{d\_s} = 175$ ns). **d** E ($L_m = 48$ μH, $T_{d\_s} = 160$ ns)

in the point C of Fig. 3.21. Figures 3.23, 3.24 and 3.25 show the ZVS waveforms of each switch at different duty cycles and loads. Figure 3.23 shows the steady-state waveforms of the CF-DAB converter with the proposed method under no load and the duty cycle of 0.5, which is the most difficult to achieve ZVS. From the enlarged ZVS waveforms on the right, it can be seen that all switches can realize zero-voltage conduction. In addition, Figs. 3.24 and 3.25 show the experimental waveforms of the CF-DAB converter under full load ($P = 1000$ W) with the duty cycle of 0.5 and 0.7, respectively. Similarly, each switch can achieve ZVS.

The experimental waveforms of the converter working in buck mode at light load and full load are shown in Figs. 3.26 and 3.27, respectively. The good ZVS performance with the proposed method under different loads is still maintained in buck mode. All switches can achieve zero-voltage conduction well under various

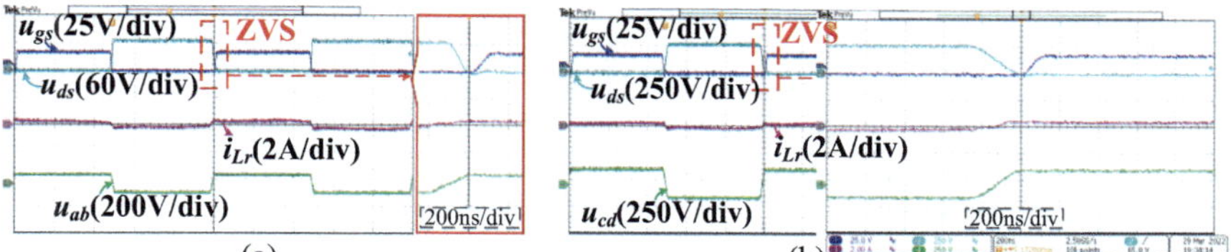

(a) (b)

Fig. 3.23 Experimental waveforms under the proposed method when $P = 0$ W and $d = 0.5$. **a** Q_{1a}&Q_{2a}. **b** S_3&S_4

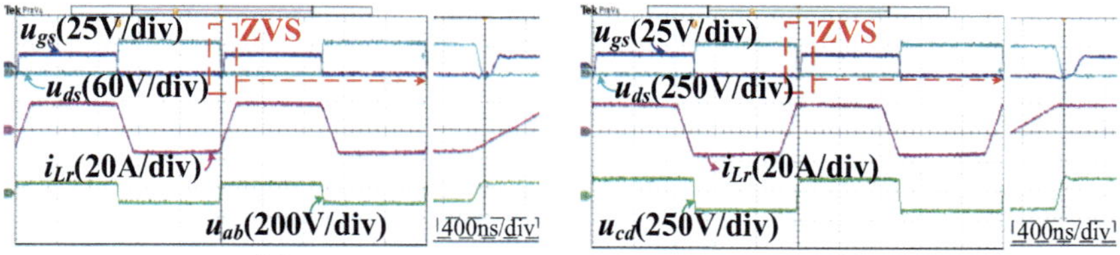

(a) (b)

Fig. 3.24 Experimental waveforms under the proposed method when $P = 1000$ W and $d = 0.5$. **a** Q_{1a}&Q_{2a}. **b** S_3&S_4

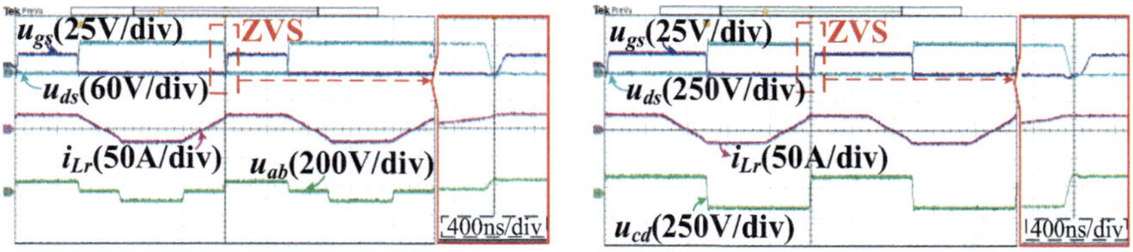

(a) (b)

Fig. 3.25 Experimental waveforms under the proposed method when $P = 1000$ W and $d = 0.7$. **a** Q_{1a}&Q_{2a}. **b** S_3&S_4

working conditions. In conclusion, it proves that the converter can achieve the full load range ZVS within a wide input voltage range after designing the appropriate magnetizing inductance and secondary side dead time.

Figure 3.28 illustrates the efficiency curves of the proposed converter at different voltage V_1 in the LVS and different loads. As the voltage V_1 decreases and the conduction loss increases, the efficiency curve shifts down. In Fig. 3.28, the efficiency of the converter under different input voltages at 100 W, 600 W, and 1000 W is marked, respectively. When the transmission power is 600 W, the highest efficiency can reach 97.4%. As the power increases, the efficiency curves of the input voltage 32 V and 27 V gradually coincide.

In addition, Fig. 3.29 compares the efficiency of the proposed method with the conventional method in detail at different input voltages. In Fig. 3.29, the points with

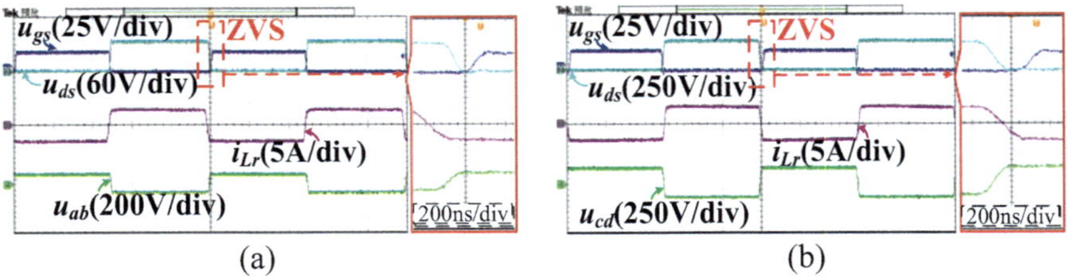

Fig. 3.26 Experimental waveforms under the proposed method when $P = -200$ W and $d = 0.5$. **a** Q_{1a}&Q_{2a}. **b** S_3&S_4

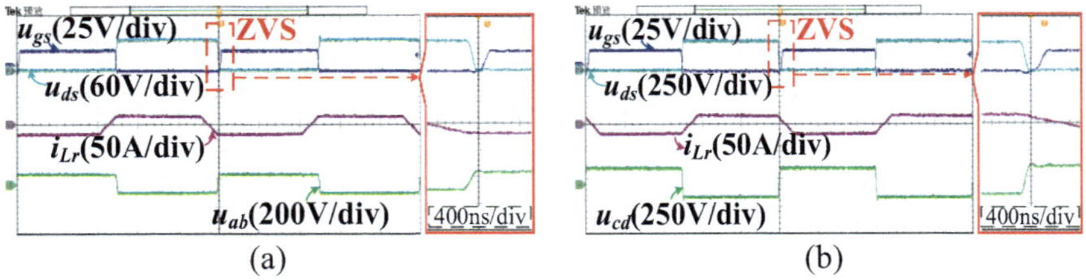

Fig. 3.27 Experimental waveforms under the proposed method when $P = -1000$ W and $d = 0.5$. **a** Q_{1a}&Q_{2a}. **b** S_3&S_4

Fig. 3.28 The efficiency curves with the proposed converter under different voltages V_1

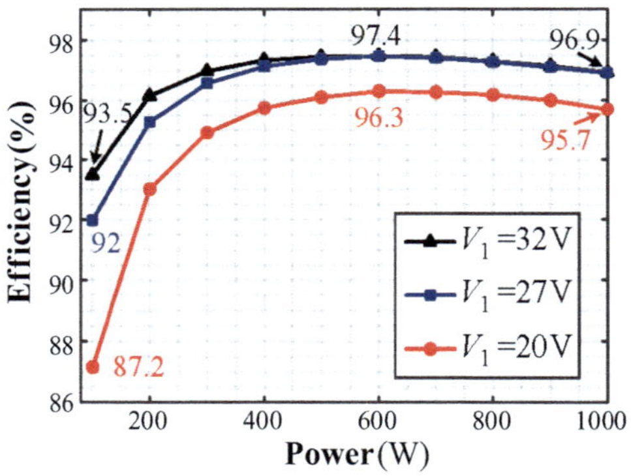

a big difference in efficiency are circled by dotted lines, and the corresponding efficiency values are marked. From Fig. 3.29a, b, the proposed method can significantly improve the efficiency under light load (such as $d = 0.5$ and $V_1 = 32$ V, $\eta = 93.5\%$ when $P = 100$ W). Although it is slightly lower than the conventional method under heavy load, the difference is small, by a margin less than 0.2%, and the curves are close to coincidence. As the input voltage decreases, the efficiency of the proposed method is close to that of the conventional method, but the efficiency of the proposed method is high in the whole voltage range.

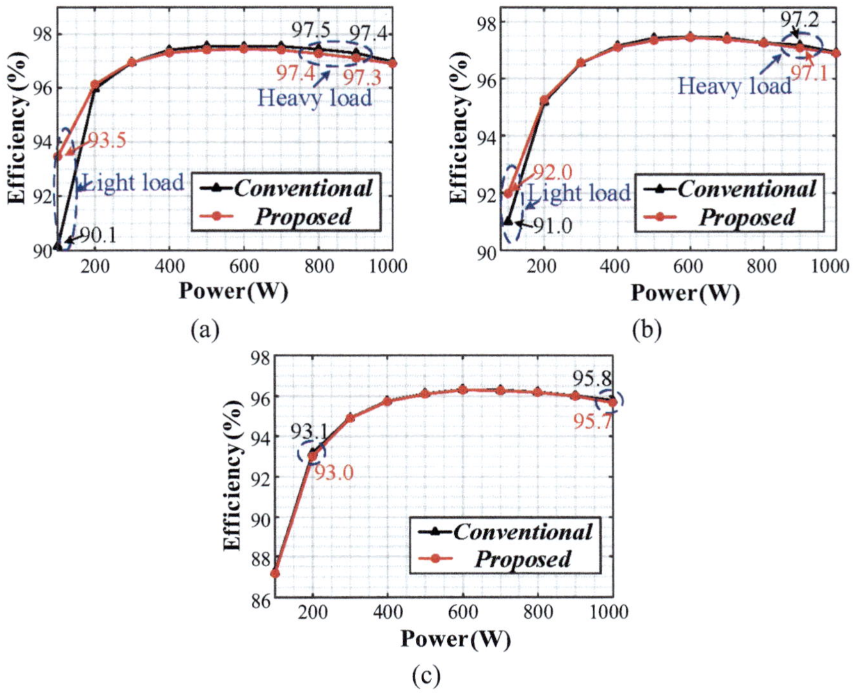

Fig. 3.29 Efficiency comparison between the conventional method and the proposed method. **a** $V_1 = 32$ V. **b** $V_1 = 27$ V. **c** $V_1 = 20$ V

Figure 3.30a, b are the comparisons of loss breakdown between the proposed method and the conventional method under light load and heavy load, respectively. As shown in Fig. 3.30a, the proposed method has less total loss than the conventional method at light load because of the implementation of ZVS, which greatly reduces the switching turn-on loss. While under heavy load, due to the introduction of the magnetizing inductance, the secondary side current of the transformer increases, which will bring a certain amount of transformer loss and switches conduction loss. And the conventional method can also achieve ZVS under heavy load, and its total loss is slightly less than the proposed method, as shown in Fig. 3.30b. However, this method optimizes the design of magnetizing inductance and dead time as much as possible to improve efficiency. It can be seen from Fig. 3.31 that the secondary side RMS current of the transformer is at most 0.53 A higher than that of the conventional method, despite the introduction of magnetizing inductance. And the efficiency difference is less than 0.2% of the total power at 1000 W. In addition, full load range ZVS will also reduce the EMI noise caused by hard switching turn-on.

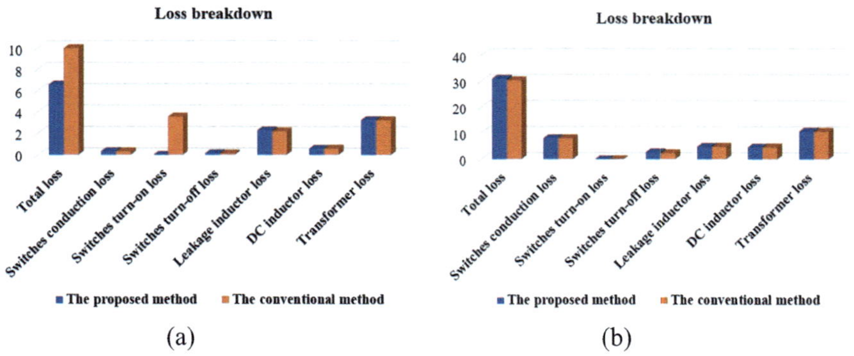

Fig. 3.30 Loss breakdown comparisons(V_1 = 32 V). **a** At light load (100 W). **b** At heavy load (1000 W)

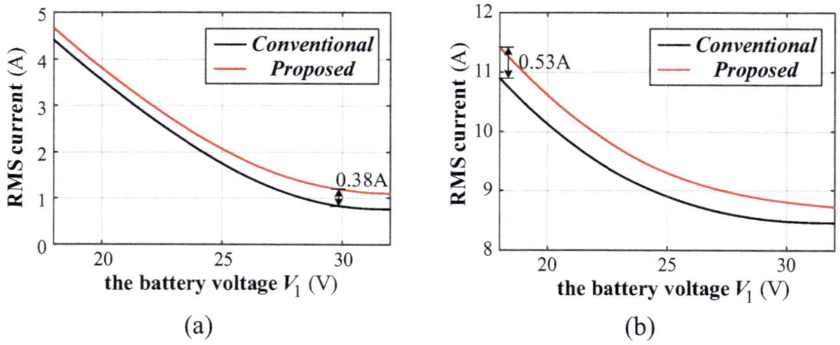

Fig. 3.31 The secondary side RMS currents of the transformer under different loads. **a** At light load (100 W). **b** At heavy load (1000 W)

3.6 Conclusion

In this chapter, to achieve ZVS of the CF-DAB converter under full load range, a dead-time optimization and magnetizing current design are proposed. First, the new ZVS analysis for CF-DAB converters under PPS control, considering the junction capacitors and the dead time, reveals the phenomenon that soft switching is difficult to satisfy under light load conditions. Then, to expand the ZVS range, a magnetizing inductance is introduced. The influence of magnetizing inductance and dead time on the ZVS range is discussed in detail, and the corresponding relationship between the magnetizing inductance and dead time is presented. Furthermore, the design of DC inductance and leakage inductance is explained to obtain low RMS currents with the precondition of realizing maximum power transmission capability. In the end, a 1000 W prototype is built to validate the effectiveness of the proposed solutions.

References

1. Peng, F. Z., Li, H., Su, G.-J., & Lawler, J. S. (2004). A new ZVS bidirectional DC–DC converter for fuel cell and battery application. *IEEE Transactions on Power Electronics, 19*(1), 54–65.
2. Xiao, H., & Xie, S. (2008). A ZVS bidirectional DC–DC converter with phase-shift plus PWM control scheme. *IEEE Transactions on Power Electronics, 23*(2), 813–823.
3. Zhu, L., Bai, H., Brown, A., & McAmmond, M. (2020). Transient analysis when applying GaN + Si hybrid switching modules to a zero-voltage-switching EV onboard charger. *IEEE Transactions on Transportation Electrification, 6*(1), 146–157.
4. Yan, Y., Gui, H., & Bai, H. (2021). Complete ZVS analysis in dual active bridge. *IEEE Transactions on Power Electronics, 36*(2), 1247–1252.
5. Sha, D., Wang, X., Liu, K., & Chen, C. (2019). A Current-fed dual-active-bridge DC–DC converter using extended duty cycle control and magnetic-integrated inductors with optimized voltage mismatching control. *IEEE Transactions on Power Electronics, 34*(1), 462–473.

Part II
Dual-Active-Bridge Converter with Modified Active Bridge

Chapter 4
Hybrid-Bridge-Based DAB Converter with Modified Voltage Matching Control and Inductor Current Injections

Abstract In this chapter, a modified voltage matching control and current injections method are proposed for the hybrid-bridge-based dual active bridge (DAB) converter, which is aimed at achieving full load range zero voltage switching (ZVS) with consideration of all the switch junction capacitors. With the proposed modified voltage matching control, the non-ZVS region under backward power transfer (even without the junction capacitors) is eliminated. The analysis of the operation principle, ZVS model analysis considering junction capacitors and current injection method are discussed in detail. In addition, the problem of discontinues ZVS region with consideration of the switch junction capacitors is solved through injecting two current sources with paralleling one inductor to the transformer. Finally, the effectiveness of the proposed solution is verified by experimental results on a 100 kHz, 1 kW converter prototype.

Keywords Auxiliary inductors · Hybrid-bridge-based dual active bridge · Wide zero voltage switching (ZVS) range

4.1 Topology and Conventional VM Control

The multi-level single-stage DAB converter can be regulated as a more appropriate scheme for wide voltage applications. The DAB converter can realize multi-level by adding dual transformers [1–3], or using the neutral point clamped (NPC) three-level full bridge [4], or adding the auxiliary half bridge network [5]. The dual-transformer based multi-level DAB converter generates three-level wave forms by modulating the waveforms at both ends of the two transformers respectively. According to voltage matching (VM), optimized modulation is proposed in [1, 2] which is applied to wide voltage gain and reduce leakage current peak. However, ZVS of some switches is limited in [1]. Three bridge arms replace two bridge arms at [3], and the whole domain ZVS is realized by utilizing segment control. Nevertheless, the use of two transformers and three bridge arms increases the volume and complexity of the device.

G. Xu and X. Chen, *Topology Deduction and Control for Dual-Active-Bridge Converters*, https://doi.org/10.1007/978-981-97-4856-3_4

The NPC three-level full-bridge converter [4] needs four additional active switching devices comparing with the conventional two-level converter. The analytical solution of minimize RMS current under various gains is solved by global optimization (GOC) in [4]. However, the optimization of four control variables brings complex calculation and classification. For simplifying control, the hybrid-bridge-based DAB converter [5] uses two control variables to realize wide voltage operation. "$V_1/2$" level is generated by adding an auxiliary half bridge on the conventional two-level full bridge in [5]. Compared with dual-transformer based multi-level DAB converter [3] and NPC three-level DAB converter [4], hybrid-bridge-based DAB converter [5] generates multi-level with a simpler structure. And the ZVS of most switches can be realized with the use of VM for a wide input voltage range by simple calculation and a PI closed loop [5].

However, under backward power transfer, ZVS will be lost when the load increase or the voltage is close to the voltage matching point. In addition, the junction capacitors of the switches are not considered in [5], and only the theoretical ZVS constraint is considered to analyze the ZVS region. While, the influence of junction capacitors on DAB converter has been investigated [6, 7]. These researches show that the converter performance will be affected by the junction capacitors, which is also worthy of study for the hybrid bridge DAB converter under VM control.

In this Chapter, the switch junction capacitors are added to the analysis of ZVS region for the hybrid bridge DAB converter, and a more accurate ZVS model is obtained. To increase the ZVS region under backward power transfer, a modified voltage matching control is proposed. In addition, an auxiliary inductor [8] and the magnetizing inductance of the transformer are utilized to generate enough induced current to help achieve full load ZVS.

4.1.1 Topology and Operation Principle of VM Control

The hybrid-bridge-based DAB converter is shown in Fig. 4.1. The DAB converter consists of the full bridge composed of switches S_1–S_4, the auxiliary half-bridge composed of switches S_5, S_6 and capacitors C_{11}, C_{12} at the V_1 side, and the active half bridge composed of switches Q_1, Q_2 and capacitors C_{21}, C_{22} at the V_2 side. L_k and n denote the leakage inductor and the ratio of transformer T, respectively. v_{AB} and v_{CD} are high frequency voltages of input and output sides respectively, and v_{EF} denotes the voltage across points E and F.

The operation waveform of conventional VM control [5] is shown in Fig. 4.2 under the condition of $0 < D_\varphi < D_y$ and $0.5 < m < 1$, where the item $m = nV_2/(2V_1)$. In this figure, T_s is the switch period, D_φ is the duty cycle corresponding to the phase-shift angle between Q_1 and S_1, and D_y represents the duty cycle of S_3 and S_4. Switches S_1, S_2 and Q_1, Q_2 work alternately at a fixed 50% duty. It is worth noting

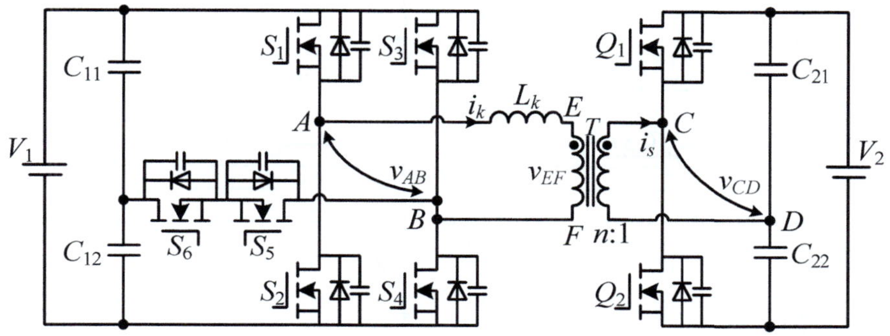

Fig. 4.1 Topology of the hybrid-bridge-based DAB converter [5]

that the turn- on instant for S_2 is the same as that of S_3, and the turn-on ins tant for S_4 and S_1 are also the same as shown. And D_y constant satisfies the following equation under the conventional VM control:

$$D_y = m - 0.5. \tag{4.1}$$

As shown in Fig. 4.2, the value of D_y is limited to (0, 0.5), and m is in the range of (0.5, 1).

The working mode of the converter can be divided into four cases according to the range of D_φ: mode A ($0 < D_\varphi < D_y$), mode B ($D_y < D_\varphi < 0.5$), mode C ($D_y - 0.5$

Fig. 4.2 Conventional working waveform under VM control when $0 < D_\varphi < D_y$ and $0.5 < m < 1$

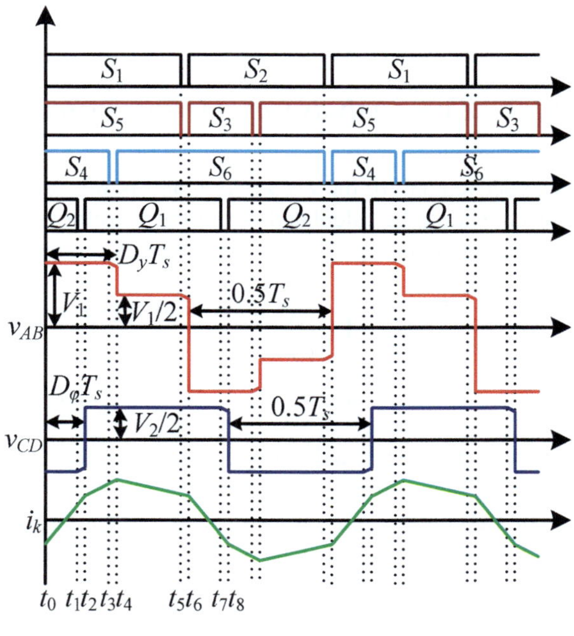

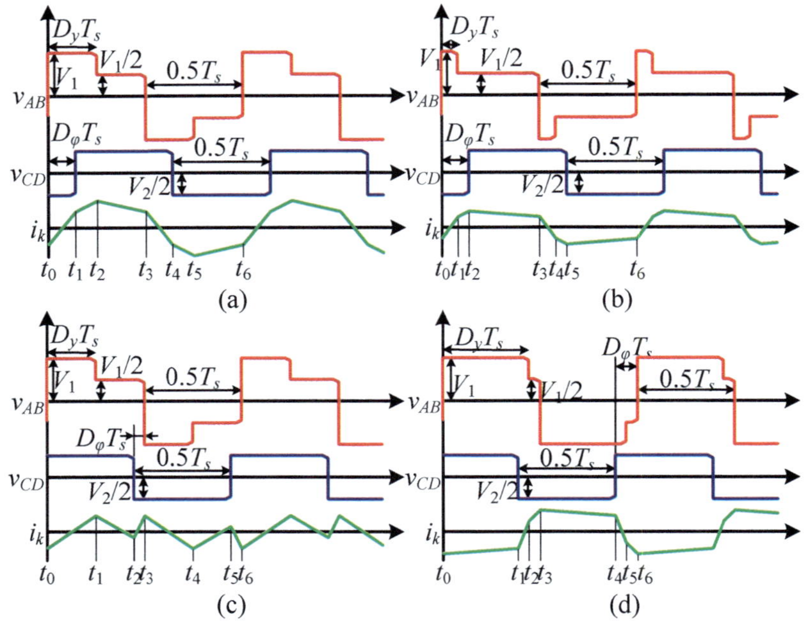

Fig. 4.3 Four operation modes. **a** Mode A: $0 < D_\varphi < D_y$. **b** Mode B: $D_y < D_\varphi < 0.5$. **c** Mode C: $D_y - 0.5 < D_\varphi < 0$. **d** Mode D: $-0.5 < D_\varphi < D_y - 0.5$

$< D_\varphi < 0$), and mode D ($-0.5 < D_\varphi < D_y - 0.5$). Figure 4.3 depicts the theoretical operation waveforms of these patterns, respectively.

4.1.2 ZVS Analysis Considering Switch Junction Capacitors

In order to analyze the ZVS region and power transfer characteristics, the leakage current should be calculated. From the waveforms of v_{AB} and v_{CD} in Fig. 4.3a, the expression of leakage current i_k of mode A can be obtained as follows:

$$i_k(t) = \begin{cases} i_k(t_0) + (V_1 + nV_2/2)(t - t_0)/L_k & t_0 < t < t_1 \\ i_k(t_1) + (V_1 - nV_2/2)(t - t_1)/L_k & t_1 < t < t_2 \\ i_k(t_2) + (V_1/2 - nV_2/2)(t - t_2)/L_k & t_2 < t < t_3 \end{cases}. \tag{4.2}$$

where, $i_k(t_0)$, $i_k(t_1)$, and $i_k(t_2)$ represent the instantaneous current value of L_k at t_0, t_1 and t_2, respectively.

In a complete cycle, the current waveform at t_3–t_6 is symmetrical with that of t_0-t_3. Meanwhile, the average value of i_k is zero in the steady state. The expressions of $i_k(t_0)$, $i_k(t_1)$, and $i_k(t_2)$ can be obtained. Then, based on (4.1), these expressions of mode A can be simplified as shown in Table 4.1. Based on the operation waveform of other working modes in Fig. 4.3, the leakage current expressions at the instants when the switches are turned on are summarized in Table 4.1.

The ZVS equivalent circuits at t_0, t_1 and t_2 of mode A are illustrated in Fig. 4.4. The transformer is equivalent to a voltage source in these ZVS equivalent circuits. Assuming that the value of the V_1 side switch junction capacitor is equal to C_p, and the value of the V_2 side switch junction capacitor is C_s. Where, C_{s1}–C_{s6}, C_{q1} and C_{q2} represent the value of junction capacitor corresponding to each switch. Taking mode A as an example, the ZVS constraints considering the charge and discharge of switch junction capacitors are analyzed in detail.

$$\begin{cases} C_{s1} = C_{s2} = C_{s3} = C_{s4} = C_{s5} = C_{s6} = C_p \\ C_{q1} = C_{q2} = C_s \end{cases} \tag{4.3}$$

At t_0, S_2, S_6 are turned off. The drain-source voltage of S_2 is zero, the drain-source voltages of S_3 and S_4 are both equal to $V_1/2$, and the voltage across junction capacitor of S_1 is equal to V_1. i_k starts to discharge C_{s1} and C_{s4}, while charge C_{s2} and C_{s3}, as shown in Fig. 4.4a. The expressions of i_k and the drain-source voltage of S_4 can be calculated in (4.4) during the discharge of switch junction capacitors.

Table 4.1 Instant leakage current expressions at four working modes under conventional VM control

	$i_k(t_0)$	$i_k(t_1)$	$i_k(t_2)$
A: $0 < D_\varphi < D_y$	$-\frac{D_\varphi nV_2T_s}{2L_k}$	$\frac{D_\varphi V_1T_s}{L_k}$	$\frac{(D_\varphi nV_2+2D_yV_1-D_ynV_2)T_s}{2L_k}$
B: $D_y < D_\varphi < 0.5$	$-\frac{D_\varphi nV_2T_s}{2L_k}$	$\frac{(D_ynV_2+2D_yV_1-D_\varphi nV_2)T_s}{2L_k}$	$\frac{(D_\varphi+D_y)V_1T_s}{2L_k}$
C: $D_y - 0.5 < D_\varphi < 0$	$\frac{D_\varphi nV_2T_s}{2L_k}$	$\frac{(D_\varphi nV_2+2D_yV_1-D_ynV_2)T_s}{2L_k}$	$\frac{D_\varphi V_1T_s}{2L_k}$
D: $-0.5 < D_\varphi < D_y - 0.5$	$\frac{D_\varphi nV_2T_s}{2L_k}$	$\frac{(4D_\varphi-2D_y+1)V_1T_s}{4L_k}$	$\frac{(D_ynV_2-D_\varphi nV_2-V_1)T_s}{2L_k}$

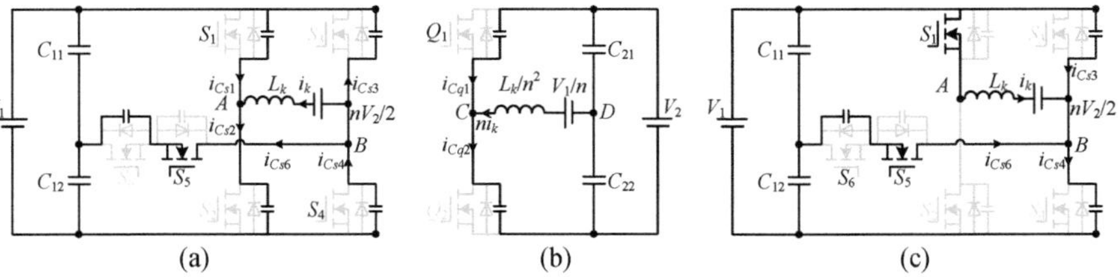

Fig 4.4 Equivalent ZVS circuits of mode A. **a** Time t_0. **b** Time t_1. **c** Time t_2

$$\begin{cases} i_k(t) = C_p \dfrac{du_{Cs4}(t)}{dt} - C_p \dfrac{du_{Cs3}(t)}{dt} - C_p \dfrac{du_{Cs6}(t)}{dt} \\ = C_p \frac{du_{Cs2}(t)}{dt} - C_p \frac{du_{Cs1}(t)}{dt} \\ u_{Cs4}(t) = V_1 - u_{Cs1}(t) - L_k \dfrac{di_k(t)}{dt} + \dfrac{nV_2}{2} \end{cases} \tag{4.4}$$

Since C_{11} and C_{12} are large enough for the switch junction capacitors, the variation law of the switch junction capacitors satisfies the following equation:

$$\begin{cases} C_p \dfrac{du_{Cs4}(t)}{dt} = -C_p \dfrac{du_{Cs3}(t)}{dt} = -C_p \dfrac{du_{Cs6}(t)}{dt} \\ C_p \dfrac{du_{Cs2}(t)}{dt} = -C_p \dfrac{du_{Cs1}(t)}{dt} \end{cases}. \tag{4.5}$$

From (4.4) and (4.5), the expression of u_{Cs4} can be represented by $i_k(t_0)$.

$$u_{Cs4}(t) = \sqrt{\frac{2L_k}{15C_p}} i_k(t_0) \sin\left(\sqrt{\frac{5}{6L_k C_P}} t\right) + \left(\frac{V_1}{5} - \frac{nV_2}{5}\right) \cos\left(\sqrt{\frac{5}{6L_k C_P}} t\right) + \frac{3V_1}{10} + \frac{nV_2}{5}. \tag{4.6}$$

To ensure that the voltage of C_{s4} can be zero to meet ZVS conditions, the following equation should be satisfied.

$$\left(\frac{V_1}{5} - \frac{nV_2}{5}\right)^2 + \frac{2L_k}{15C_p} i_k^2(t_0) \geq \left(\frac{3V_1}{10} + \frac{nV_2}{5}\right)^2. \tag{4.7}$$

From (4.7), the ZVS condition of S_4 can be obtained:

$$i_k(t_0) \leq -V_1 \sqrt{\frac{3C_p(8m+1)}{8L_k}}. \tag{4.8}$$

Similarly, according to (4.4) and (4.5), the ZVS condition of S_1 can be derived as:

$$i_k(t_0) \leq -V_1 \sqrt{\frac{4C_p(3m+1)}{3L_k}} \leq -V_1 \sqrt{\frac{3C_p(8m+1)}{8L_k}}. \tag{4.9}$$

At t_1, Q_2 is turned off, the u_{Cq2} is zero, and the u_{Cq1} is V_2. C_{q1} begins to be discharged and C_{q2} begins to be charged, as shown in Fig. 4.4b. The energy released. and absorbed by C_{q1} and C_{q2} is the same. Then, the ZVS condition of Q_1 is

$$i_k(t_1) \geq 0. \tag{4.10}$$

At t_2, S_4 is turned off. The drain-source voltage of S_1 and S_4 are zero, the drain-source voltage of S_6 is equal to $V_1/2$, and the drain-source voltage of S_2 and S_3 are both equal to V_1. As shown in Fig. 4.4c, the C_{s6} and C_{s3} begin to be discharged and C_{s4} begins to be charged. the expression of u_{Cs6} is derived in (4.11) during the discharge of switch junction capacitors:

$$u_{Cs6}(t) = -\sqrt{\frac{L_k}{3C_p}} I_k(t_2) \sin\left(\sqrt{\frac{1}{3L_k C_p}} t\right) + \left(V_1 - \frac{nV_2}{2}\right) \cos\left(\sqrt{\frac{1}{3L_k C_p}} t\right) - \frac{V_1}{2} + \frac{nV_2}{2}. \tag{4.11}$$

Then, the ZVS condition of S_6 can be calculated.

$$\begin{cases} I_k(t_2) \geq 0 \\ (V_1 - \frac{nV_2}{2})^2 + \frac{L_k}{3C_p} I_k^2(t_2) \geq (\frac{V_1}{2} - \frac{nV_2}{2})^2 \end{cases}. \tag{4.12}$$

In one switching period, the positive half-cycle and negative half-cycle waveforms are symmetrical, and the working state is also symmetrical. The ZVS conditions of S_2, S_3, S_5 and Q_2 are consistent with that of S_1, S_4, S_6 and Q_1. By replacing $i_k(t_0)$, $i_k(t_1)$, $i_k(t_2)$ in (4.9), (4.10) and (4.12) with the current expressions in Table 4.1, the ZVS constraint conditions at mode A can be obtained. On the other hand, the ZVS restrictions of other modes can also be obtained and be summarized in Table 4.2 together with that of mode A, where $\tau_a = (L_k C_p)^{1/2}/T_s$, $\tau_b = (L_k C_s)^{1/2}/T_s$.

Table 4.2 ZVS conditions of all power switches at four work modes with considering junction capacitors

	ZVS for S_1-S_4	ZVS for Q_1&Q_2	ZVS for S_5&S_6
A: $0 < D_\varphi < D_y$	$D_\varphi \geq \frac{\tau_a}{m}\sqrt{\frac{3(8m+1)}{8}}$	$D_\varphi \geq 0$	$D_\varphi \geq 0$
B: $D_y < D_\varphi < 0.5$	$D_\varphi \geq \frac{\tau_a}{m}\sqrt{\frac{3(8m+1)}{8}}$	$D_\varphi \geq 0$	$(D_\varphi - D_y)nV_2 + 2D_yV_1 \geq 0$
C: $D_y - 0.5 < D_\varphi < 0$	$D_\varphi \leq 0$	$D_\varphi \leq -\frac{4\tau_b}{n}\sqrt{m}$	$(D_y - D_\varphi)nV_2 + 2D_yV_1 \geq 0$
D: $-0.5 < D_\varphi < D_y - 0.5$	$D_\varphi \leq 0$	$D_\varphi \leq 0.5m - 0.5 - \frac{2\tau_b}{n}\sqrt{2m}$	$(D_y - D_\varphi)nV_2 - V_1 \geq 0$

4.1.3 ZVS Region Under Conventional VM Control with Considering Switch Junction Capacitors

According to the operation waveform in Fig. 4.3a and the instantaneous current expression in Table 4.1, the transferred power of mode A is

$$
\begin{aligned}
P_{o\_A}(D_y, D_\varphi) &= \frac{2}{T_s}\left[\int_{t_0}^{t_1} V_1 i_k(t)dt + \int_{t_1}^{t_2} V_1 i_k(t)dt + \int_{t_2}^{t_3} 0.5V_1 i_k(t)dt\right] \\
&= \frac{nV_2V_1T_s}{8L_k}(8D_\varphi^2 - 2D_\varphi - 4D_\varphi D_y - D_y + 2D_y^2)
\end{aligned} \tag{4.13}
$$

Similarly, the transferred power functions of the other three modes and formula (4.13) are summarized in (4.14). Where, P_{o_B}, P_{o_C} and P_{o_D} represent the power transfer functions of mode B, mode C and mode D respectively. The power transfer is normalized by $P_{base} = (nV_2)^2/(16L_kf_s)$.

Based on (4.14) and the ZVS constraints conditions listed in Table 4.2, the working areas of different powers under different modes are normalized, and the ZVS region of the converter under conventional VM control is presented in Fig. 4.5. Figure 4.5a describes the theoretical ZVS region ignoring switch junction capacitors, and Fig. 4.5b illustrates the ZVS region considering the charge and discharge of switch junction capacitors. It can be seen that some switches in the transition area between modes A and C cannot realize ZVS with considering switch junction capacitors. Moreover, ZVS cannot be realized in most of the backward power transmission region under the conventional VM controlling.

$$
\begin{cases}
P_{o\_A} = \dfrac{-nV_2V_1T_s}{8L_k}(8D_\varphi^2 - 2D_\varphi - 4D_\varphi D_y - D_y + 2D_y^2) \\
P_{o\_B} = \dfrac{-nV_2V_1T_s}{8L_k}(4D_\varphi^2 - 2D_\varphi + 4D_\varphi D_y - D_y - 2D_y^2) \\
P_{o\_C} = \dfrac{nV_2V_1T_s}{8L_k}(4D_\varphi^2 + 2D_\varphi + 4D_\varphi D_y + D_y - 2D_y^2) \\
P_{o\_D} = \dfrac{nV_2V_1T_s}{8L_k}(8D_\varphi^2 + 6D_\varphi - 4D_\varphi D_y - 3D_y + 2D_y^2 + 1)
\end{cases}. \tag{4.14}
$$

4.2 Magnetizing Inductor to Eliminate the Effect of Junction Capacitor

Due to the influence of switch junction capacitors, the leakage inductor needs enough energy to realize ZVS of switches. However, in the transition interval between modes A and C, i_k is small, which leads to the failure of ZVS as shown in Fig. 4.5b.

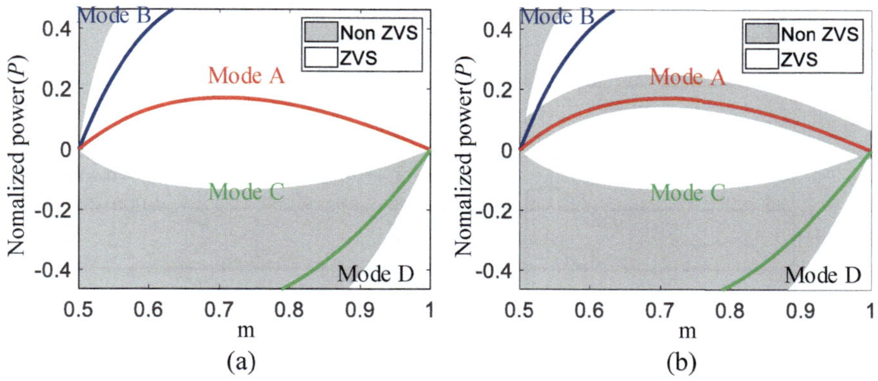

Fig. 4.5 ZVS Region under conventional VM control. **a** Ignoring switch junction capacitors. **b** Considering switch junction capacitors

This chapter proposes a current injections method to solve this problem is depicted in Fig. 4.6. Figure 4.7 shows the working waveform of the improved converter. Compared with the conventional converter in Fig. 4.1, the auxiliary inductor L_1 and the magnetizing inductance L_m are added to both sides of the transformer. The waveform of the current $i_{L1}(t)$ in mode A is shown in Fig. 4.7a, b presents the waveform of the current $i_{Lm}(t)$ in mode C. These inductors provide additional energy to charge and discharge switch junction capacitors for realizing ZVS. According to Table 4.2, switch junction capacitors limit the realization of ZVS of switches S_1-S_4, Q_1 and Q_2. The following will take modes A and C as examples to analysis the influences of L_1 and L_m on ZVS, where $i(t) = i_{L1}(t) + i_k(t)$ and $i_s(t) = ni_k(t)–i_{Lm}(t)$.

At mode A, since the switching state changes at t_0 and t_2 during the positive half-cycle on the V_1 side, ZVS analysis is only carried out at these moments.

According to the symmetry of the voltage waveform of L_1 and the principle that the average value of i_{L1} in each steady switching cycle must be equal to zero, the

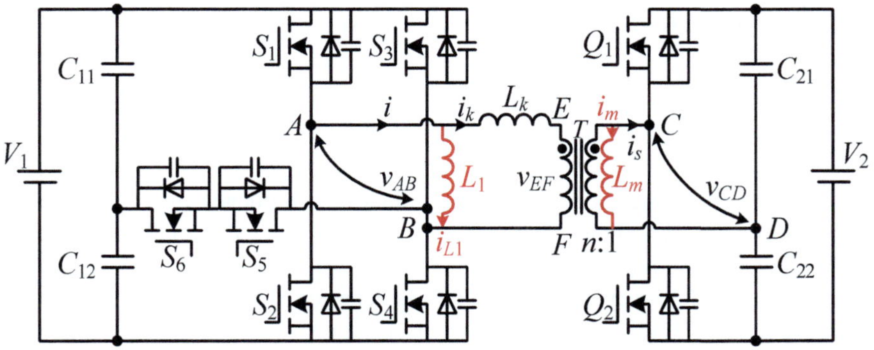

Fig. 4.6 Hybrid-bridge-based DAB converter with the auxiliary inductor L_1 and the magnetizing inductance L_m

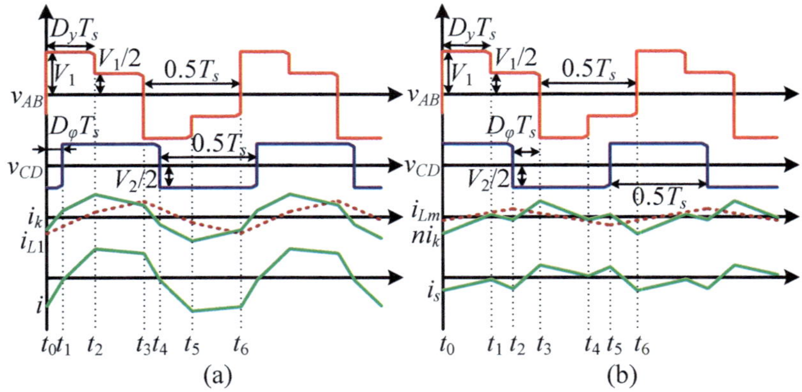

Fig. 4.7 **a** Mode A current waveforms considering L_1 at V_1 side. **b** Mode C current waveforms considering L_m at V_2 side

values of i_{L1} at t_0 and t_2 can be derived as:

$$\begin{cases} i_{L1}(t_0) = -\dfrac{V_1 T_s(1+2D_y)}{8L_1} \\ i_{L1}(t_2) = \dfrac{V_1 T_s(6D_y-1)}{8L_1} \end{cases}. \tag{4.15}$$

Based on current constraints of mode A at t_0 and t_2 listed in (4.9) and (4.12), the ZVS conditions of switches S_1–S_4 after adding the auxiliary inductor L_1 can be calculated as follows:

$$\begin{cases} i(t_0) = i_k(t_0) + i_{L1}(t_0) \le -V_1\sqrt{\dfrac{4C_p(3m+1)}{3L_k}} \\ i(t_2) = i_k(t_2) + i_{L1}(t_2) \ge 0 \end{cases}. \tag{4.16}$$

Then the range of L_1 is obtained by (1) and (16), where $L_a = T_s L_k/(L_k C_p)^{1/2}$:

$$L_1 \le \sqrt{\frac{3}{64(3m+1)}} m L_a. \tag{4.17}$$

Reducing the inductance of L_1 can eliminate the influence of switch junction capacitors on the ZVS of switches S_1-S_4. On the other hand, the smaller the inductor L_1, the larger the current flowing through it, which can lead unnecessary conduction losses. As a result, the optimal value of L_1 is determined as $(3/64(3m+1))^{1/2} m L_a$. In practice, some margin can be designed accordingly.

When $D_y < 1/6$, $i_{L1}(t_2)$ is less than zero which has an adverse effect on the ZVS implementations of S_5, S_6. The ZVS constraints can be calculated as follows from

(4.9) and (4.16):

$$D_\varphi \geq \frac{L_k - 6D_yL_k - 8D_yL_1 + 8D_yL_1m}{8L_1m}. \tag{4.18}$$

At mode C, the charging and discharging of switch junction capacitors occur at the time t_2 on the V_2 side. According to the symmetry of the voltage waveform of L_m and the average value of its current is zero in one switching period. The value of $i_{Lm}(t_2)$ is:

$$i_{Lm}(t_2) = \frac{V_2T_s}{8L_m}. \tag{4.19}$$

Similarly, the current limiting conditions at mode C can be obtained. After adding magnetizing inductance L_m, the ZVS conditions of Q_1 and Q_2 are as follows:

$$i_s(t_2) = \frac{nD_\varphi V_1T_s}{2L_k} - \frac{V_2T_s}{8L_m} \leq -V_1\sqrt{\frac{4C_sm}{L_k}}. \tag{4.20}$$

The range of L_m can be obtained, where $L_b = T_sL_k/(L_kC_s)^{1/2}$:

$$L_m \leq \frac{\sqrt{m}}{8n}L_b. \tag{4.21}$$

Under the same analysis, the ZVS conditions of the other two modes are also summarized in Table 4.3. Combining the four mode power expressions listed in (4.14) with the ZVS conditions in Table 4.3, the ZVS region of S_5 and S_6 after adding L_1 and L_m is depicted in Fig. 4.8.

It is obvious from Figs. 4.5b and 4.8 that adding L_1 and L_m can effectively eliminate the influence of the switch junction capacitors in the mode transition, and the ZVS region of the converter is extended. Noted that, only one additional inductor is needed for the hardware and L_m is inductance formed by the magnetizing inductance of the transformer. However, the problem that the ZVS region of the converter is limited when operating in the backward state has not been well solved as shown in Fig. 4.8.

Table 4.3 ZVS conditions after adding L_1 and L_m under conventional VM control

	ZVS for S_1-S_4	ZVS for Q_1&Q_2	ZVS for S_5&S_6
A: $0 < D_\varphi < D_y$	$D_\varphi \geq 0$	$D_\varphi \geq 0$	$D_\varphi \geq \frac{L_k-6D_yL_k-8D_yL_1+8D_yL_1m}{8L_1m}$
B: $D_y < D_\varphi < 0.5$	$D_\varphi \geq 0$	$D_\varphi \geq 0$	$D_\varphi \leq \frac{6D_yL_k-L_k+8D_yL_1+8D_yL_1m}{8L_1m}$
C: $D_y - 0.5 < D_\varphi < 0$	$D_\varphi \leq 0$	$D_\varphi \leq 0$	$D_\varphi \geq \frac{L_k-6D_yL_k-8D_yL_1+8D_yL_1m}{8L_1m}$
D: $-0.5 < D_\varphi < D_y - 0.5$	$D_\varphi \leq 0$	$D_\varphi \leq 0$	$D_\varphi \leq -\frac{L_k+4L_1-6D_yL_k-8D_yL_1m}{8L_1m}$

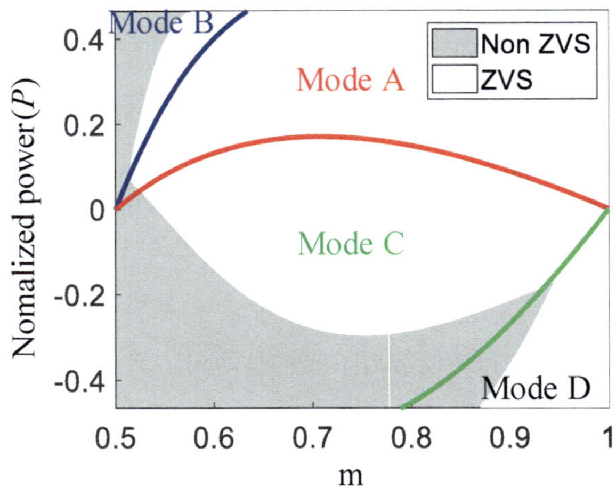

Fig. 4.8 ZVS Region of S_5 and S_6 after adding inductors L_1 and L_m

4.3 Modified VM Control to Achieve ZVS Under Backward Power Conditions

4.3.1 Principle of Modified VM Control

Combined with the above analysis, a modified VM control is also proposed to extend the ZVS region under backward power. Figure 4.9 shows the operation waveforms under forward and backward power of the proposed control scheme. The definition of each variable in the modified VM control is consistent with that in Fig. 4.2, and D_y still satisfies (4.1). As shown in Fig. 4.9, the modification of the control is focused on the waveform of v_{AB}. For example, under forward power transmission, when S_1, S_4 and S_5 are turned on, the voltage level of v_{AB} is V_1. While, under backward power transmission, the voltage of v_{AB} is $V_1/2$ instead of V_1, which is achieved by changing the control scheme.

Figure 4.10 presents the four working modes of modified VM control under backward power, in which the range of mode A is $0 < D_\varphi < 0.5 - D_y$, the range of mode B is $0.5 - D_y < D_\varphi < 0.5$, the range of mode C is $-D_y < D_\varphi < 0$, and the range of mode D is $-0.5 < D_\varphi < -D_y$. Based on these working waveforms, the leakage inductor current i_{Lk} expressions are given in Table 4.4.

From (4.13) and leakage current value in Table 4.4, the power expressions of the modified VM control under backward power transfer are

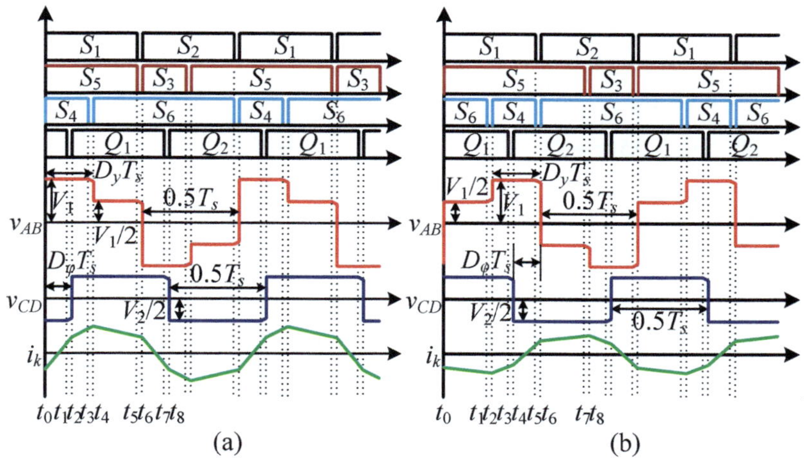

Fig. 4.9 Working waveforms of modified VM scheme. **a** Under forward power. **b** Under backward power

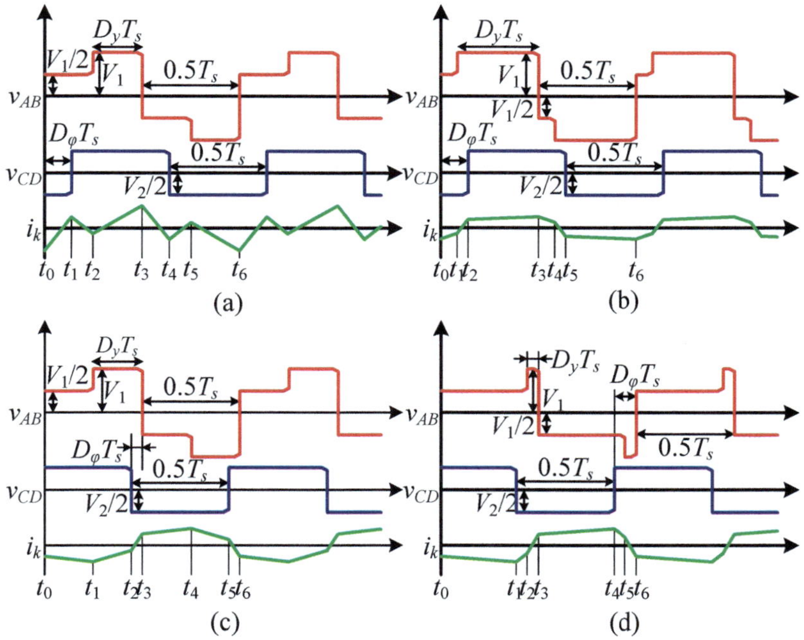

Fig. 4.10 Four operation modes of modified VM control under backward power transfer. **a** Mode A: $0 < D_\varphi < 0.5\text{-}D_y$. **b** Mode B: $0.5\text{-}D_y < D_\varphi < 0.5$. **c** Mode C: $\text{-}D_y < D_\varphi < 0$. **d** Mode D: $-0.5 < D_\varphi < \text{-}D_y$

Table 4.4 Instant leakage current expressions of modified VM control under backward power

	$i_k(t_0)$	$i_k(t_1)$	$i_k(t_2)$
A: $0 < D_\varphi < 0.5 - D_y$	$-\frac{D_\varphi nV_2T_s}{2L_k}$	$\frac{D_\varphi V_1T_s}{2L_k}$	$\frac{(D_\varphi nV_2+D_ynV_2-2D_yV_1)T_s}{2L_k}$
B: $0.5 - D_y < D_\varphi < 0.5$	$-\frac{D_\varphi nV_2T_s}{2L_k}$	$\frac{(V_1-D_\varphi nV_2-D_ynV_2)T_s}{2L_k}$	$\frac{(4D_\varphi V_1-2V_1+nV_2)T_s}{4L_k}$
C: $-D_y < D_\varphi < 0$	$\frac{D_\varphi nV_2T_s}{2L_k}$	$\frac{(D_\varphi nV_2+D_ynV_2-2D_yV_1)T_s}{2L_k}$	$\frac{D_\varphi V_1T_s}{L_k}$
D: $-0.5 < D_\varphi < -D_y$	$\frac{D_\varphi nV_2T_s}{2L_k}$	$\frac{(D_\varphi-D_y)V_1T_s}{2L_k}$	$-\frac{(D_\varphi nV_2+D_ynV_2+2D_yV_1)T_s}{2L_k}$

$$\begin{cases} P_{o\_A} = \frac{-nV_2V_1T_s}{8L_k}(8D_\varphi^2 - 2D_\varphi - 4D_\varphi D_y - D_y + 2D_y^2) \\ P_{o\_B} = \frac{-nV_2V_1T_s}{8L_k}(4D_\varphi^2 - 2D_\varphi + 4D_\varphi D_y - D_y - 2D_y^2) \\ P_{o\_C} = \frac{nV_2V_1T_s}{8L_k}(4D_\varphi^2 + 2D_\varphi + 4D_\varphi D_y + D_y - 2D_y^2) \\ P_{o\_D} = \frac{nV_2V_1T_s}{8L_k}(8D_\varphi^2 + 6D_\varphi - 4D_\varphi D_y - 3D_y + 2D_y^2 + 1) \end{cases} \quad (4.22)$$

where, P_{o2_A}, P_{o2_B}, P_{o2_C} and P_{o2_D} are the power transfer functions of mode A, mode B, mode C and mode D respectively.

Considering the interaction of the junction capacitors, the auxiliary inductor L_1 and the magnetizing inductance L_m, the ZVS constraints of the modified VM control under backward power can be obtained in Table 4.5. The ZVS region of the modified VM control is depicted in Fig. 4.11b by (4.22), constraint conditions of ZVS in Table 4.5 and the ZVS region in forward power in Fig. 4.8. Compared with the ZVS region under the conventional VM control in Fig. 4.11a, it is obviously seen that the full range ZVS of all switches are realized under the modified VM control when $m > 0.58$.

Table 4.5 ZVS constraints of modified VM control under backward power

	ZVS for $S_1, S_2\&S_5, S_6$	ZVS for $Q_1\&Q_2$	ZVS for $S_3\&S_4$
A: $0 < D_\varphi < 0.5 - D_y$	$D_\varphi \geq 0$	$D_\varphi \geq 0$	$D_\varphi \leq \frac{1+2m^2-3m-\sqrt{3}\tau_a}{2m} - \frac{2-3m}{4km}$
B: $0.5 - D_y < D_\varphi < 0.5$	$D_\varphi \geq 0$	$D_\varphi \geq 0$	$D_\varphi \leq \frac{1-2m^2+m+\sqrt{3(4m+3)}\tau_a}{2m} + \frac{2-3m}{4km}$
C: $-D_y < D_\varphi < 0$	$D_\varphi \leq 0$	$D_\varphi \leq 0$	$D_\varphi \leq \frac{1+2m^2-3m-\sqrt{3}\tau_a}{2m} - \frac{2-3m}{4km}$
D: $-0.5 < D_\varphi < -D_y$	$D_\varphi \leq 0$	$D_\varphi \leq 0$	$D_\varphi \leq \frac{1-2m^2-m+\sqrt{3(4m+3)}\tau_a}{2m} + \frac{2-3m}{4km}$

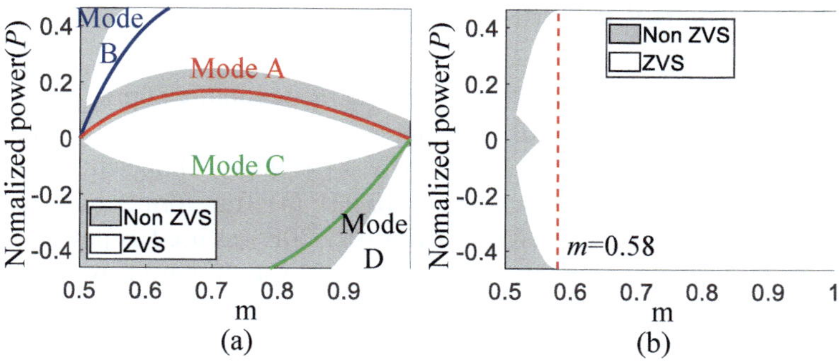

Fig. 4.11 ZVS Restrictions. **a** Conventional VM control. **b** Modified VM control

4.3.2 Implementation of the Modified VM Control

Figure 4.12 shows the control block diagram of the control proposed above. V_1 and V_2 are the actual input voltage and output voltage respectively, and V_{ref} is the output reference voltage. The error between V_2 and V_{ref} is used to obtain D_φ through the PI regulator. Due to the power limitation of the converter, D_φ is limited between $D_{\varphi max}$ and $D_{\varphi min}$. It is easy to get D_y through V_1 and V_2 as well as the Eq. (4.1). By the output value D_φ of the PI control loop and D_y of the D_y generation loop as well as the power expressions (4.14) and (4.22), output power estimated value P_{o_e} can be obtained. The power estimated value is used to evaluate the magnitude and direction of the output power. Then, the switching modulation is changed to extend the ZVS region according to the direction of power flow. The use of switch modulation changes has a negative impact on the dynamic performance of the system. However, the steady-state stability of the converter is improved due to the realization of full range ZVS with the proposed solution.

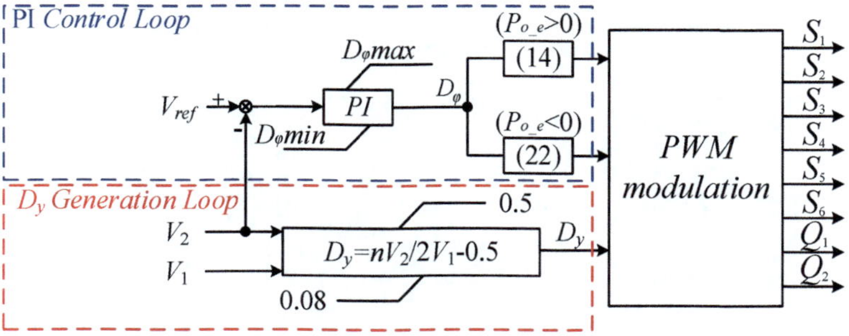

Fig. 4.12 Control block diagram of modified VM control

4.4 Experimental Results

A laboratory prototype was established to verify the feasibility of proposed solution scheme in the hybrid-bridge DAB converter is presented in Fig. 4.13. The converter was designed as follows: (1) $V_1 = 120$–200 V, $V_2 = 96$ V; (2) Switching frequency $f_s = 100$ kHz; (3) Leakage inductor $L_k = 10$ μH; (4) Transformer turns ratio $n =$ 20:8; (5) Rated power $P_e = 1$ kW; (6) IXFK102N30P was used as power switches S_1-S_6, and its junction capacitor is 920 pF. IRFP4321 was the power switch on the V_2 side, and its junction capacitor is 308 pF. According to (4.17) and (4.21), the effect of junction capacitors can be eliminated by reducing inductance L_1 and L_m. Then, L_1 and L_m were designed as 70 μH and 40 μH respectively to extend ZVS region under the minimum circulating current. Under the parameters of the laboratory prototype, the adding of L_1 and L_m has little effect on the converter gain, which can be ignored.

According to the AP method, ferrite core EE42/20 is selected as the transformer core and the primary and secondary side turn ratio N_p:N_s = 20:8 can be calculated from the input voltage and output voltage range. The related equations are as follows,

$$n = \frac{2V_{1\min}}{V_2}. \tag{4.23}$$

$$N_s = \frac{V_2 T_s}{4A_e \Delta B}. \tag{4.24}$$

$$N_p = nN_s. \tag{4.25}$$

Fig. 4.13 The laboratory prototype of hybrid-bridge DAB converter with current injections

where, A_e is the area for flux flow, ΔB is the flux swing and is designed as a constant value of 0.125 T. The magnetizing inductance L_m can be designed by adding appropriate air gap in the transformer (the core reluctance is rather small compared with the air gap and can be ignored):

$$L_m = \frac{\mu_0 n^2 A_e}{l_g}. \tag{4.26}$$

where μ_0 is free air magnetic permeability, A_e is area for flux flow, l_g is the length of air gap.

In addition, the V_1 side capacitors C_{11} and C_{12} were designed as 10 μF, and the values capacitors C_{21} and C_{22} on the V_2 side were designed as 20 μF.

When input voltages $V_1 = 200$ V and 120 V respectively, the ZVS waveforms of S_1, S_3 and Q_1 in the mode transition area are shown in Figs. 4.14 and 4.15 under conventional control [5] and proposed solution. u_{ds}, u_{gs} are respectively drain-source voltage and drive signal of the corresponding switch. The current i, i_s are the bridge injection current between the primary and secondary sides of the converter, and the voltage u_{AB}, u_{CD} are the voltage at both ends of the transformer. It can be seen that the phase shift between u_{CD} and u_{ds} of the switch S_3 is close to zero, and the converter operates in the transition region between mode A and mode C. Compared with conventional control [5], these switches always can realize ZVS under the proposed solution scheme in the transition region. The leakage current at the boundary of mode A and mode C is small with conventional control [5], which is not enough to support the charge and discharge of switch junction capaci tance. The current injections method by adding L_1 and L_m can effectively solution the problem.

The working waveforms of forward heavy load (1000 W), forward light load (100 W), backward light load (−100 W), and backward heavy load (-1000 W) are shown in Figs. 4.16, 4.17, and 4.18 under different input voltages of 200 V, 150 V, and 120 V with the proposed solution. These figures show the waveform of voltage u_{AB}, u_{CD} and the current i, i_s. In order to directly determine whether ZVS is reached, the instantaneous current values are measured at time the corresponding switch is turned on. Meanwhile, the current limiting conditions of ZVS are summarized in Table 4.6 through the same analysis method in Sect. 4.1.

In Table 4.6, i_{ZVS_S} represents the value of i_s at the rising time of u_{CD} to judge whether Q_1 and Q_2 can achieve ZVS. i_{ZVS_1} and i_{ZVS_2} have different meanings under different ope rating conditions. In the case of forward power, i_{ZVS_1} and i_{ZVS_2} refer to the i values at the rising time and falling time of u_{AB}, respectively. In the case of backward power, i_{ZVS_1} and i_{ZVS_2} indicate the values of i when the voltage u_{AB} rises from '-V_1' to '$V_1/2$' and from '$V_1/2$' to 'V_1', respectively. i_{ZVS_1} and i_{ZVS_2} decide whether switches S_1-S_6 can realize ZVS. Noted that $i_{ZVS\_1} = i_{ZVS\_2}$ when $m = 1.0$ under backward power, so the implementation of ZVS only be judged by the value of i_{ZVS_1}.

From Table 4.6, the current limiting conditions under different input voltages can be obtained. Then, the measured current values in the experiment were compared with

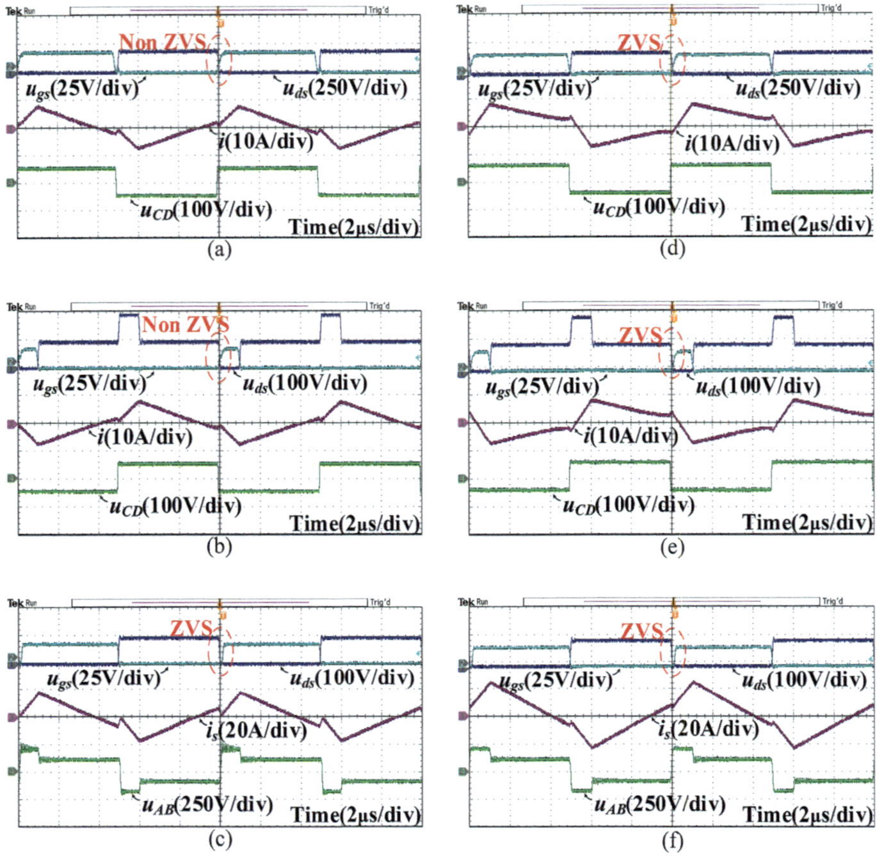

Fig. 4.14 Experimental waveforms under $V_1 = 200$ V with conventional control [5] **a** $S_1\&S_2$. **b** $S_3\&S_4$. **c** $Q_1\&Q_2$. Experimental waveforms with proposed solution **d** $S_1\&S_2$. **e** $S_3\&S_4$. **f** $Q_1\&Q_2$

the theoretically calculated currents which are the constraint conditions of ZVS, and all currents meet to the inequalities in Table 4.6. Hence, it is proved that all switches can realize ZVS at these cases.

The efficiency curves under the conventional control [5] and the proposed solution which uses the current injections method under the modified VM control are depicted in Fig. 4.19. Figure 4.19a shows the efficiency curves of two control schemes under forward power when $m = 0.65$ and 1.0. When $m = 0.65$, the efficiency under proposed solution is significantly higher than that under conventional control [5] in the medium power range. Due to the influence of junction capacitors, there are more switching losses under conventional control as shown in Fig. 4.20b. However, the addition of the auxiliary inductor L_1 and the magnetizing inductance L_m also brings more additional loss when each switch can realize ZVS under the two control schemes, which slightly reduces the efficiency. The loss breakdown under light load ($P = 100$ W), half load ($P = 500$ W) and heavy load ($P = 1000$ W) at $m = 0.65$ are shown in Fig. 4.20 with

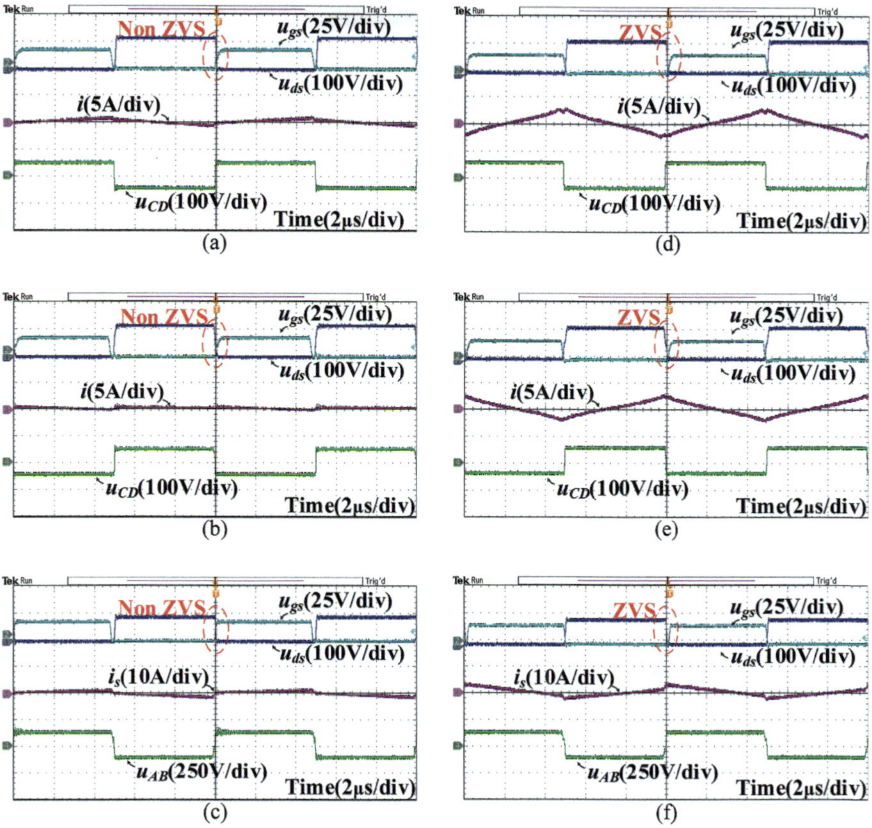

Fig. 4.15 Experimental waveforms under $V_1 = 120$ V with conventional control [5] **a** S_1&S_2. **b** S_3&S_4. **c** Q_1&Q_2. Experimental waveforms with proposed solution **d** S_1&S_2. **e** S_3&S_4. **f** Q_1&Q_2

two schemes. P_T, P_{Lk}, P_{L1}, P_{sw}, P_{sw_c} and P_{Total} respectively are transformer loss, leakage inductor loss, auxiliary inductor loss, switching loss, switches conduction loss and total loss of the converter. It can be seen that the switches conduction loss P_{sw_c} and the additional core loss P_{L1} of the proposed solution lower is large than the convention control. For this reason, the total loss of the converter is large under the proposed solution in Fig. 4.20a, c Compared with the total loss, these additional losses are small at full load in Fig. 4.20c.

This phenomenon is also obvious when $m = 1.0$, as shown in Fig. 4.19a. Figure 4.19b shows the efficiency comparisons of the two control schemes at $m = 0.6$ and 0.9 under backward power. It can be seen that the efficiency under modified VM control is higher than that under conventional control in heavy load, because ZVS of the converter is difficult to realize at this time under conventional control.

Furthermore, the comparison between this work and the DAB under different modulation schemes is shown in the Table 4.7. These converters with proposed modulation have wide voltage gain range. The main characteristics of these converters

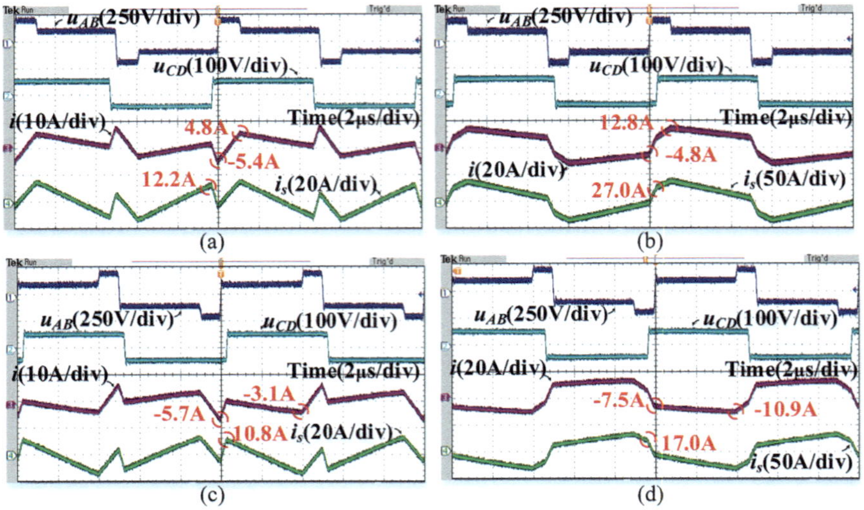

Fig. 4.16 Experimental waveforms under $V_1 = 200$ V. **a** P = 100 W. **b** P = 1000 W. **c** P = −100 W. **d** P = −1000 W

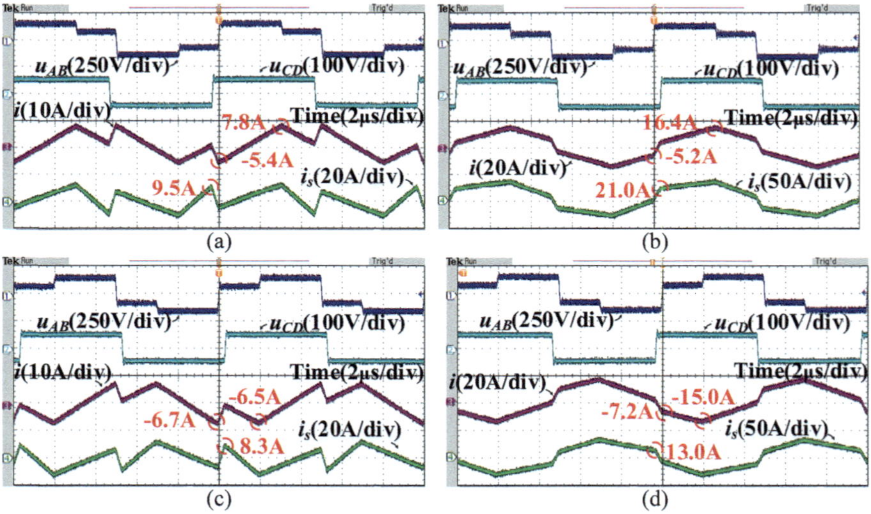

Fig. 4.17 Experimental waveforms under $V_1 = 150$ V. **a** P = 100 W. **b** P = 1000 W. **c** P = −100 W. **d** P = −1000 W

are summarized in the table, including structure, modulation strategy, soft switching range, working range and peak efficiency.

The DAB converter in [9–11] is studied based on the traditional two-level full bridge structure. References [3–5] propose multi-level DAB converters which are

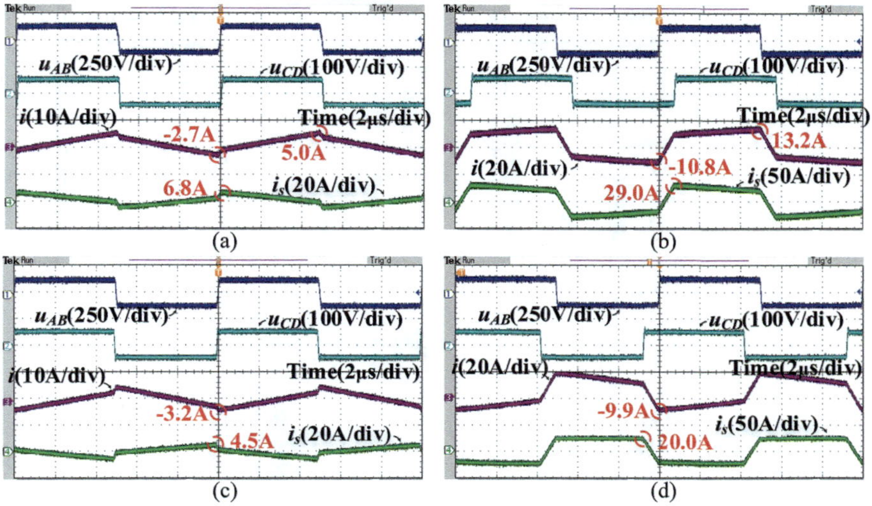

Fig. 4.18 Experimental waveforms under $V_1 = 120$ V. **a** P = 100 W. **b** P = 1000 W. **c** P = −100 W. **d** P = −1000 W

Table 4.6 Instant leakage current limit for prototype

	i_{ZVS_1}	i_{ZVS_2}	i_{ZVS_S}
Mode A of $P > 0$	$i < \frac{-1.329\sqrt{3m+1}}{m}$	$i > 0$	$i_s > 0$
Mode C of $P > 0$	$i < 0$	$i > 0$	$i_s > \frac{0.528}{\sqrt{m}}$
Mode A of $P < 0$	$i < \frac{-0.940\sqrt{6m-1}}{m}$	$i < \frac{-0.996\sqrt{4m-3}}{m}$	$i_s > 0$
Mode C of $P < 0$	$i < 0$	$i < \frac{-0.996\sqrt{4m-3}}{m}$	$i_s > \frac{0.756}{\sqrt{m}}$

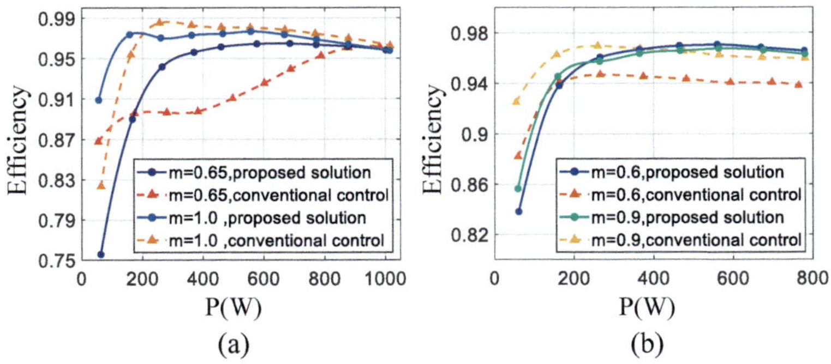

Fig. 4.19 Efficiency comparison. **a** $m = 0.65$ and 1.0 at forward power. **b** $m = 0.6$ and 0.9 at backward power

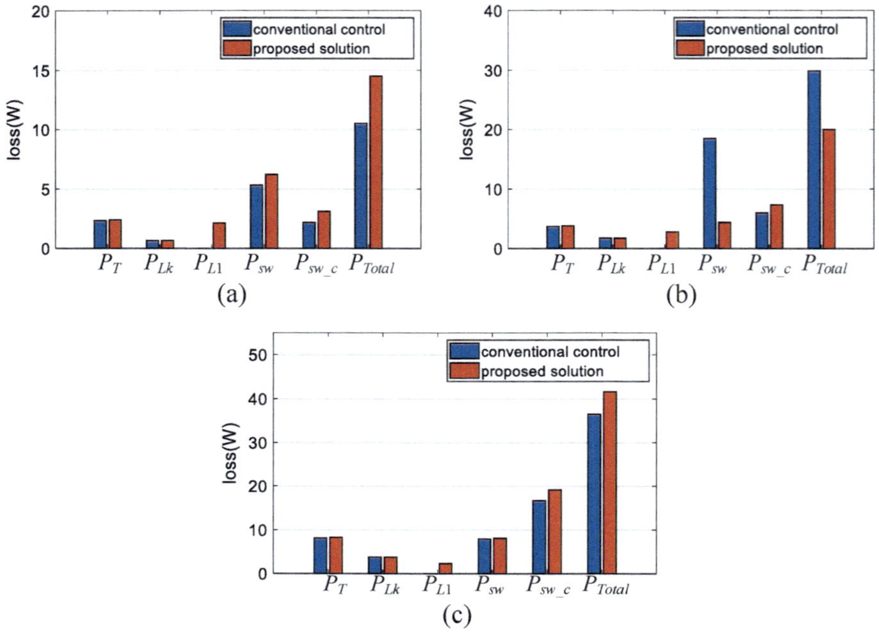

Fig. 4.20 Loss breakdown comparison at $m = 0.65$. **a** under light load (100 W). **b** under half load (500 W). **c** under full load (1000 W)

applied to the wide voltage range. In terms of the number of switches, the hybrid-bridge-based DAB converter [5] has the same number of switches as the conventional DAB converter [10, 11], while the DT DAB converter [3] and the NH3L-DAB converter [4] use more switches. For the DAB converters [5, 10, 11], additional components are added to generate multilevel. Compared with the conventional DAB converter [10, 11], DT DAB converter [3] and NH3L-DAB converter [4], the DAB converter with blocking capacitor [9] and the hybrid-bridge-based DAB converter [5] use two control variables to achieve a wide range of working. However, the ZVS range of these converter [5, 9] is limited. The proposed solution base on the converter [5] in this chapter realizes the full range soft switching under 2-variable decoupling control.

Table 4.7 Comparison between this work and existing DAB converter with different modulation schemes

Converter features	DAB [10]	DAB [11]	DAB with blocking capacitor [9]	DT DAB [3]	NH3L-DAB [4]	Hybrid-bridge-based DAB [5]	This work
Number of MOSFETs	8	8	6	10	10	8	8
Additional component	No	No	A capacitor	A Transformer	2 diodes and 2 capacitors	2 capacitors	2 capacitors and an inductor
Number of control variable	3	5	2	4	4	2	2
ZVS for MOSFETs	Narrow	Narrow	Narrow	Full range	Full range	Narrow	Full range
Modulation	FDM	PWM + PSM	APWM + PSM	PSM	PSM	PWM + PSM	PWM + PSM
Input	200 V	380 V	100–200 V	100–400 V	200–450 V	120–240 V	120–200 V
Output	200–400 V /0.5–1.0 kW	95–190 V /1.2 kW	300 V/500 W	400 V /1.0 kW	20–28.8 V /1.0 kW	96 V /1.0 kW	96 V /1.0 kW
Peak efficiency	-	96.8%	97.9%	97.8%	98.8%	–	97.7%

4.5 Conclusion

This chapter proposed a modified VM control and current injections method for hybrid-bridge-based DAB converter. In the proposed solution, the sequence of half-bridge state and full-bridge state of the converter is changed according to the transmission direction of power. The ZVS region of the converter is extended in backward power and the switches realize the full range ZVS under wide voltage input. In addition, the influence of the switch junction capacitors is considered to analyze ZVS in this chapter, and a more accurate ZVS model is obtained. This chapter proposed to add the auxiliary inductor L_1 and the magnetizing inductance L_m, which provide additional current for realization for charging and discharging junction capacitance. A 1 kW converter prototype was built to verify the effectiveness of the proposed solution. The steady-state waveforms and the ZVS waveforms under the two controls are given.

Compared with the conventional control, ZVS is easier to realize with the proposed solution in mode conversion area. From the comparison of efficiency and loss breakdown, the converter efficiency under the proposed solution is higher when the power switch is difficult to realize ZVS with the conventional control. The proposed solution realizes the power switching ZVS in wide input voltage range in this work. Furthermore, as for the future work, the impact of the control on dynamic performance is worthy of further study.

References

1. Xu, G., Sha, D., Xu, Y., & Liao, X. (2018). Dual-transformer-based DAB converter with wide ZVS range for wide voltage conversion gain application. *IEEE Transactions on Industrial Electronics, 65*(4), 3306–3316.
2. Sha, D., Zhang, J., & Liu, K. (2020). Leakage inductor current peak optimization for dual-transformer current-fed dual active bridge DC–DC converter with wide input and output voltage range. *IEEE Transactions on Power Electronics, 35*(6), 6012–6024.
3. Chen, D., Deng, J., Li, M., Wang, Z., & Wang, S. (2021). An enhanced dual active bridge converter with full domain ZVS by utilizing a simple segment control for wide voltage range applications. *IEEE Transactions on Industrial Electronics, 69*(7), 6817–6827.
4. Zhang, Z., Hang, L., Tong, A., et al. (2022). Optimized modulation strategy of NH3L-DAB converter to minimize RMS current for wide voltage range applications. *IEEE Transactions on Power Electronics, 37*(7), 7789–7808.
5. Xu, G., Sha, D., Xu, Y., & Liao, X. (2018). Hybrid-bridge-based DAB converter with voltage match control for wide voltage conversion gain application. *IEEE Transactions on Power Electronics, 33*(2), 1378–1388.
6. Yan, Y., Gui, H., & Bai, H. (2021). Complete ZVS analysis in dual active bridge. *IEEE Transactions on Power Electronics, 36*(2), 1247–1252.
7. Xu, G., Li, L., Chen, X., Liu, Y., Sun, Y., & Su, M. (2021). Optimized EPS control to achieve full load range ZVS with seamless transition for dual active bridge converters. *IEEE Transactions on Industrial Electronics, 68*(9), 8379–8390.
8. Everts, J., Krismer, F., Keybus, J. V.d., Driesen, J., & Kolar, J. W. (2014). Optimal ZVS modulation of single-phase single-stage bidirectional DAB AC–DC converters. *IEEE Transactions on Power Electronics, 29*(8), 3954–3970.

9. Deng, J., & Wang, H. (2021). A hybrid-bridge and hybrid modulation-based dual-active-bridge converter adapted to wide voltage range. *IEEE Journal of Emerging and Selected Topics in Power Electronics, 9*(1), 910–920.
10. Choi, W., Rho, K., & Cho, B. (2016). Fundamental duty modulation of dual-active-bridge converter for wide-range operation. *IEEE Transactions on Power Electronics, 31*(6), 4048–4064.
11. Li, J., Luo, Q., Mou, D., Wei, Y., Sun, P., & Du, X. (2022). A hybrid five-variable modulation scheme for dual-active-bridge converter with minimal RMS current. *IEEE Transactions on Industrial Electronics, 69*(1), 336–346.

Part III
Dual-Active-Bridge Converter with Modified Transformer

Chapter 5
Three-Winding-Coupled-Inductor-Based Dual Active Bridge DC–DC Converter with Full Load Range ZVS Under Wide Voltage Range

Abstract In this chapter, a three-winding-coupled -inductor-based dual active bridge (TWCI DAB) DC-DC converter is proposed. Similar to the current-fed DAB converter, the proposed TWCI DAB has small input current ripple and wide input voltage range. In addition, the zero-voltage-switching (ZVS) of all switches in the full load range under wide input voltage variation can be realized. With the magnetic integration structure, the size of the total magnetics can be reduced, and the efficiency is also increased compared with the discrete magnetic case. To better analyze the working principle of the three-winding-coupled-inductor, a general analysis method based on magnetic reluctance model is firstly introduced to obtain the expressions of high frequency current. Furthermore, ZVS analysis and inductor design to obtain low inductor root mean square (RMS) current and full load range ZVS based on the ZVS conditions, maximum power transfer and RMS current are presented. Finally, the validity of the proposed topology and design are verified by experimental results from a 1-kW laboratory prototype.

Keywords Dual active bridge (DAB) · Three winding coupled inductor · Wide voltage gain range · Zero voltage switching (ZVS)

5.1 Converter Topology and Operation Principle

In CF-DAB, although the improved modulation strategies and topologies can improve the performance. However, the dual boost topology in the above-mentioned CF-DAB has two DC inductors, one leakage inductor and a transformer, which would increase the number of magnetic elements, thus increasing the occupying volume and related loss. In [1] and [2], the two DC inductors on the LVS are coupled and integrated. The power density has been improved to a certain extent. However, the design of coupled inductors is not discussed. The high voltage gain ratio can be obtained with series connection of the outputs of an interleaved bidirectional buck-boost converter and a dual-active half-bridge converter, using the windings of a coupled inductor [3–5]. However, the advantages of using a coupled inductor in such topologies are

G. Xu and X. Chen, *Topology Deduction and Control for Dual-Active-Bridge Converters*, https://doi.org/10.1007/978-981-97-4856-3_5

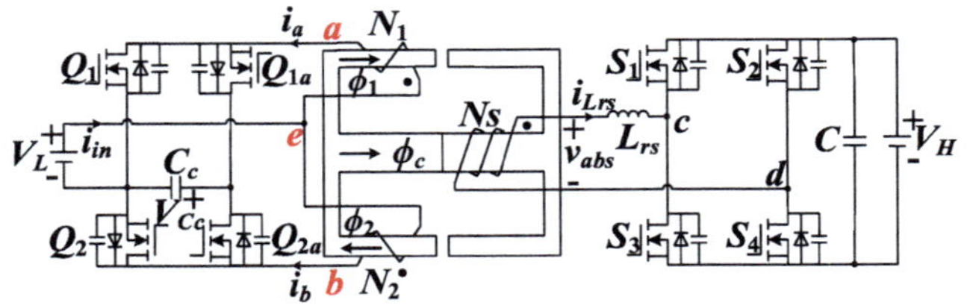

Fig. 5.1 The topology of proposed TWCI DAB

not clearly articulated. A coupled inductor is integrated into two LVS full bridges for ZVS operation throughout full-load range [6]. However, the coupled inductors are only used to integrate the DC inductors [1, 2] or worked as transformer [3–5, 7] of CF-DAB. This means that magnetic integration is only applied to part of the magnetic elements in the converter. Though, the power density is improved, the overall integration of converter is still low.

In this chapter, a new three-winding-coupled-inductor-based dual active bridge (TWCI DAB) DC–DC converter is proposed, the integrated magnetic structure can significantly improve the power density and efficiency (Fig. 5.1).

5.1.1 Topology and Modulation Schedule Using Phase-Shift Control

The topology of TWCI DAB converter is depicted in Fig. 5.1. In the low voltage side (LVS), V_L represents the wide range voltage. C_c is the clamping capacitor. N_1 and N_2 are turns number of the two LVS windings of coupled inductors, respectively. And N_s is the turns number of high voltage side (HVS) winding. The inductor L_{rs} denotes the leakage inductor on the secondary side of the TWCI. C is the HVS filter capacitor. The converter works in the boost mode as the power flows from the LVS to the HVS; otherwise, it works in the buck mode.

Figure 5.2 shows PWM modulation signals and the steady-state waveforms of the converter operating in boost mode. The boost mode is mainly analyzed since the working principles of the two modes are symmetrical. The duty ratio d of the switches Q_1 and Q_2 makes the voltage clamp of clamping capacitor C_c to be V_{Cc}. v_{abs} is an induced electromotive force generated by v_{ab}. S_1–S_4 are driven by pulse signals with 50% duty cycles. And the phase-shift control is applied between the LVS and the HVS.

To analyze the TWCI magnetic technique, the magnetic reluctance model of TWCI is shown in the Fig. 5.3.

The core reluctance is ignored since it is rather small compared with the air gap, R_1 and R_2 are the air gap reluctances of two outer legs respectively. In this chapter, it is designed that $N_1 = N_2 = N_p$ and $R_1 = R_2 = R_p$ in order to simplify the working modes and analysis. N_p is the turns of the LVS winding, and R_p is the reluctance.

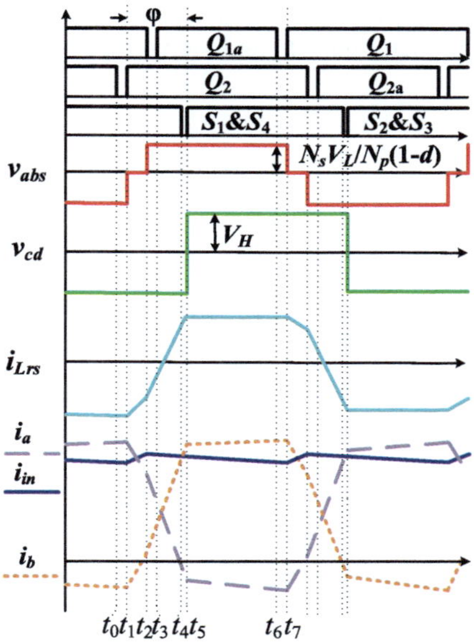

Fig. 5.2 Steady-state waveforms for proposed TWCI DAB in boost mode. (Interval A: $\varphi \in [(2d\text{-}1)\,\pi, \pi])$

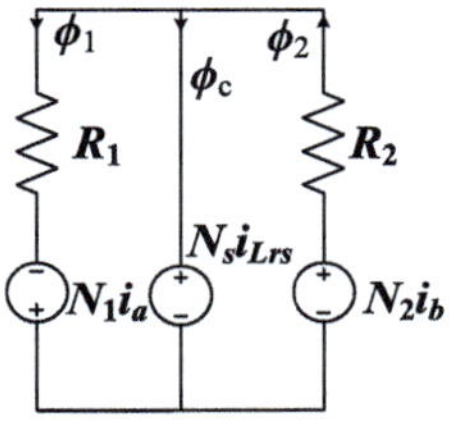

Fig. 5.3 Magnetic reluctance model of TWCI

According to the superposition theorem, the magnetic flux in each core leg is derived as

$$\begin{cases} \phi_1 = \dfrac{N_p i_a + N_s i_{Lrs}}{R_P} \\ \phi_c = \dfrac{-N_p(i_a - i_b) - 2N_s i_{Lrs}}{R_P} \\ \phi_2 = \dfrac{-N_s i_{Lrs} + N_p i_b}{R_P} \end{cases} \tag{5.1}$$

According to Faraday's law of electromagnetic induction, the voltages of the three windings can be written as

$$v_{ae} = -N_1 \frac{d\phi_1}{dt},\ v_{be} = -N_2 \frac{d\phi_2}{dt},\ v_{abs} = N_s \frac{d\phi_c}{dt} \tag{5.2}$$

At the same time, according to the relationship between the slope of the leakage inductor current and the voltage across the leakage inductor, v_{abs} can be written as

$$v_{abs} = L_{rs}\frac{di_{Lrs}}{dt} + v_{cd} \tag{5.3}$$

5.1.2 Working Stages of the Converter

The circuit working modes are shown in Fig. 5.4, and the detailed circuit analysis is as follows.

Stage 1 (before t_0): During this stage, Q_1, Q_{2a}, S_2 and S_3 are turned on as shown in Fig. 5.4a. The N_1, N_2, N_s windings are all conducted. Consequently, v_{ae} and v_{be}, which are the voltages across N_1 and N_2 respectively, and the mid-point voltage v_{cd} of the two arms of the HVS are expressed as

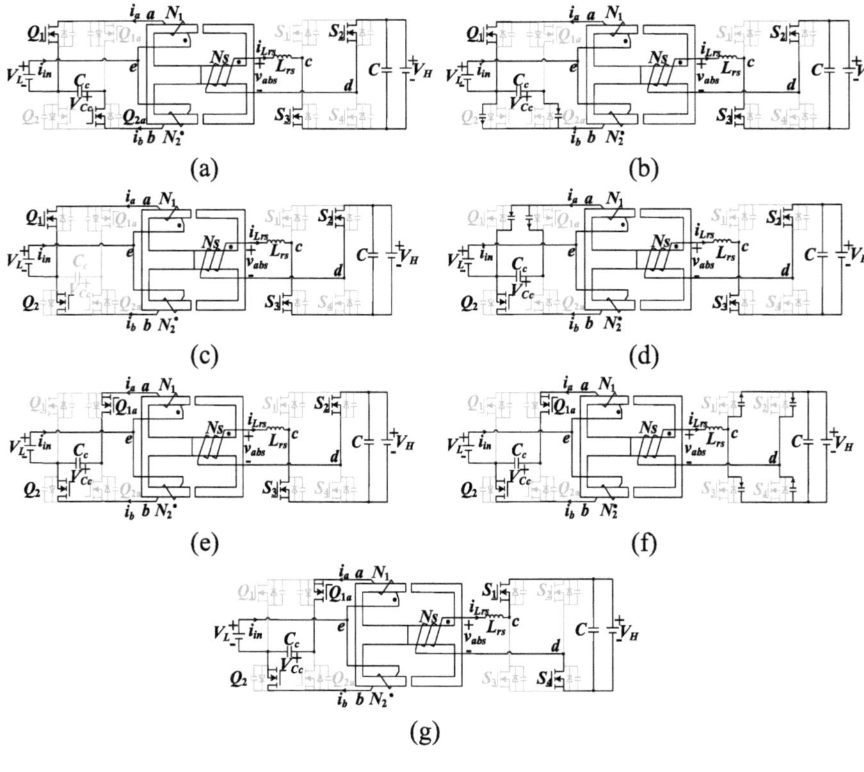

Fig. 5.4 Operation modes of the converter in the boost mode. **a** *Before* t_0. **b** t_0–t_1. **c** t_1-t_2. **d** t_2–t_3. **e** t_3–t_4. **f** t_4–t_5. **g** t_5–t_6

$$v_{ae} = -V_L, v_{be} = V_{Cc} - V_L, v_{cd} = -V_H \tag{5.4}$$

Then, according to Faraday's Law, the relation between the voltages and the magnetic flux of the TWCI can be obtained as

$$\begin{cases} \dfrac{d\phi_1}{dt} = \dfrac{V_L}{N_p} \\ \dfrac{d\phi_2}{dt} = \dfrac{-V_{Cc} + V_L}{N_p} \\ \dfrac{d\phi_c}{dt} = \dfrac{L_{rs}}{N_s}\dfrac{di_{Lrs}}{dt} - \dfrac{V_H}{N_s} \end{cases} \tag{5.5}$$

According to $\begin{cases} \phi_1 = \dfrac{N_p i_a + N_s i_{Lrs}}{R_P} \\ \phi_c = \dfrac{-N_p(i_a - i_b) - 2N_s i_{Lrs}}{R_P} \\ \phi_2 = \dfrac{-N_s i_{Lrs} + N_p i_b}{R_P} \end{cases}$ (5.1) and (5.5), the current slope of three windings and v_{abs} can respectively be derived as

$$\begin{cases} \dfrac{di_a}{dt} = \dfrac{V_{Cc}N_s^2 - N_pN_sV_H + L_{rs}R_pV_L}{L_{rs}N_p^2} \\ \dfrac{di_{Lrs}}{dt} = \dfrac{N_pV_H - N_sV_{Cc}}{L_{rs}N_p} \\ \dfrac{di_b}{dt} = \dfrac{-V_{Cc}N_s^2 + N_pN_sV_H + L_{rs}\left(R_pV_L - R_pV_{Cc}\right)}{L_{rs}N_p^2} \\ v_{abs} = -\dfrac{N_s}{N_p}V_{Cc} \end{cases} \tag{5.6}$$

Stage 2 (t_0–t_1): At t_0, Q_{2a} is turned off. At this time, i_b charges C_{2a} and discharges C_2 until the body diode of Q_2 is conducted. It can be deduced that the condition for Q_2 to achieve ZVS conduction is $i_b\ (t_0) < 0$.

Stage 3 (t_1–t_2): At t_1, Q_2 is turned on under ZVS. Therefore, v_{ae}, v_{be} and v_{cd} are expressed as

$$v_{ae} = -V_L, v_{be} = -V_L, v_{cd} = -V_H \tag{5.7}$$

Similarly, the current slope and v_{abs} can be written as

$$\begin{cases} \dfrac{di_a}{dt} = \dfrac{-N_pN_sV_H + L_{rs}R_pV_L}{L_{rs}N_p^2} \\ \dfrac{di_{Lrs}}{dt} = \dfrac{V_H}{L_{rs}} \\ \dfrac{di_b}{dt} = \dfrac{N_pN_sV_H + L_{rs}R_pV_L}{L_{rs}N_p^2} \end{cases}, v_{abs} = 0 \tag{5.8}$$

Stage 4 (t_2–t_3): Q_1 is turned off at t_2. i_a charges C_1 and discharges C_{1a} until the body diode of Q_{1a} is conducted. It can be deduced that the condition for Q_{1a} to achieve ZVS conduction is $i_a\ (t_2) > 0$.

Stage 5 (t_3–t_4): At t_3, Q_{1a} is turned on under ZVS. At this stage, v_{ae}, v_{be} and v_{cd} are expressed as

$$v_{ae} = V_{Cc} - V_L, v_{be} = -V_L, v_{cd} = -V_H \tag{5.9}$$

Likewise, the current slope and v_{abs} can be calculated as

$$\begin{cases} \dfrac{di_a}{dt} = \dfrac{-N_s^2V_{Cc} - N_pN_sV_H + L_{rs}\left(R_pV_L - R_pV_{Cc}\right)}{L_{rs}N_p^2} \\ \dfrac{di_{Lrs}}{dt} = \dfrac{N_pV_H + N_sV_{Cc}}{L_{rs}N_p} \\ \dfrac{di_b}{dt} = \dfrac{N_s^2V_{Cc} + N_pN_sV_H + L_{rs}R_pV_L}{L_{rs}N_p^2} \end{cases}, v_{abs} = \frac{N_s}{N_p}V_{Cc} \tag{5.10}$$

Stage 6 (t_4–t_5): At time t_4, S_2 and S_3 are turned off. The leakage inductor current charges C_{s2} and C_{s3} and discharges C_{s1} and C_{s4} until the body diodes of S_1 and S_4 are conducted. The condition for S_1 and S_4 to achieve ZVS is $i_{Lrs}\ (t_4) > 0$.

Stage 7 (t_5–t_6): At time t_5, S_1 and S_4 are turned on under ZVS. Accordingly, v_{ae}, v_{be} and v_{cd} are expressed as

$$v_{ae} = V_{Cc} - V_L, v_{be} = -V_L, v_{cd} = V_H \tag{5.11}$$

Correspondingly, the current slope and v_{abs} can be written as

$$\begin{cases} \dfrac{di_a}{dt} = \dfrac{-N_s^2V_{Cc} + N_pN_sV_H + L_{rs}\left(R_pV_L - R_pV_{Cc}\right)}{L_{rs}N_p^2} \\ \dfrac{di_{Lrs}}{dt} = \dfrac{-N_pV_H + N_sV_{Cc}}{L_{rs}N_p} \\ \dfrac{di_b}{dt} = \dfrac{N_s^2V_{Cc} - N_pN_sV_H + L_{rs}R_pV_L}{L_{rs}N_p^2} \\ v_{abs} = \dfrac{N_s}{N_p}V_{Cc} \end{cases} \tag{5.12}$$

The previous analysis provides a general analysis method of TWCI based on magnetic reluctance model. The current slope in each mode can be obtained, which is an important basis for analysis of power characteristics and ZVS.

5.2 Power Characteristics and ZVS Analysis

5.2.1 Transmission Power with PPS Control

As Fig. 5.5 shown, the phase shift angle between the rising edge of v_{abs} and v_{cd} is defined as φ. The total volt-seconds applied to the N_1 or N_2 winding of LVS over one switching period are

$$-V_L d + (-V_L + V_{Cc})(1-d) = 0 \tag{5.13}$$

Consequently, the duty cycle of Q_1 and Q_2 is given by

$$d = 1 - \frac{V_L}{V_{Cc}} \tag{5.14}$$

where V_{Cc} is the voltage across the clamping capacitor C_c and is a fixed value.

V_{Cc} is controlled to match the output voltage, and current slope of leakage inductor is zero during the power transmission stage 1 and 7. The turns number relationship of the TWCI can be obtained as

$$\frac{N_p}{N_s} = \frac{V_{Cc}}{V_H} \tag{5.15}$$

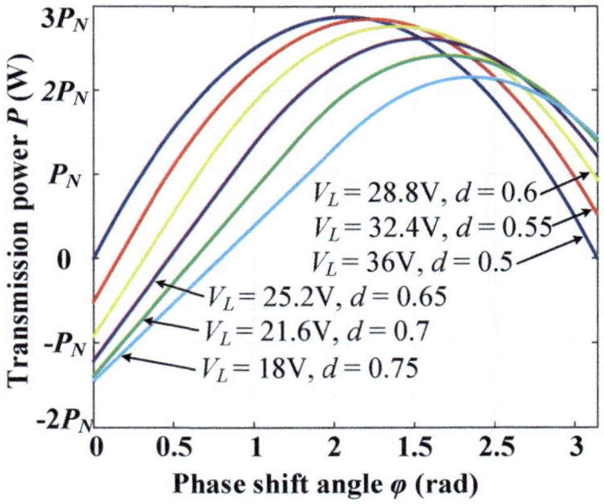

Fig. 5.5 Curves of transmission power and phase shift angle

Through integration of the transformer voltage and current under PPS control, the power can be calculated as

$$P = \begin{cases} \frac{V_H^2(1-d)[\varphi-(d-0.5)\pi]T_s}{\pi L_{rs}} & \varphi \subset [0, (2d-1)\pi] \\ \frac{V_H^2[-\varphi^2+2d\pi\varphi-d(2d-1)\pi^2]T_s}{2\pi^2 L_{rs}} & \varphi \subset [(2d-1)\pi, \pi] \end{cases} \tag{5.16}$$

Figure 5.5 illustrates the relationship between transmission power and phase shift angle under PPS control. Under different voltages V_L, when the voltage matches and the phase shift angle φ range is $[0, 0.5\pi]$, the transmission power increases monotonically as the phase shift angle increases.

$$I_{ZVS\_Q_1\&Q_2} = i_b(t_0)$$
$$= \begin{cases} \frac{R_p T_s(V_L-V_{Cc})V_L}{2N_p^2 V_{Cc}} & \frac{\varphi}{\pi} \subset [0, (2d-1)] \\ \frac{P}{2V_L} + \frac{R_p T_s V_L^2}{2N_p^2 V_{Cc}} - \frac{R_p T_s V_L}{2N_p^2} - \frac{V_H^2 T_s}{4V_{Cc}L_{rs}} + \frac{V_H}{2V_{Cc}L_{rs}}\sqrt{\frac{V_H^2 T_s^2 V_L}{V_{Cc}}\left(1-\frac{V_L}{V_{Cc}}\right) - 2L_{rs}T_s P} & \frac{\varphi}{\pi} \subset [(2d-1), 1] \end{cases} \tag{5.17}$$

5.2.2 ZVS Range Analysis

Based on mode analysis and calculation of instantaneous current, the current limit for charging/discharging the junction capacitor C_{2a}/C_2 to achieve ZVS of Q_1 and Q_2 is written as (5.17).

The current limit for charging/discharging the junction capacitor C_{1a}/C_1 to achieve ZVS of Q_{1a} and Q_{2a} can be obtained in the same way, which is written as

$$I_{ZVS\_Q_{1a}\&Q_{2a}} = i_a(t_2)$$
$$= \begin{cases} -\frac{R_p T_s(V_L-V_{Cc})V_L}{2N_p^2 V_{Cc}} & \frac{\varphi}{\pi} \subset [0, (2d-1)] \\ \frac{P}{2V_L} - \frac{R_p T_s V_L^2}{2N_p^2 V_{Cc}} + \frac{R_p T_s V_L}{2N_p^2} - \frac{V_H^2 T_s}{4V_{Cc}L_{rs}} + \frac{V_H^2 T_s V_L}{V_{Cc}^2 L_{rs}} - \frac{V_H}{2V_{Cc}L_{rs}}\sqrt{\frac{V_H^2 T_s^2 V_L}{V_{Cc}}\left(1-\frac{V_L}{V_{Cc}}\right) - 2L_{rs}T_s P} & \frac{\varphi}{\pi} \subset [(2d-1), 1] \end{cases} \tag{5.18}$$

And the ZVS current limit for charging/discharging the junction capacitor of all the HVS switches is written as

$$I_{ZVS\_S_1\&S_2\&S_3\&S_4} = i_{Lrs}(t_4)$$

$$=\begin{cases}\frac{V_H T_s(V_{Cc}-2V_L)}{4V_{Cc}L_{rs}} & \frac{\varphi}{\pi}\subset[0,(2d-1)]\\ \frac{V_H T_s}{4L_{rs}}-\frac{1}{2L_{rs}}\sqrt{\frac{V_H^2T_s^2V_L}{V_{Cc}}\left(1-\frac{V_L}{V_{Cc}}\right)-2L_{rs}T_sP} & \frac{\varphi}{\pi}\subset[(2d-1),1]\end{cases} \tag{5.19}$$

From (5.17)–(5.19) of the ZVS analysis above, it can be seen that after the specification parameters are determined, such as transmission power and winding turns ratio, etc., the ZVS of the converter switches are only related to ${N_p}^2/R_p$ and leakage inductor L_{rs}. To make the expression simple, it is defined that $L_{eq}={N_p}^2/R_p$.

5.3 Design Considerations

5.3.1 ZVS Design Related to Leq and Leakage Inductor

In this section, the parameter considerations for L_{eq} and leakage inductor L_{rs} are discussed with the goal of achieving full load range ZVS. The related parameters are: V_L = 18~36 V, V_H = 360 V, P_N = 1 kW. According to (5.14), the duty cycle range is 0.5~0.75.

According to the ZVS current limit formula of each switch, Fig. 5.6a and b show ZVS of the LVS switches Q_1 and Q_2 at different d. When d is equal to 0.5, from no load to the rated power point P_N, the current is less than zero; when d is equal to 0.75, the current is less than zero from no load to P_N. Consequently, despite the change in d and leakage inductor value L_{rs}, within the range of [2 μH, 10 μH] for the equivalent inductors L_{eq} of outer legs, the current value $I_{ZVS\_Q1\&Q2}$ is always less than zero. It means that the ZVS of the LVS switches Q_1 and Q_2 can be achieved in the full load range with different L_{eq}. Similarly, Fig. 5.6c–f show that ZVS of other switches can also be realized within the full load range. Therefore, the appropriate value range of L_{eq} and leakage inductor L_{rs} can be preliminarily selected by ZVS design. However, the inductor RMS current should also be considered to achieve a smaller inductor loss.

5.3.2 Inductor RMS Current Design Considerations

In this section, the design considerations for leakage inductor L_{rs} and L_{eq} are discussed with the goal of obtaining low RMS current.

For the leakage inductor design, from the relationship between the transmission power and the phase shift angle φ in Fig. 5.2, when the phase shift angle $\varphi = 0.5\pi$, the relationship between the transmission power versus the leakage inductor L_{rs} and the duty cycle is written as (5.20).

Figure 5.7 shows that the corresponding leakage inductor is different under different d when the rated power is transmitted.

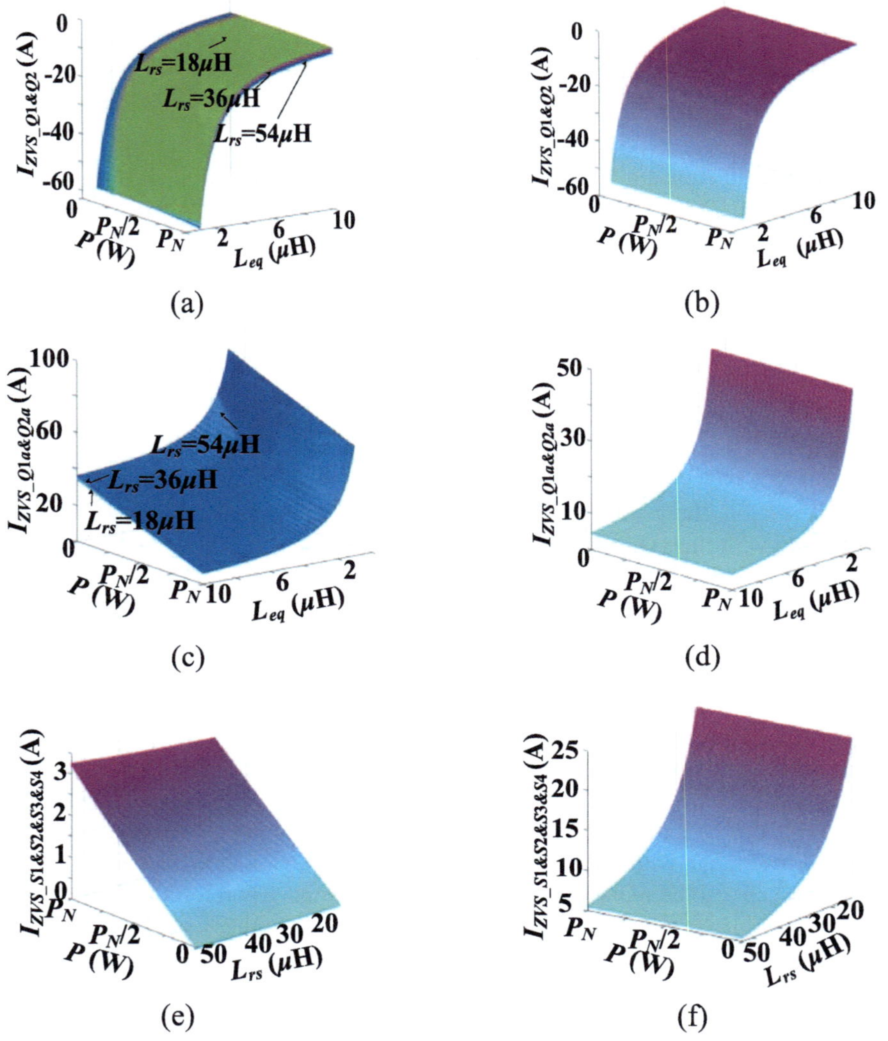

Fig. 5.6 ZVS of switches. **a** Q_1 & Q_2 when $V_L = 36$ V. **b** Q_1 & Q_2 when $V_L = 18$ V. **c** Q_{1a} & Q_{2a} when $V_L = 36$ V. **d** Q_{1a} & Q_{2a} when $V_L = 18$ V. **e** S_1 & S_2 & S_3 & S_4 when $V_L = 36$ V. **f** S_1 & S_2 & S_3 & S_4 when $V_L = 18$ V. ($V_L = 18$~36 V, $V_H = 360$ V, $P_N = 1$ kW)

$$P(L_{rs}, d) = \frac{N_s^2 V_c^2 \left[-8d^2 + 8d - 1\right] T_s}{8 N_p^2 L_{rs}} \tag{5.20}$$

Consequently, taking Fig. 5.7 as an example, $L_{rs} \leq 54$ μH should be satisfied to achieve maximum power transfer at the extremely case.

Under rated power P_N, the RMS current of the leakage inductor can be calculated as

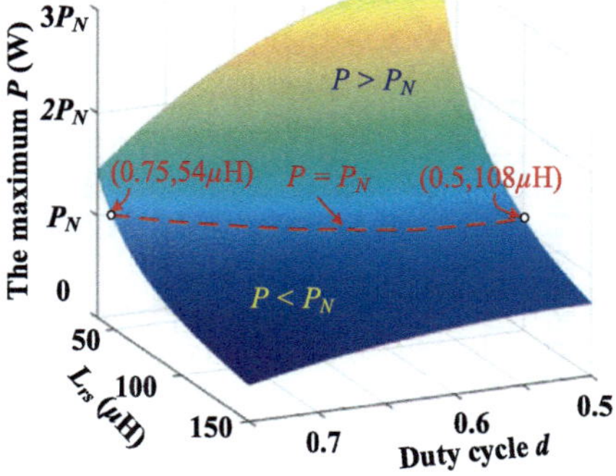

Fig. 5.7 Transmission power versus leakage inductor and duty cycle ($\varphi = 0.5\pi$, $P_N = 1$ kW)

$$I_{L_{rs}\_RMS} = \begin{cases} \frac{V_H T_s}{2\pi L_{rs}}\sqrt{2\phi(1-d)[\phi-(2d-1)\pi]-\frac{\pi^2}{12}(2d-1)^2(4d-5)} & \left\{\begin{array}{c} V_{L\_\min} \le V_L \le V_{L\_div} \\ \phi \subset [0, (2d-1)\pi] \end{array}\right\} \\ \frac{V_H T_s}{\pi L_{rs}}\sqrt{\frac{\phi[\phi(\pi d-\phi)+3\pi^2 d(1-2d)]}{6\pi}+\frac{\pi^2(2d-1)^2(4d-1)}{16}} & \left\{\begin{array}{c} V_{L\_div} \le V_L \le V_{L\_\max} \\ \phi \subset [(2d-1)\pi, \pi] \end{array}\right\} \end{cases} \quad (5.21)$$

L_{eq} is the equivalent inductor, but its physical meaning is uncertain. To deal with this, the study finds that when the TWCI meets certain conditions, it can be completely equivalent to the discrete structure of two DC inductors plus a transformer. The detailed analysis is as follows: it is assumed that the leakage flux through the air is negligible, through the dual transformation method [8], that is, according to the reluctance model of the TWCI shown in Fig. 5.3, the dual transformation is performed to derive the equivalent circuit of the magnet circuit model as shown in Fig. 5.8. And the inductor parameters of the equivalent circuit can be written as

$$L_{dc1} = \frac{N_1^2}{R_1} = \frac{N_p^2}{R_p} = L_{eq},\ L_{dc2} = \frac{N_2^2}{R_2} = \frac{N_p^2}{R_p} = L_{eq} \quad (5.22)$$

Figure 5.9 illustrates the curves for the inductor RMS current. Figure 5.9a illustrates the relationship between the leakage inductor RMS current and the voltage for different leakage inductances. When the leakage inductance L_{rs} is equal to 18 μH, the

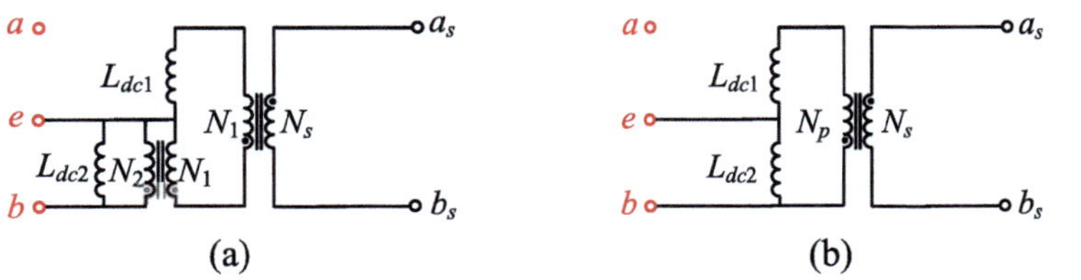

Fig. 5.8 Equivalent circuit for the TWCI. **a** $N_1 \neq N_2$; **b** $N_1 = N_2 = N_p$

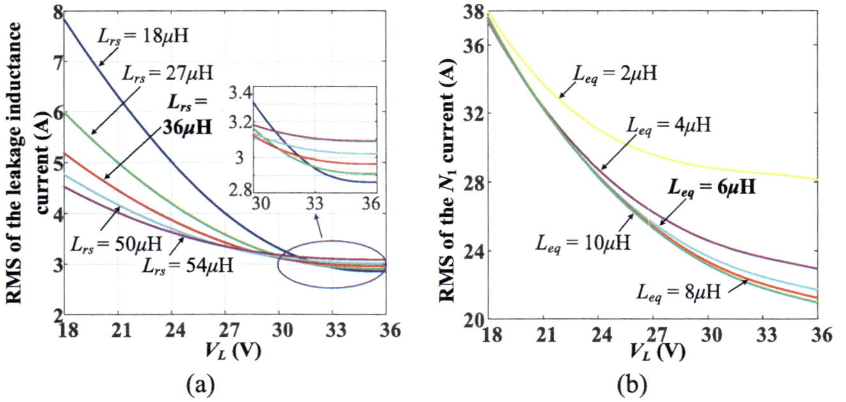

Fig. 5.9 Curves for the RMS current. **a** RMS currents under different values of leakage inductor L_{rs}. **b** RMS currents under different L_{eq} values

RMS current increases significantly if the voltage is low; when L_{rs} is larger than 50 μH, the leakage currents change slightly. As can be seen, the RMS current decreases with the value of leakage inductance increases near V_L = 18–33 V. However, the trend of leakage inductor RMS current is opposite near V_L = 33–36 V. That is the RMS current increases with the leakage inductance increases. Considering that more windings will be needed to increase the inductance, the leakage inductance is designed to be 36 μH to make a tradeoff between RMS current and copper loss.

Based the transferred power, the L_{eq} (DC inductor) value is around 10 μH when the current ripple is set as 15% of rated average current according to the general engineering design. And when the ripple is 100%, the L_{eq} value is around 2 μH. Consequently, Fig. 5.9b illustrates the RMS current curves of the outer-leg windings when L_{eq} changes from 2 to 10 μH. Similar to leakage inductor, considering the magnetic loss of the outer legs and the RMS current, L_{eq} = 6 μH is finally selected for the equivalent inductor of the two outer legs.

5.3.3 Inductor RMS Current Design Considerations

The physical topology of TWCI is shown in Fig. 5.10. The standard EE cores and the Litz wire windings constitute the TWCI structure. According to Fig. 5.8, the two windings on the outer legs function as the equivalent inductors as well as the transformer primary. It can not only reduce the volume of the magnetic element, but also reduce the input current ripple. The winding of the center leg can be used as the transformer secondary winding.

Different TWCI structures will produce different inherent leakage inductance. The structure A of TWCI is shown in Fig. 5.11a. Since the windings are all separated, it will produce a large leakage inductance, and reduce the maximum transmission

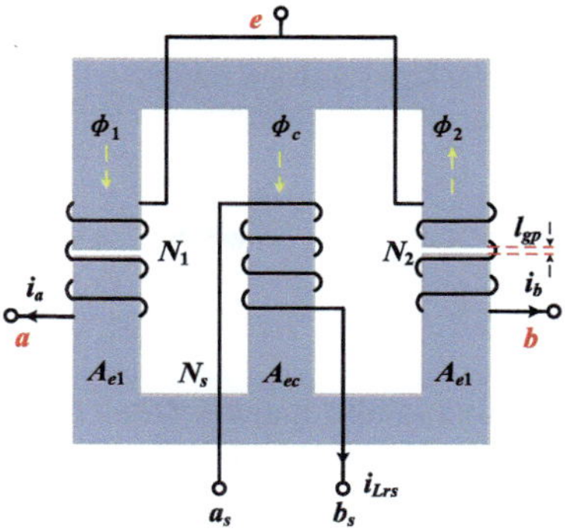

Fig. 5.10 Physical topology of TWCI using EE cores

power of the converter. In order to reduce the leakage inductance, according to source transfer equivalent transformation method, the center leg winding can be split into two outer leg windings, forming the structure B of TWCI, as Fig. 5.11b shown.

By the measurement of impedance analyzer, the leakage inductance parameters of structure A and structure B of TWCI are shown in Table 5.1.

It can be seen that the leakage inductance of the TWCI with winding interleaving structure can be effectively reduced. In order to decrease the leakage inductance, the structure B is finally selected for the prototype in this manuscript.

Then, the required leakage inductance is obtained by series connection of an external inductor, that is the leakage inductance L_{rs_TWCI} of the TWCI is compensated by an external series inductance L_{rs_series} to realize the required leakage inductance L_{rs}, as Fig. 5.12 shows.

It should be noted that the TWCI central leg does not need the air gap to achieve decoupled-integration of the two outer leg windings. Meanwhile, dc component of

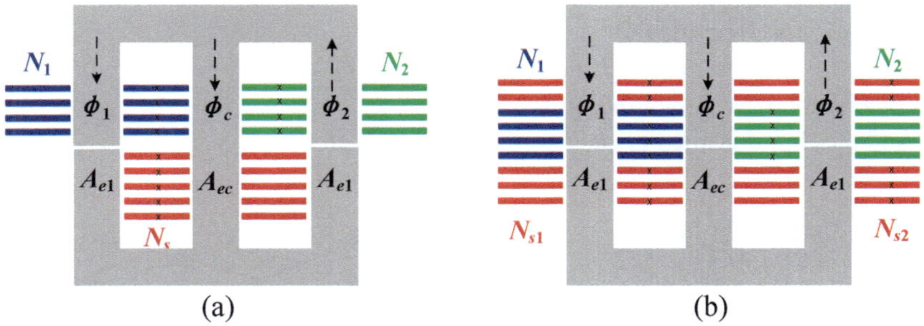

Fig. 5.11 Different structures of TWCI. **a** Structure A of TWCI. **b** Structure B of TWCI

Table 5.1 Measured leakage inductance of different structures of TWCI

Item	Leakage inductance
Structure A (μH)	41.6
Structure B (μH)	11.2

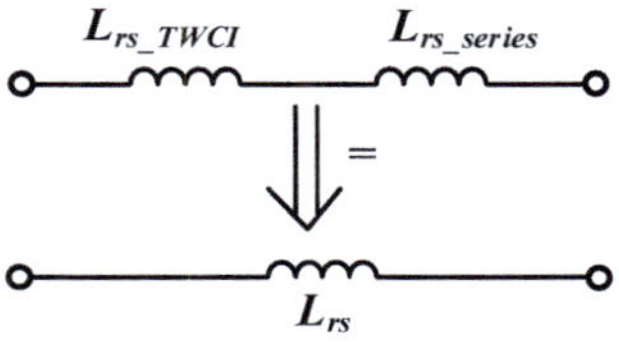

Fig. 5.12 The diagram of external series leakage inductor

the flux is eliminated in the center leg, and the magnetic loss can be eliminated. For the proposed TWCI, parameters including the core reluctance R_p and the numbers of turns of the windings N_p, N_s need to be determined. The detailed design process of TWCI is as follows.

In order to limit the core loss of a specified material, the maximum flux density change of the core is set as 150 mT. And then the turns number N_p ($N_1 = N_2 = N_p$) of the two LVS windings of TWCI can be calculated as

$$N_P = \frac{dV_L T_s}{\Delta B A_e} \tag{5.23}$$

where ΔB is the change of the magnetic flux density, A_e is the cross-sectional area of the two outer legs of EE42 cores, and d is the duty cycle. And according to (4.15), the turns number N_s of the HVS winding of TWCI can be obtained.

The self-inductances of two outer legs of TWCI are DC equivalent inductances L_{eq}, and $L_{eq} = 6\ \mu\text{H}$ has been designed to obtain low RMS current. According to the (5.22), the air gap reluctances of two outer legs can be derived, and the air gap length of two outer legs can be written as

$$l_{gp} = R_p \mu_0 A_e \tag{5.24}$$

where μ_0 is space permeability.

According to the magnetic flux expression (5.1), the magnetic flux waveforms can be obtained, as shown in Fig. 5.13.

And the flux densities in each core leg are estimated as

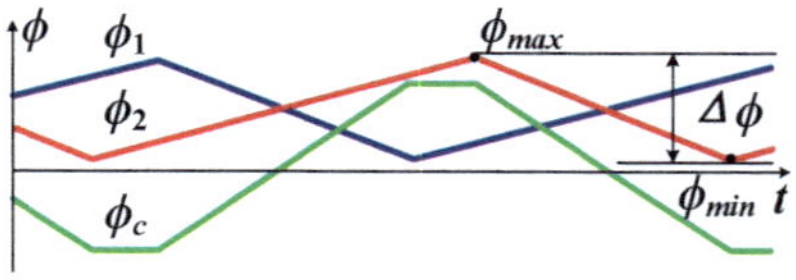

Fig. 5.13 The magnetic flux waveforms of the TWCI

$$\begin{cases} B_1 = \dfrac{N_p i_a + N_s i_{Lrs}}{R_P A_{e1}} \\ B_c = \dfrac{-N_p(i_a - i_b) - 2N_s i_{Lrs}}{R_P A_{ec}} \\ B_2 = \dfrac{-N_s i_{Lrs} + N_p i_b}{R_P A_{e2}} \end{cases} \quad (5.25)$$

where A_{e1}, A_{e2} and A_{ec} are cross-sectional areas of the two outer legs and the center post respectively.

5.3.4 *Inductor RMS Current Design Considerations*

Figure 5.6b describes the structure of discrete magnetic parts with two inductors and a transformer. The parameters for comparisons of TWCI and discrete magnetics are shown in Table 5.2. As previously discussed, $N_1 = N_2$, and $R_1 = R_2$ are adopted for the TWCI. The core type is selected to ensure that the change of magnetic flux density to be the same. The proposed TWCI using a single EE42 can reduce the magnetic core occupied volume and achieve a higher power density compared with the two UF25 boost DC inductors and one EE42 transformer in the CF-DAB

The loss breakdown comparisons between the discrete magnetics case of CF-DAB and the proposed TWCI DAB at different V_L (V_L = 18 V, 27 V and 36 V) and different loads (P = 100W and 1000W) are shown in Fig. 5.14. P_{TWCI_DAB} and $P_{CF\text{-}DAB}$ are total losses of TWCI DAB converter and CF-DAB converter, respectively. P_{TWCI_copper} and P_{TWCI_core} are copper loss and core loss of the TWCI DAB converter,

Table 5.2 Comparison of different magnetic parts designs

Magnetic category	Discrete magnetics		TWCI DAB
	Transformer design	2 × inductor design	
Core material	Mn-Zn power ferrite DMR40		
Core model	EE42	UF25	EE42
Length (mm)	42	2 × 25	42
Width (mm)	20	2 × 13	20
Height (mm)	42	2 × 40	42
Magnetic flux density change (T)	$\Delta B = 0.15$	$B_{max} = 0.39$ ($\Delta B = 0.158$)	$\Delta B = 0.15$
Effective *Ve* (mm^3)	22,601	2 × 9264.4	22,601
Effective *Ae* (mm^2)	233	106	233
Turns	N_1:N_2 = 5:25	2 × 5	N_p:N_s = 5:25
Wire diameter (mm)	$d = 0.1$		
Window factor	0.197	0.15	0.246

respectively. And $P_{TWCI\_Lrs\_copper}$ and $P_{TWCI\_Lrs\_core}$ are copper loss and core loss of the leakage inductor in TWCI DAB converter, respectively. P_{Ldc_copper} and P_{Ldc_core} are copper loss and core loss of the two DC inductors in CF-DAB converter, respectively. P_{T_copper} and P_{T_core} are copper loss and core loss of the transformer in CF-DAB converter, respectively. And P_{Lrs_copper} and P_{Lrs_core} are copper loss and core loss of the leakage inductor in CF-DAB converter, respectively. $P_{turn\text{-}on}$ and $P_{turn\text{-}off}$ are turn-on loss and turn-off loss of all the switches, respectively. And P_{con} is conduction loss for switches. P_{dri} is driving loss. Under the PWM modulation strategy, the driving signals for the two switches in any bridge are complementary. The conduction of the antiparallel diode only exists in the dead time, and it lasts for a very short time. Thus, the conduction loss of antiparallel diode is ignored in this chapter to simplify the analysis. Except for the difference of two DC inductors and transformers, other experimental parameters are the same, such as MOSFETs, PCB circuit boards, etc. In TWCI DAB, all the MOSFETs are turned on with ZVS. Hence, the turn-on losses are zero. The total losses consist of four parts, namely driving loss, switches loss, external inductor loss, and TWCI loss. As Fig. 5.14a shown, when the converter works at $V_L = 18$ V and light load (100 W), the total loss of CF-DAB is higher than that of the proposed TWCI DAB about 6 W, which is mainly caused by hard turning off loss, leading to low efficiency at light load. When $V_L = 36$ V, the difference is reduced to 3W. As the input voltage increases, the current decreases, then the turn-off loss and the copper loss of the magnetic elements decrease. As Fig. 5.14b shown, when the converter works at $V_L = 18$ V and heavy load condition (1000 W), the power loss with the proposed CF-DAB is higher than that of TWCI DAB about 6.6 W, which is mainly caused by the increased core and copper losses. Based on the theory of Steinmetz Equation [9], the product of the magnetic loss density of each leg and the effective volume is used to obtain the magnetic core loss of TWCI.

At low input voltages condition, the conduction time and input current increases as the input voltage decreases. The corresponding switching loss and magnetic element loss increase. Magnetic element loss and hard turn off loss dominate at light loads, leading to low efficiency at light load. As can be seen, the total loss of the integrated

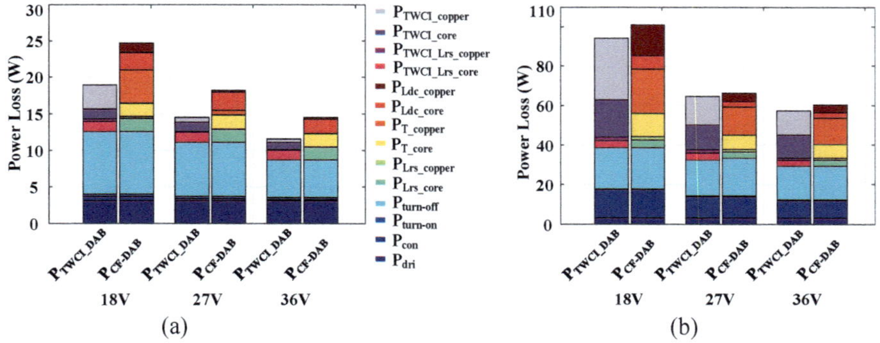

Fig. 5.14 Loss breakdown at different V_L and different loads. **a** At light load ($P = 100$ W). **b** At full load ($P = 1000$ W)

magnetic TWCI is reduced according to Fig. 5.14. Therefore, TWCI DAB efficiency is higher than CF-DAB.

In addition to the benefits of easy to manufacture and leakage inductance reduction, the proposed TWCI can be modeled and analyzed directly with combination of the magnetic circuit and the electrical circuit, which can simplify the analysis process compared with analyzing them separately. On this basis, a general analysis method of TWCI is provided. In addition, according to Table 5.2, when transmitting the same power, the power density of TWCI DAB can also be increased due to the reduction of core number and volume. Furthermore, efficiency of TWCI DAB is higher than CF-DAB with discrete magnetics according to the loss breakdown.

5.4 System Parameters Design

In order to verify the effectiveness of the design method, a 1kW experimental prototype is built. And the experimental prototype is shown in Fig. 5.15. The system specifications are illustrated in Table 5.3. All the design parameters are designed taking into consideration the safe operation of converter for highest LVS current, i.e., 18 V LVS voltage and 1kW output power.

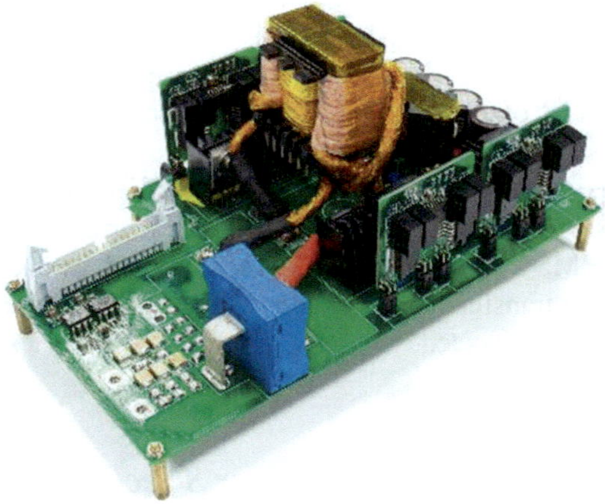

Fig. 5.15 The experimental prototype

Table 5.3 System specifications

P	1 kW	L_{rs}	36 μH
f	150 kHz	L_{eq}	6 μH
V_L	18~36 V	n	5
V_H	360 V	C_c	60 μF
Q_1,Q_2,Q_{1a},Q_{2a}	IPP023N10N5	S_1-S_4	UJ3C065030K3S

5.4.1 Equivalent Inductance Leq Design

In practice, the first consideration of the equivalent inductance L_{eq} is the ZVS performance of the LVS switches. According to Fig. 5.5.6a–d, the appropriate value range [2 μH, 10 μH] of L_{eq} can be preliminarily selected. On the other hand, the RMS current should be discussed based on the allowed current ripple to obtain the optimized inductance L_{eq}. Taking the ZVS design of the LVS switches, current ripple and RMS current into consideration, L_{eq} is designed to be 6 μH.

5.4.2 Turns Ratios of TWCI Design

The turns ratio of TWCI should be designed to ensure the voltage matching control, that is the voltage of clamping capacitor V_{Cc} matches the output voltage V_H. The clamping voltage $V_{Cc} = 72$ V and the output voltage $V_H = 360$ V. According to (5.15), the turns ratio of TWCI is $n = N_s/N_p = V_H/V_{Cc} = 5$.

5.4.3 Leakage Inductance Lrs Design

The ZVS performance of the HVS switches is affected by the value of the leakage inductance L_{rs}. According to Fig. 5.5.6e and f, the appropriate value range [18 μH, 54 μH] of L_{rs} can be preliminarily selected. In addition, the leakage inductor is closely related to the output power. Thus, $Lrs \leq 54$ μH should be satisfied to achieve maximum power transfer form Fig. 5.7. Besides, the optimized leakage inductor is designed by the tradeoff between RMS current and copper loss based on the analysis of RMS current from Fig. 5.9a. Based on considering the ZVS range, maximum power transfer capability and the RMS current, the leakage inductance $L_{rs} = 36$ μH is finally selected.

5.4.4 Clamping Capacitor Selection

The selection of the clamping capacitor C_c should ensure that the voltage ripple across C_c is confined within an allowed range. Due to the complementary of the LVS modulation, only half a switching cycle $[t_0, t_6)$ is analyzed. The clamping voltage $v_{Cc}(t)$ can be calculated by

$$v_{Cc}(t) = \frac{1}{C_c}\int_{t_3}^{t} i_a(t)dt + v_{Cc}(t_3) \tag{5.26}$$

The current i_a will charge/discharge Cc. During $[t_3, t_4)$, i_a is larger than zero, so Cc will be charged and v_{Cc} will rise; while during (t_4, t_6), i_a is smaller than zero, so Cc is discharged and v_{Cc} will fall. So, the maximum of v_{Cc} is at t_4, and minimum one is either at t_3 or t_6. So the voltage ripple ΔV_{Cc} across Cc can be illustrated as

$$\Delta V_{Cc} = v_{Cc}(t_4) - \min\{v_{Cc}(t_3), v_{Cc}(t_6)\} \tag{5.27}$$

ΔV_{Cc} is designed to satisfy the following criteria

$$\Delta V_{Cc}/V_{Cc} \leq 1\% \tag{5.28}$$

Calculation for the case of 18 V input and 1 kW load, the critical value for Cc can be obtained.

$$C_c \geq 42.3\ \mu\text{F} \tag{5.29}$$

A near 40% margin is designed. Thus, the clamping capacitor is selected to be 60 µF.

5.4.5 *Clamping Capacitor Selection*

Considering the maximum voltage across the LVS switches $V_{Q1\_Q1a\_Q2\_Q2a}$ is the clamping voltage V_{Cc}, that is $V_{Q1\_Q1a\_Q2\_Q2a} = 72$ V. And the maximum voltage across the HVS switches $V_{S1\_S2\_S3\_S4}$ is the output voltage V_H, that is $V_{S1\_S2\_S3\_S4} =$ 360 V. The current stresses of LVS switches $I_{Q1} = i_a(t_1) = 60.67$ A. And the current stresses of HVS switches $I_{S1} = i_{Lrs}(t_5) = 10.71$ A. According to the voltage, current stresses and margin should be considered, device type IPP023N10N5 from Infineon company is selected for the LVS switches, and device type UJ3C065030K3S from United SiC company is selected for the HVS switches.

5.5 Experimental Results

Figure 5.16 shows the steady-state waveforms of the proposed TWCI DAB in boost mode under different voltages and loads. As shown, v_{ab}, v_{cd}, V_H, i_{Lrs} are the voltages and current of the proposed TWCI DAB, as described in Fig. 5.1. Figure 5.16a–c describe the steady-state waveforms at light load with $V_L = 18, 27, 36$ V, respectively. The corresponding duty cycle of Fig. 5.16a–c is $d = 0.75$, $= 0.625$, and $= 0.5$, respectively. It is seen that the pulse width of the midpoint voltage v_{cd} between the HVS bridges is always fixed at 0.5 in these figures, which means the duty cycle of HVS switches is always equal to 0.5. The waveforms of the LVS voltage v_{ab} are similar when $d > 0.5$. When $d = 0.5$, the operation waveforms of the PPS control is exactly the same as a SPS controlled converter. In addition, v_{cd} lags behind v_{ab} in phase to adjust the power among all waveforms. Meanwhile, the current slope is zero when both v_{ab} and v_{cd} have been regulated to be stabilized values with the same polarity. Similarly, Fig. 5.16d–i depict steady-state waveforms at half load and full load, respectively. These demonstrate the working principle of the proposed TWCI DAB. And it is clear that the experimental waveforms coincide with the theoretical analysis pretty well.

The ZVS experimental waveforms of switches are shown in Figs. 5.17 and 5.18. Compared to heavy load, it is most difficult to achieve ZVS under light load. Meanwhile, compared with the top switch Q_{1a} of LVS and the bottom switch Q_2 of LVS, S_1 of HVS are more difficult to realize ZVS. Consequently, Fig. 5.17 illustrates the ZVS experimental waveforms of the bottom switch Q_2 of LVS and switch S_1 of HVS under different voltages at light load. The results show that ZVS can be achieved despite the voltage fluctuations at light load.

Figure 5.18 illustrates the ZVS experimental waveforms of the bottom switch Q_2 of LVS and S_1 of HVS at different voltages at full load. It can be seen that ZVS can be achieved when input voltage changes.

Figure 5.19 illustrates the load-step experimental result from half load to full load with $V_L = 27$V. V_H can keep stable and can return to the steady state value within 20ms.

Figure 5.20 shows the measured efficiency at different loads and different voltages. In order to verify the effectiveness of the TWCI DAB, the discrete magnetic case of CF-DAB is also measured. The overall conversion efficiency is high, and the highest efficiency of the TWCI DAB reaches 95.7%. The proposed TWCI DAB achieves higher efficiency than the discrete magnetic elements of CF-DAB with the same magnetic material. The power loss at light load can be reduced for TWCI DAB, because the core loss makes up the majority of the total loss when converter operates at light load condition.

5.6 Conclusion

In this chapter, a TWCI DAB converter is proposed. Similar to the CF-DAB, the proposed TWCI DAB can achieve small input current ripple and wide input voltage range. In addition, the integrated magnetic structure can make the converter more efficient and improve the power density significantly. With the adopted PWM modulation, the proposed TWCI DAB can achieve ZVS of the primary and secondary side switches within the full load range under wide voltage range. In addition, the design of equivalent inductor L_{eq} and leakage inductor are introduced considering the maximum power transmission, soft switching conditions and RMS current. The experimental results of the proposed prototype verified the effectiveness of the proposed solution.

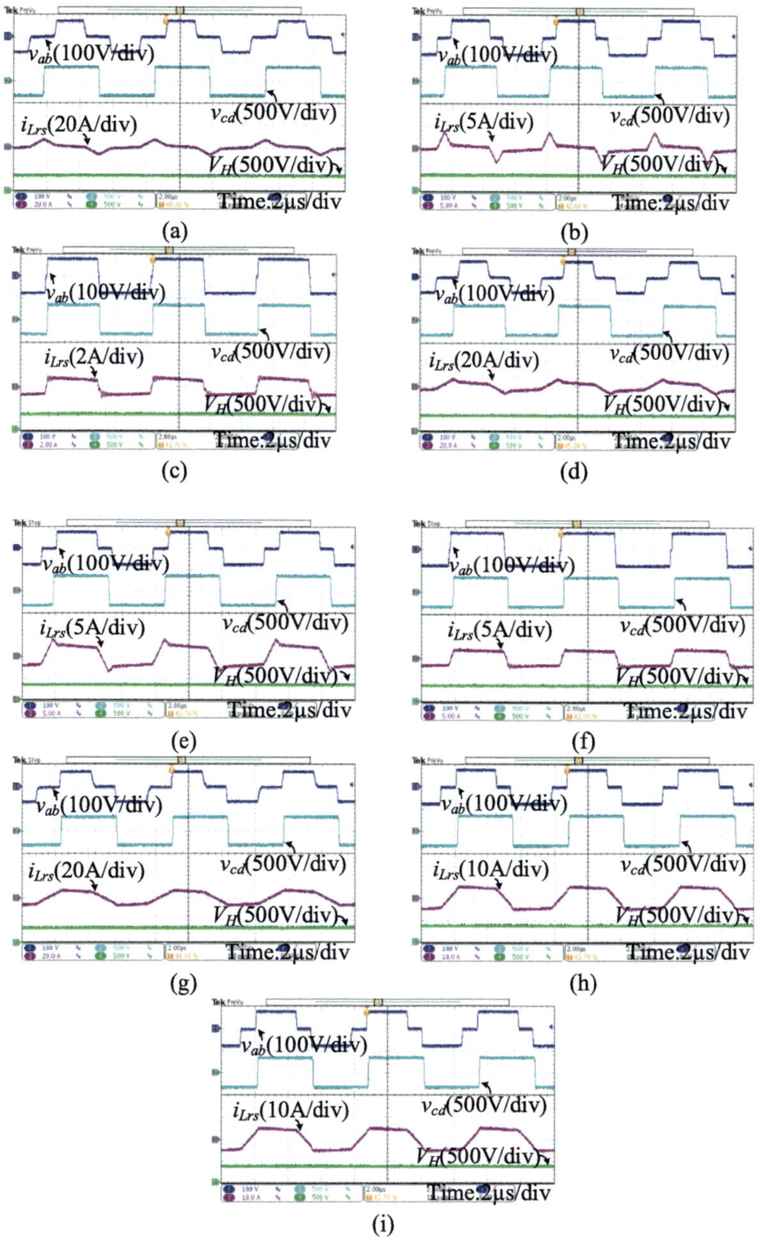

Fig. 5.16 Steady-state waveforms with the proposed TWCI DAB. **a** At light load when $V_L = 18$ V. **b** At light load when $V_L = 27$ V. **c** At light load when $V_L = 36$ V. **d** At half load when $V_L = 18$ V. **e** At half load when $V_L = 27$ V. **f** At half load when $V_L = 36$ V. **g** At full load when $V_L = 18$ V. **h** At full load when $V_L = 27$ V. **i** At full load when $V_L = 36$ V

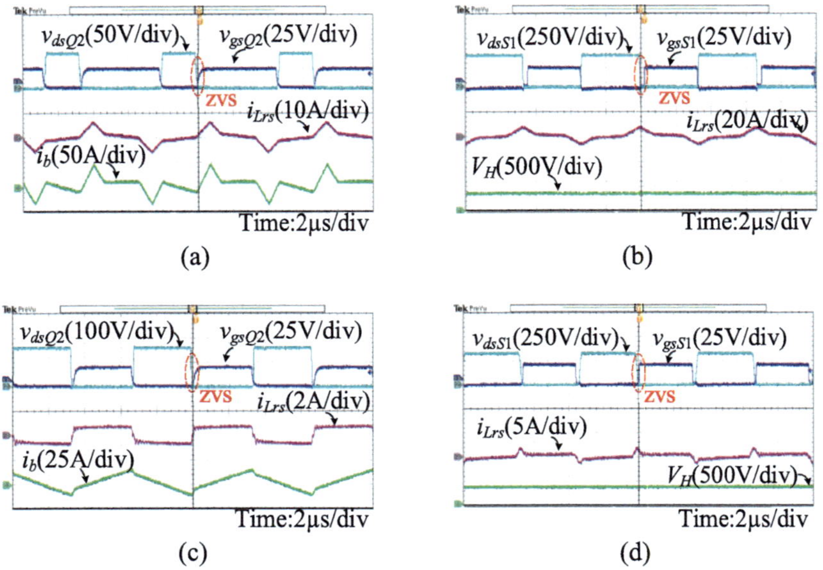

Fig. 5.17 ZVS experimental waveforms of switches with light load. **a** Q_2 when $V_L = 18$ V. **b** S_1 when $V_L = 18$ V. **c** Q_2 when $V_L = 36$ V. **d** S_1 when $V_L = 36$ V

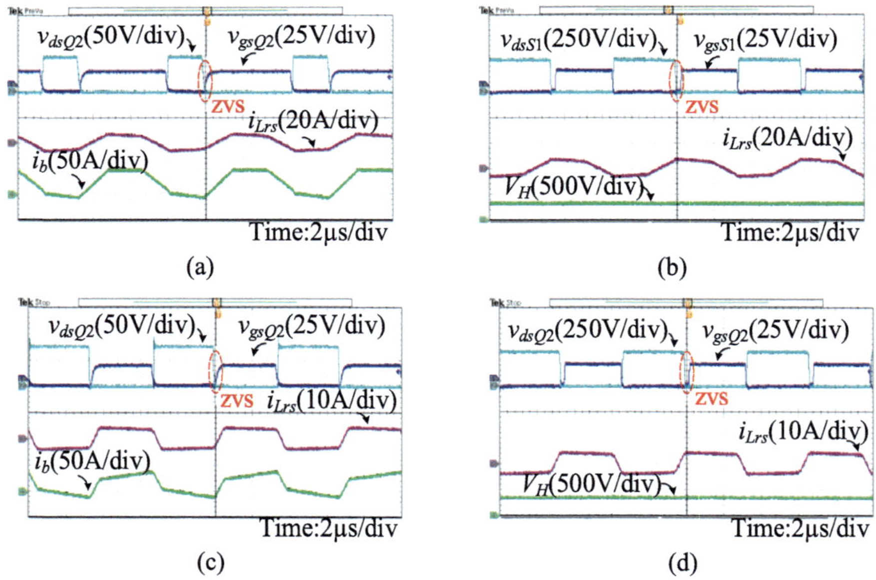

Fig. 5.18 ZVS experimental waveforms of switches with full load. **a** Q_2 when $V_L = 18$ V. **b** S_1 when $V_L = 18$ V. **c** Q_2 when $V_L = 36$ V. **d** S_1 when $V_L = 36$ V

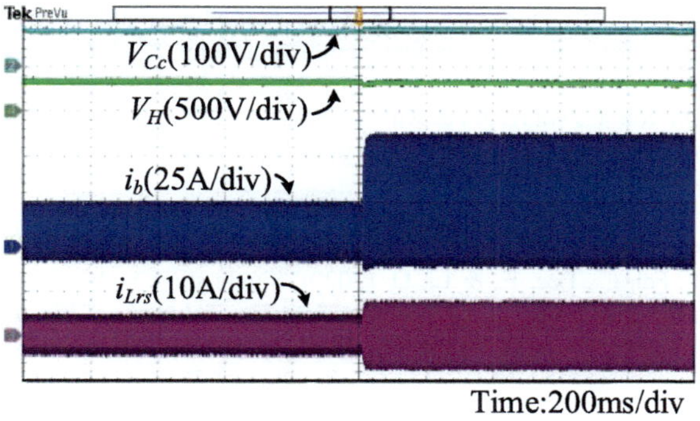

Fig. 5.19 Load-step experimental result from half load to full load with $V_L = 27$ V

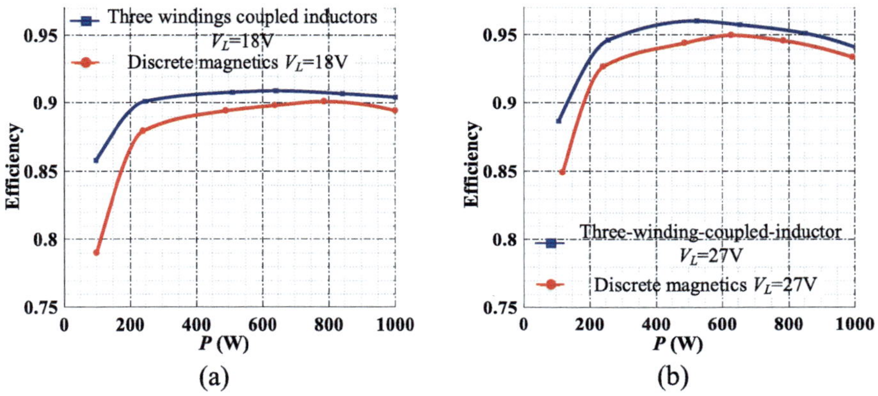

Fig. 5.20 Efficiency comparisons between the proposed TWCI DAB and CF-DAB at different loads and different voltages. **a** $V_L = 18$ V. **b** $V_L = 27$ V. **c** $V_L = 36$ V

References

1. Sha, D., Wang, X., Liu, K., & Chen, C. (2019). A current-fed dual-active-bridge DC–DC converter using extended duty cycle control and magnetic-integrated inductors with optimized voltage mismatching control. *IEEE Transactions on Power Electronics, 34*(1), 462–473.
2. Jiang, C., & Liu, H. (2021). A novel interleaved parallel bidirectional dual-active-bridge DC–DC converter with coupled inductor for more-electric aircraft. *IEEE Transactions on Industrial Electronics, 68*(2), 1759–1768.
3. Bahrami, H., Farhangi, S., Iman-Eini, H., & Adib, E. (2018). A new interleaved coupled-inductor nonisolated soft-switching bidirectional DC–DC converter with high voltage gain ratio. *IEEE Transactions on Industrial Electronics, 65*(7), 5529–5538.
4. Bahrami, H., Farhangi, S., Iman-Eini, H., & Adib, E. (2019). Analysis, design, and implementation of DC–DC IBBC-DAHB converter with voltage matching to improve efficiency. *IEEE Transactions on Industrial Electronics, 66*(7), 5209–5219.
5. Yan, Z., Zeng, J., Lin, W., & Liu, J. (2020). A novel interleaved nonisolated bidirectional DC–DC converter with high voltage-gain and full-range ZVS. *IEEE Transactions on Power Electronics, 35*(7), 7191–7203.

6. Guo, Z., & Sha, D. (2019). Dual-active-bridge converter with parallel-connected full bridges in low-voltage side for ZVS by using auxiliary coupling inductor. *IEEE Transactions on Industrial Electronics, 66*(9), 6856–6866.
7. Sha, D., Lin, Q., You, F., Wang, X., & Xu, G. (2016). A ZVS bidirectional three-level DC–DC converter with direct current slew rate control of leakage inductance current. *IEEE Transactions on Industry Applications, 52*(3), 2368–2377.
8. Bloom & Gordon. (1985). *Modern DC-to-DC switchmode power converter circuits* (pp. 262–325). Van Nostrand Reinhold Inc.
9. Ahmed, M. H., Fei, C., Lee, F., & Li, Q. (2020). Single-stage high-efficiency 48/1 V sigma converter with integrated magnetics. *IEEE Transactions on Industrial Electronics, 67*(1), 192–202.

Chapter 6
Modulated Coupled Inductor Based ISOP Dual-Active-Bridge Converter with Hybrid SPS Control

Abstract The Input-serial-output-parallel (ISOP) dual-active-bridge (DAB) converter is an attractive solution to be used as a DC transformer (DCT) in the DC power distribution system. However, due to zero-voltage-switching (ZVS) limitation and large circulating current, the efficiency of the converter will be reduced. To improve the efficiency, an ISOP DAB converter based on a modulated coupled inductor, and a corresponding hybrid SPS control are proposed in this chapter. With the hybrid SPS control strategy, the equivalent leakage inductance of modulated coupled inductor can be controllable. The large inductance is adopted at light load and the small inductance for the transformer is applied at heavy load to reduce circulating current and extend the ZVS range. By optimizing the root-mean-square (RMS) current of two operation modes of the converter, a control boundary of the hybrid SPS control can be obtained to modulate the coupled inductor. Meanwhile, the additional magnetizing inductance is utilized to achieve ZVS for all switches. The analysis based on the RMS current optimization and ZVS condition is given. Based on the conditions of the full load range ZVS and small RMS current, the parameters of the modulated coupled inductor, the turn ratio of the transformer, and magnetizing inductance are designed. Finally, a 250 kHz, 800/400 V, 2000 W ISOP DAB prototype is built to verify the effectiveness of the proposed method. Meanwhile, a comparison experiment with the conventional individual leakage inductors solution is given, which shows performance improvements of the proposed solution.

6.1 Converter Topology and Operation Principle

6.1.1 Topology and Working Principle of the Proposed Converter

The schematic circuit of the proposed converter is shown in Fig. 6.1. As seen, the converter consists of two half-bridge DAB modules connected with the ISOP structure, and the leakage inductors of two DAB modules are coupled. And the magnetizing inductance is specially addressed instead of ignoring it, since it is utilized to

G. Xu and X. Chen, *Topology Deduction and Control for Dual-Active-Bridge Converters*, https://doi.org/10.1007/978-981-97-4856-3_6

help achieve ZVS in this chapter. The parameters of the two DAB modules, including the isolation transformer, the half-bridge capacitors, the leakage inductor, and the switch device, are designed to be the same to ensure power balance. The two ports of the converter are connected to two DC buses.

The hybrid SPS control method is adopted in the proposed converter, whose key waveforms are illustrated in Figs. 6.2 and 6.3. As seen, the phase-shift ratio btween primary side switches and secondary side switches is defined as φ. The phase between two DAB modules, namely the phase between the voltages of two windings of the coupled inductor, is defined as φ_{ab}.

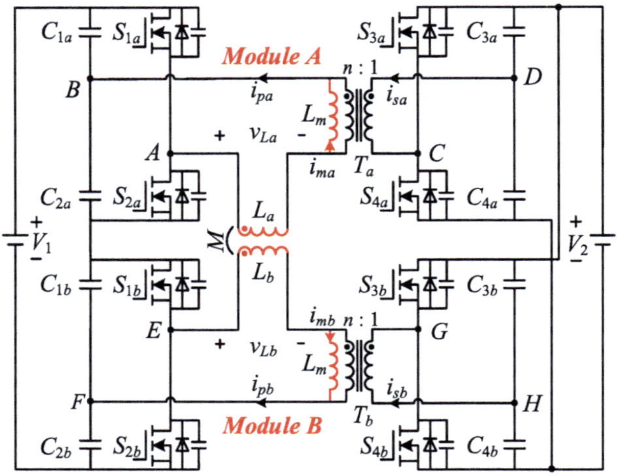

Fig. 6.1 The circuit of the proposed converter

Fig. 6.2 The key waveforms when the converter operates in Mode I

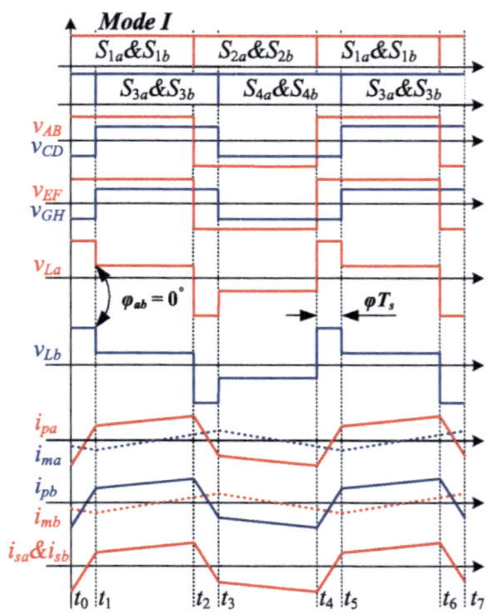

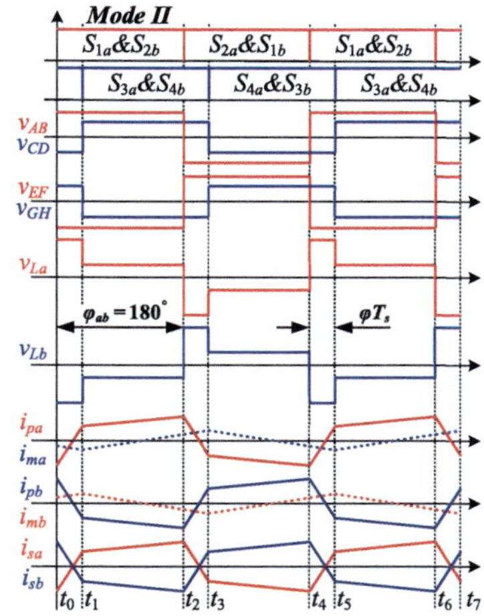

Fig. 6.3 The key waveforms when the converter operates in Mode II

There are two working modes for the proposed converter. When the converter works in Mode I, φ_{ab} is operated at 0°, whose key operating waveforms are shown in Fig. 6.2 In this mode, the phase between two DAB modules is controlled to be 0° by modulating the PWM signals. The driving signal of S_{1a} and S_{1b} is the same, and S_{3a} and S_{3b} is the same. And the phase between v_{La} and v_{Lb} is 0°, namely $v_{La} = v_{Lb}$.

When the converter works in Mode II, φ_{ab} is operated at 180°, whose key operating waveforms are shown in Fig. 6.3. In this mode, the phase between two DAB modules is controlled to be 180° by modulating the PWM signals. The driving signal of S_{1a} and S_{2b} is the same, and S_{3a} and S_{4b} is the same. And the phase between v_{La} and v_{Lb} is 180°, namely $v_{La} = -v_{Lb}$.

6.1.2 Model Analysis of the Modulated Coupled Inductor

As for the coupled inductor shown in Fig. 6.1, its mathematical model can be derived by the following equation according to the basic circuit theory.

$$\begin{cases} v_{La} = L_a \dfrac{di_{pa}}{dt} + M \dfrac{di_{pb}}{dt} \\ v_{Lb} = M \dfrac{di_{pa}}{dt} + L_b \dfrac{di_{pb}}{dt} \end{cases} \tag{6.1}$$

where L_a, L_b, and M are the self-inductance of each winding and the mutual inductance between two windings, respectively.

Equation (6.1) can be rearranged as

$$\begin{cases} (1-k_c^2)L_a\dfrac{di_{pa}}{dt} = v_{La} - k_c v_{Lb} \\ (1-k_c^2)L_b\dfrac{di_{pb}}{dt} = v_{Lb} - k_c v_{La} \end{cases} \tag{6.2}$$

where k_c is the coupling coefficient, which is defined as

$$k_c = \frac{M}{\sqrt{L_a L_b}} \tag{6.3}$$

When the converter operates in Mode I, the relation of the two voltages across the two windings of the coupled inductor can be expressed as

$$v_{La} = v_{Lb} \tag{6.4}$$

To achieve the power balance of two DAB modules, the parameters of two DAB converters are designed to be the same. Then, the self-inductance of the coupled inductor is designed to be the same that $L_k = L_a = L_b$. Therefore, according to (6.4), (6.2) can be simplified as

$$\begin{cases} (1+k_c)L_k\dfrac{di_{pa}}{dt} = v_{La} \\ (1+k_c)L_k\dfrac{di_{pb}}{dt} = v_{Lb} \end{cases} \tag{6.5}$$

Then, the equivalent leakage inductance when φ_{ab} is 0° can be obtained as

$$L_{eq0} = (1+k_c)L_k \tag{6.6}$$

When the converter operates in Mode II, the relation of the two voltages across the two windings of the coupled inductor can be expressed as

$$v_{La} = -v_{Lb} \tag{6.7}$$

According to (6.7), (6.2) can be simplified as

$$\begin{cases} (1-k_c)L_k\dfrac{di_{pa}}{dt} = v_{La} \\ (1-k_c)L_k\dfrac{di_{pb}}{dt} = v_{Lb} \end{cases} \tag{6.8}$$

Then, the equivalent leakage inductance can be obtained as

$$L_{eq180} = (1-k_c)L_k \tag{6.9}$$

As analyzed above, when the converter operates in different modes, which can be controlled by modulating the PWM signals of the converter, the equivalent leakage inductance can be exchanged between L_{eq0} and L_{ea180}.

6.1.3 Power Transfer Characteristics of the Proposed Converter

Because of the symmetrical current characteristics of two DAB modules, only module A is analyzed in detail. When the converter works in Mode I, the instantaneous current of the coupled inductor can be calculated as

$$\begin{cases} i_{pa}(t_1) = i_{pa}(t_0) + \dfrac{V_1 + 2nV_2}{4(1+k_c)L_k}\dfrac{\varphi}{f_s} \\ i_{pa}(t_2) = i_{pa}(t_1) + \dfrac{V_1 - 2nV_2}{4(1+k_c)L_k}\dfrac{1-2\varphi}{2f_s} \end{cases} \tag{6.10}$$

where f_s is the switching frequency of the converter, φ is the phase-shift ratio of each DAB module. According to the volts-second theory, the relation between $i_{pa}(t_0)$ and $i_{pa}(t_2)$ is

$$i_{pa}(t_0) = -i_{pa}(t_2) \tag{6.11}$$

Then, according to (6.10) and (6.11), $i_{pa}(t_0)$ can be solved as

$$i_{pa}(t_0) = -\frac{V_1 - 2nV_2 + 8n\varphi V_2}{16(1+k_c)L_k f_s} \tag{6.12}$$

After that, the transferred power of the converter can be calculated as

$$P_{\text{I}}(\varphi) = 2f_s \int_{t_0}^{t_2} V_1 i_{pa}(t)dt = \frac{nV_1V_2\varphi(1-2\varphi)}{4(1+k_c)L_k f_s} \tag{6.13}$$

To normalize the transferred power, the base power can be set when the converter works with $\varphi = 0.25$, $k_c = 0$, and $V_1 = 2nV_2$, which can be expressed as

$$P_{base} = P(0.25)_{|V_1=2nV_o,k_c=0} = \frac{V_1^2}{64L_k f_s} \tag{6.14}$$

Then the normalized power under Mode I can be obtained as

$$P_{\text{I}}^*(\varphi) = \frac{P_{\text{I}}(\varphi)}{P_{base}} = \frac{8G\varphi(1-2\varphi)}{1+k_c} \tag{6.15}$$

where G is the voltage ratio of the converter, which is defined as

$$G = \frac{2nV_2}{V_1} \tag{6.16}$$

When the converter operates in Mode II, the transferred power can be calculated with the same approach, which is expressed as

$$P_{\text{II}}(\varphi) = \frac{nV_1V_2\varphi(1-2\varphi)}{4(1-k_c)L_kf_s} \tag{6.17}$$

And the normalized power under Mode II can be obtained as

$$P_{\text{II}}^*(\varphi) = \frac{P_{\text{II}}(\varphi)}{P_{base}} = \frac{8G\varphi(1-2\varphi)}{1-k_c} \tag{6.18}$$

By comparing (6.13) and (6.18), it can be concluded that the transferred power when the converter operates in Mode II is larger than that when the converter operates in Mode I with the same φ, since the range of k_c is $0 < k_c < 1$.

6.2 ZVS Conditions Analysis and Control Boundary of the Hybrid SPS Control

6.2.1 ZVS Analysis Considering Junction Capacitance and Magnetizing Current Injection

As for the proposed converter, the working principles of the two DAB modules are symmetrical. Therefore, in this section, only module A is analyzed in detail. To achieve ZVS for switches, the energy stored in the inductive components should be high enough to charge/discharge the switch junction capacitors sufficiently.

For the primary side switches, the equivalent ZVS circuit considering the junction capacitors is illustrated in Fig. 6.4a. When S_{1a} is turned off, whose waveforms are shown in Figs. 6.2 and 6.3, i_{pa} will approach its maximum value. Then, i_{pa} will charge C_{s1} and C_{2a}, and discharge C_{s2} and C_{1a}. During the charging/discharging process, there are three parts of energy including the energy stored in L_{eq}, energy in all capacitors, and the energy flowing into V_2. According to the law of energy conservation, the sum of all energy stored in the capacitors is a constant value, while the energy stored in L_{eq} is decreased and fed back to V_2. The relation among these energies can be calculated as,

$$E_{Leq}(t_{2\_1}) = E_{Leq}(t_{2\_2}) + E_{feedback} = \frac{1}{2}L_{eq}i_p(t_{2\_1})^2 \tag{6.19}$$

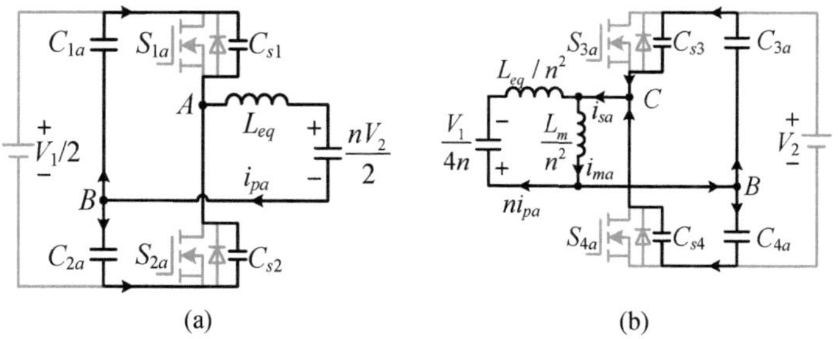

Fig. 6.4 Equivalent ZVS circuit. **a** The equivalent ZVS circuit after S_{1a} is turned off. **b** The equivalent ZVS circuit after S_{3a} is turned off

where t_{2_1} and t_{2_2} are the beginning and end of the charging/discharging process, respectively.

According to Kirchhoff's current law, the relation between the leakage inductor current and the capacitor voltage can be expressed as

$$i_{pa}(t) = C_{s1}\frac{du_{s1}(t)}{dt} - C_{s2}\frac{du_{s2}(t)}{dt} = 2C_{s1}\frac{du_{s1}(t)}{dt} \tag{6.20}$$

where $u_{s1}(t)$ and $u_{s2}(t)$ are the voltage across the switch junction capacitors for S_{1a} and S_{2a}, respectively.

When the charging/discharging process is finished, the energy stored in L_{eq} is fed back to V_2, which is calculated as

$$E_{feedback} = \int_{t_{2\_1}}^{t_{2\_2}} \left[\frac{1}{2}nV_2 i_{pa}(t)\right]dt \tag{6.21}$$

After the charging/discharging process, if the switch junction capacitors are charged/discharged sufficiently, the remaining energy stored in L_{eq} should be greater than zero. Therefore, the constraint to realizing ZVS for the primary side switches can be obtained as

$$E_{Leq}(t_{2\_1}) - E_{feedback} \geq 0 \tag{6.22}$$

According to (6.19), (6.20), and (6.21), (6.22) can be rearranged as

$$i_{pa}(t_{2\_1}) \geq \sqrt{\frac{nV_1V_2C_{s1}}{L_{eq}}} \tag{6.23}$$

which is the constraint for primary side switches to achieve ZVS.

For ZVS realization of the secondary side switches, different from the conventional DAB converter, the magnetizing current participates in the charge/discharging process after the switches are turned off. The equivalent ZVS circuit considering the junction capacitors and the magnetizing current after S_{3a} is turned-off is illustrated in Fig. 6.4b. As seen, the magnetizing current will benefit to charge/discharge the junction capacitors. The magnetizing current i_{ma} will cancel the leakage inductor current i_{pa} and change the shape of i_{sa}, which will help secondary side switches to achieve ZVS. During the charging/discharging process, i_{sa} should always be greater than zero, which can be expressed as

$$i_s(t_3) = n\left[i_m(t_3) - i_p(t_3)\right] \geq 0 \tag{6.24}$$

According to (6.23) and (6.24), Fig. 6.5 plots the ZVS range versus P^* and G with different k_c and k_L. Noted that, k_L is the ratio between the leakage inductance and magnetizing inductance, which is defined as

$$k_L = \frac{L_k}{L_m} \tag{6.25}$$

Figure 6.5a shows ZVS comparison with different k_c. As seen, when $k_c = 0.5$, the ZVS range in Mode I is larger than that in Mode II. And ZVS range with the conventional individual leakage inductor method ($k_c = 0$), is smaller than that in Mode I. Therefore, the ZVS range can be extended by using the coupled inductor. However, hard switching still occurs at light load for the converter.

To achieve ZVS at light load, the magnetizing inductance can be properly designed to extend the ZVS range for the converter. Figure 6.5b shows the ZVS range comparison with different k_L. As seen, when $k_L = 0$, which means the magnetizing inductance is ignored, none ZVS region will occur at light load. When k_L is changed from 0 to 0.15, which means the magnetizing current is injected, the ZVS region is expanded, and the full load range ZVS can be realized. Therefore, with the proper

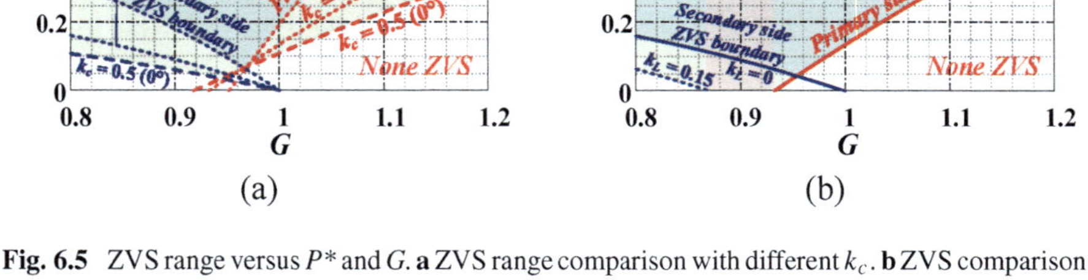

Fig. 6.5 ZVS range versus P^* and G. **a** ZVS range comparison with different k_c. **b** ZVS comparison with different k_L

design of G and k_L, full load range ZVS can be achieved for the converter under SPS control.

6.2.2 RMS Current Derivation

Because the RMS current of the two DAB modules is the same, only module A is analyzed. When the converter operates in Mode I, the normalized RMS current can be derived as

$$
\begin{aligned}
I^*_{p\_rms\_I}(\varphi) &= \frac{1}{I_{base}}\sqrt{2f_s \cdot \int_{t_0}^{t_2} i_{pa}^2(t)dt} \\
&= \frac{4\sqrt{16G\varphi^2(3-4\varphi)+(G-1)^2}}{\sqrt{3}(1+k_c)}
\end{aligned} \tag{6.26}
$$

where I_{base} is the base current, which is defined as (P_{base}/V_1). When the converter operates in Mode II, the normalized RMS current can be expressed as

$$
I^*_{p\_rms\_II}(\varphi) = \frac{4\sqrt{16G\varphi^2(3-4\varphi)+(G-1)^2}}{\sqrt{3}(1-k_c)} \tag{6.27}
$$

Substituting (6.15) and (6.18) into (6.26) and(6.27) respectively, Fig. 6.6 plots the normalized RMS current versus normalize power P^* with different coupling coefficient k_c and voltage ratio G under different operation modes in the 3D plot. As seen, when the converter operates at the voltage matching point, namely $G = 1$, the RMS current in Mode II is always smaller than that in Mode I under the full load range. However, when the converter does not operate at the voltage matching point, the RMS current in Mode I is smaller than that in Mode II at light load. Therefore, Mode I is more suitable for the converter working at light load, and Mode II is more suitable for the converter working at heavy load to reduce RMS current.

6.2.3 Control Boundary Between Mode I and Mode II

To achieve a low RMS current for the converter, the different operating modes can be adopted in different operating conditions. It can be observed from Fig. 6.6 that the converter is more suitable to operate in Mode I at light load and Mode II at heavy load. Therefore, it is necessary to propose a hybrid SPS control strategy, of which the key point is how to determine the control boundary of different operating modes. For example, Fig. 6.7a illustrates the normalized RMS current when $k_c =$ 0.5. And the control boundary can be obtained by solving the equations of (6.15),

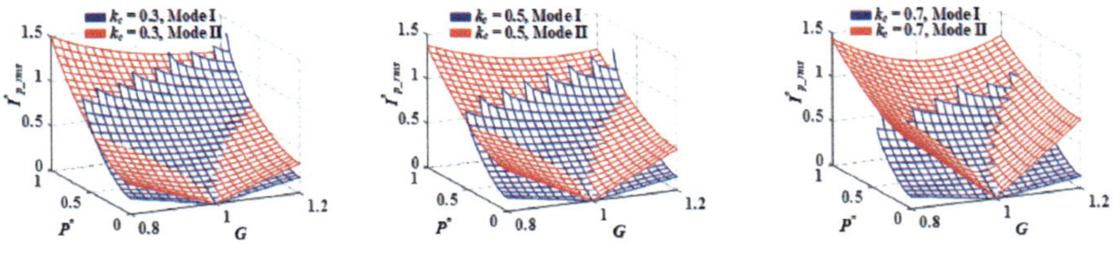

(a) (b) (c)

Fig. 6.6 The leakage inductor normalized RMS current versus normalized power P* with different G and kc under different operation modes in 3D plot. **a** $k_c = 0.3$. **b** $k_c = 0.5$. **c** $k_c = 0.7$

(6.18), (6.26) and (6.27). However, the calculation results are complicated, resulting in a large amount of resources consumed by the MCU controller.

Therefore, to simplify the expressions, the fitted boundary lines are used to approximate the original control boundary lines, shown as the dotted lines a and b in Fig. 6.7(b). Meanwhile, the control boundary with different k_c is expressed in Table 6.1. Then, the hybrid SPS control strategy can be obtained according to the control boundary in Table 6.1.

Figure 6.8 shows the comparisons of normalized RMS current between the proposed solution and the conventional solution. Noted that, the conventional solution is that the leakage inductors are not coupled, and the coupled coefficient k_c is 0. As seen, under the proposed hybrid SPS control strategy, the RMS current of the converter can be reduced in some operating conditions. But, if the coupled coefficient k_c is too large, the RMS current will be increased at the middle power point. Therefore, a suitable coupled coefficient is needed to optimize the RMS current of the converter, which will be discussed in the next Section. The control block of the proposed solution is illustrated in Fig. 6.9, of which the boundary law can be

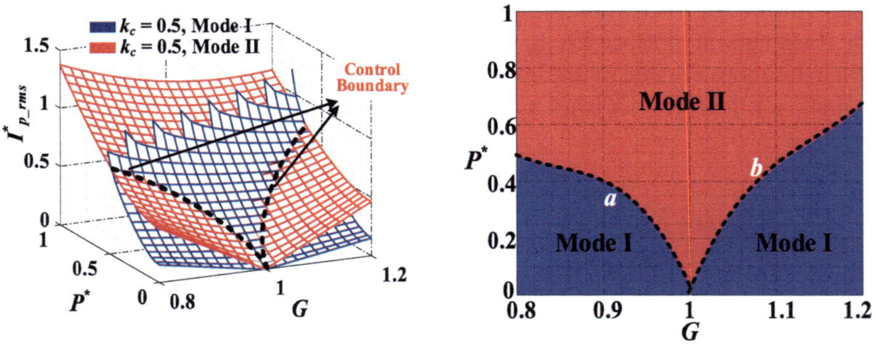

(a) (b)

Fig. 6.7 Control boundary of the converter when $k_c = 0.5$. **a** 3D plot of the normalized RMS current and control boundary. **b** The bottom view of the 3D plot and the fitted control boundary lines

Table 6.1 The fitted control boundary with different k_c

$k_c = 0.3$	a	$P^{*} = -77.732G^3 + 197.93G^2 - 168.81G + 48.637$
	b	$P^{*} = 77.607G^3 - 265.06G^2 + 303.69G - 116.22$
$k_c = 0.5$	a	$P^{*} = -89.228G^3 + 226.97G^2 - 193.13G + 55.422$
	b	$P^{*} = 86.015G^3 - 294.58G^2 + 338.22G - 129.63$
$k_c = 0.7$	a	$P^{*} = 114.45G^3 + 289.04G^2 - 243.26G + 68.71$
	b	$P^{*} = 101.09G^3 - 350.27G^2 + 406.01G - 156.79$

obtained from Table 6.1. As Fig. 6.9 shows, V_{in1} and V_{in2} are the input voltages of two DAB modules, respectively. φ_a and φ_b are the phase shift ratio of two DAB modules, respectively. V_2 is the output voltage, and V_{2_ref} is the reference value of the output voltage. To control the output voltage, the error between V_2 and V_{2_ref} is calculated through the PI controller. Then, the phase shift ratio can be obtained for the two DAB modules. To achieve voltage balance between two DAB modules, the error between V_{in1} and V_{in2} is calculated through the PI controller. Then, $\Delta\varphi$ can be obtained, which is added to φ_a. By slightly adjusting the phase shift ratio of module B, voltage balance between the two modules can be achieved. Meanwhile, the phase φ_{ab} between two DAB modules can be calculated by the control boundary law.

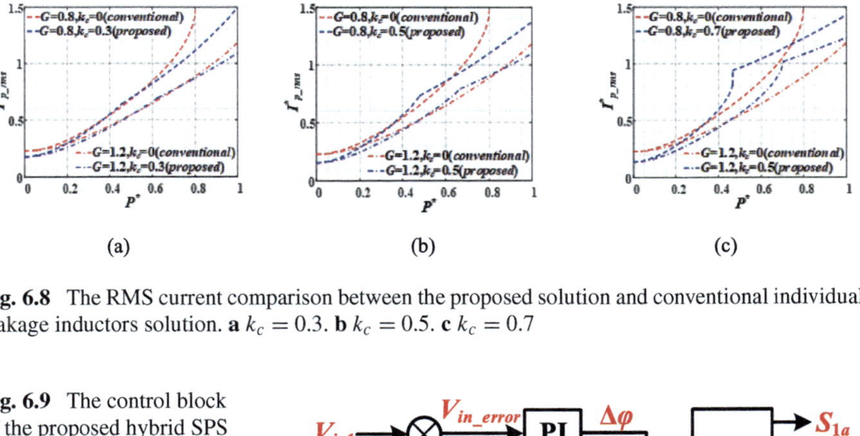

(a) (b) (c)

Fig. 6.8 The RMS current comparison between the proposed solution and conventional individual leakage inductors solution. **a** $k_c = 0.3$. **b** $k_c = 0.5$. **c** $k_c = 0.7$

Fig. 6.9 The control block of the proposed hybrid SPS control

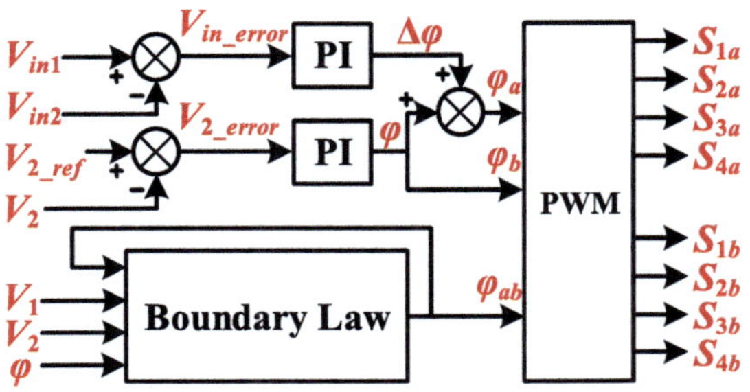

6.3 Parameters Design Consideration

As previously analyzed, when the proposed ISOP DAB converter operates under different modes, the equivalent leakage inductance can be changeable by utilizing the modulated coupled inductor, which can reduce the RMS current. However, a suitable coupling coefficient k_c of the modulated coupled inductor should be properly designed to optimize the RMS current. Meanwhile, to achieve full load range ZVS, the voltage ratio G (or the turns ratio of the transformer) of the converter and the magnetizing inductance Lm should also be designed.

In this chapter, a design example is given with the parameters of 250 kHz, 2000W, $V_{in} = 800$ V, and $V_o = 400$ V. And the voltage is considered with ±5% fluctuation under rated operating conditions. With the given parameters of the converter, the leakage inductance can be calculated according to (6.13) and shown as

$$L_k = \frac{V_1 V_2}{32 f_s P_{rated}} \tag{6.28}$$

6.3.1 *Voltage Ratio G and Coupled Coefficient k_c Design to Achieve Full Load Range ZVS for Primary Side Switches with Small RMS Current*

To achieve full load range ZVS for primary side switches, the voltage ratio G should be designed. Figure 6.10a–c plot the normalized RMS current contour lines and the ZVS range of primary side switches with different coupling coefficient kc. As seen, to achieve full load range ZVS for primary side switches, G should be designed in the full load ZVS range region. Meanwhile, G should also be designed to achieve a relatively small RMS current. As Fig. 6.10a–c show, for a certain output power P^*, the RMS current is decreased as G increases. Therefore, to achieve low RMS current, G should be designed as large as possible in the full load ZVS range.

For the design example of this chapter, considering the voltage fluctuation, the range of G is designed with ±5% fluctuation. Figure 6.10d shows the optimized operation range of G. As seen, to achieve low RMS current, the operation range of G is designed to be close to the primary side ZVS boundary. Figure 6.11a plots the operation range versus k_c. As seen, G decreases monotonically with k_c. To find a suitable G, k_c should be designed first.

Figure 6.11b plots the normalized RMS current versus P^* with different k_c at the rated voltage point. As seen, the normalized RMS current will decrease as k_c increases. However, when k_c is designed too large, the normalized RMS current will increase at the middle power point. Therefore, to make a tradeoff for the normalized RMS current over the entire load, k_c can be designed to be 0.5.

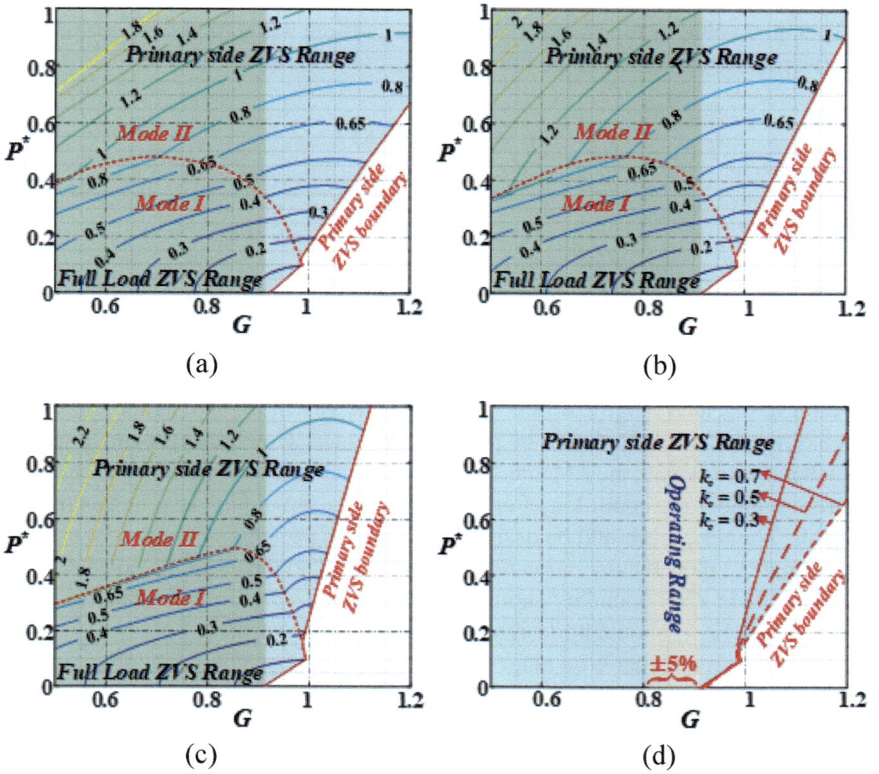

Fig. 6.10 The normalized RMS current contour lines and the ZVS range of primary side switches with different coupling coefficient kc. **a** $k_c = 0.3$. **b** $k_c = 0.5$. **c** $k_c = 0.7$. **d** The operation range for the converter

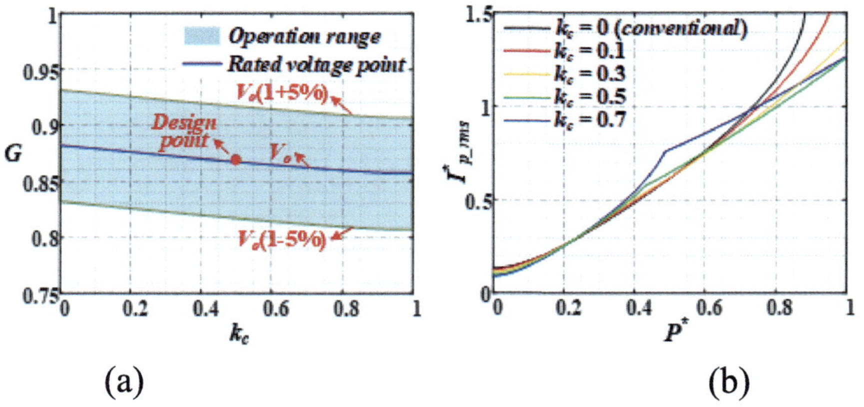

Fig. 6.11 Leakage inductor current comparison when L_k changes under rated load. **a** RMS current. **b** Peak current

Then, the voltage ratio G can be designed according to Fig. 6.11a, which is 0.857. Afterward, the turns ratio of the transformer can be calculated as

$$n = \frac{V_1 G}{2V_2} = 0.857 \quad (6.29)$$

6.3.2 Magnetizing Inductance Lm Design to Achieve Full Load Range ZVS for Secondary Side Switches

To achieve full load range ZVS for secondary side switches, the magnetizing inductance of the transformer should be properly designed. However, if the magnetizing inductance is too small, the RMS current of secondary side switches will be increased. Therefore, under the condition of ensuring ZVS, L_m should be designed as large as possible. Namely, k_L should be designed as small as possible.

Figure 6.12 plots the operation range and the ZVS boundary of the converter. The secondary side ZVS boundary should be designed close to the operating range. As seen in Fig. 6.12, k_L should be designed larger than 0.15 to achieve ZVS for secondary side switches in the operation range. Then, the range of magnetizing inductance L_m can be calculated as

$$L_m < \frac{L_k}{k_L} = 6.67L_k \quad (6.30)$$

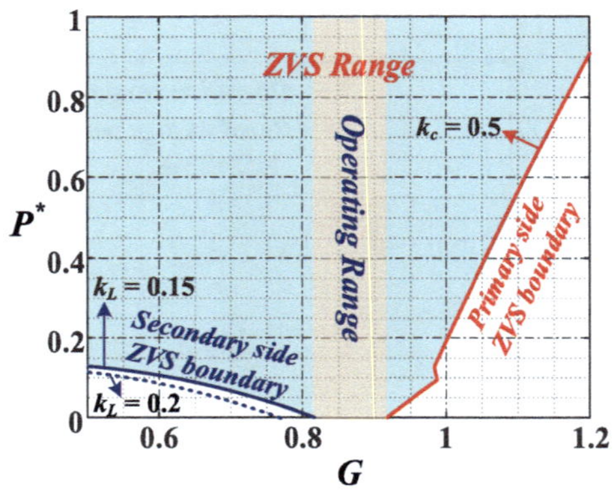

Fig. 6.12 k_L design of the converter

6.4 Comparison

The qualitative comparison between the proposed method and the different state-of-art control methods is summarized in Table III. Compared to conventional SPS control [1] and SPS control with the Tapped Transformer [2], the proposed method has two control variables to achieve full load range ZVS and relative optimal RMS current. Because of the magnetizing current injection for the converter, compared to [3], the proposed method can achieve ZVS at light load and improve the light load efficiency. Compared to the EPS control [4] and EPS control with DC-blocking capacitors [5], the optimization process of the proposed method is easier. Although the full load range soft switching achievement and easy optimization for the control method can be realized in [6], it has one more control variable than the proposed method.

6.5 Experimental Verification

A 2000W prototype of the proposed ISOP DAB converter is built to verify the effectiveness of the proposed solution, which is shown in Fig. 6.13. The detailed parameters of the converter are shown in Table 6.2. To balance the transferred power of two modules, the parameters of two DAB modules are designed to be the same, including the turns ratio of the transformer, magnetizing inductance, and leakage inductance. Besides, when producing the transformers and the coupled inductor, the tolerance of these components should be controlled as small as possible to facilitate the accuracy of power control for the hybrid SPS control scheme. Meanwhile, a comparison between the proposed solution and the conventional individual inductor solution is given. The parameters of the conventional individual inductor solution are also included in Table 6.2.

The middle point voltages of the H-bridge when the converter operates at steady state are illustrated in Fig. 6.14. As seen, when the converter operates in different modes, the operating phase between two DAB modules is different. When the converter operates at Mode I, the operating phase between two DAB modules is

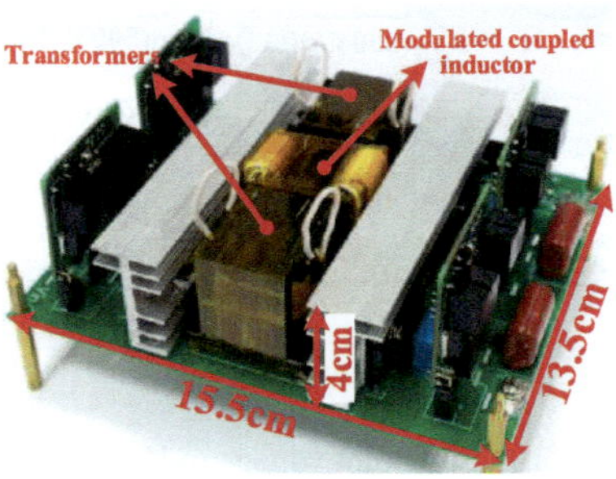

Fig. 6.13 The prototype of the proposed ISOP DAB converter

Table 6.2 Qualitative comparison of different control methods

Mode	Number of variables	Soft-sw range	Type of soft-sw.	Optimization complexity
SPS [1]	1	Partial	ZVS	None
SPS with Tapped transformer [2]	1	Partial (improved)	ZVS	Low
SPS with modulated coupled inductor [3]	2	Partial	ZVS	Low
EPS [4]	2	Partial (improved)	ZVS	Medium
Hybrid EPS with DC-blocking capacitor [5]	2	Full	ZVS	High
Hybrid trapezoidal, triangular and SPS modulation [6]	3	Full	ZVS/ZCS	None
Proposed method	2	Full	ZVS	Low

0°. As Fig. 6.14a shows, the phase between v_{AB} and v_{EF} is 0°. When the converter operates at Mode II, the operating phase between two DAB modules is 180°. As Fig. 6.14b shows, the phase between v_{AB} and v_{EF} is 180°.

Figure 6.15 illustrates the steady-state waveforms of the converter when $V_2 = 400$ V. As Fig. 6.15a and b show, when the converter operates at light load, the polarity of the primary side and secondary side current is opposite, which helps the converter to achieve full load range ZVS.

Figure 6.15c and d show the steady-state waveforms of Module A and Module B during heavy load operation. As seen, the converter can be operated in typical DAB waveforms

Figure 6.16 shows the ZVS waveforms of the converter when $V_2 = 380$ V. As seen in Fig. 6.16a and b, when the converter operates at light load, all switches can achieve ZVS. As Fig. 6.16a shows, when S_{1a} is turned on, the polarity of i_{pa} is

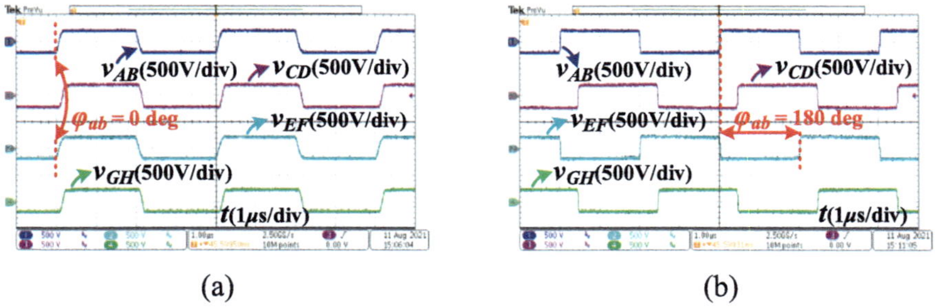

Fig. 6.14 Midpoint voltage of H-bridge with different operation modes. **a** Mode I. **b** Mode II

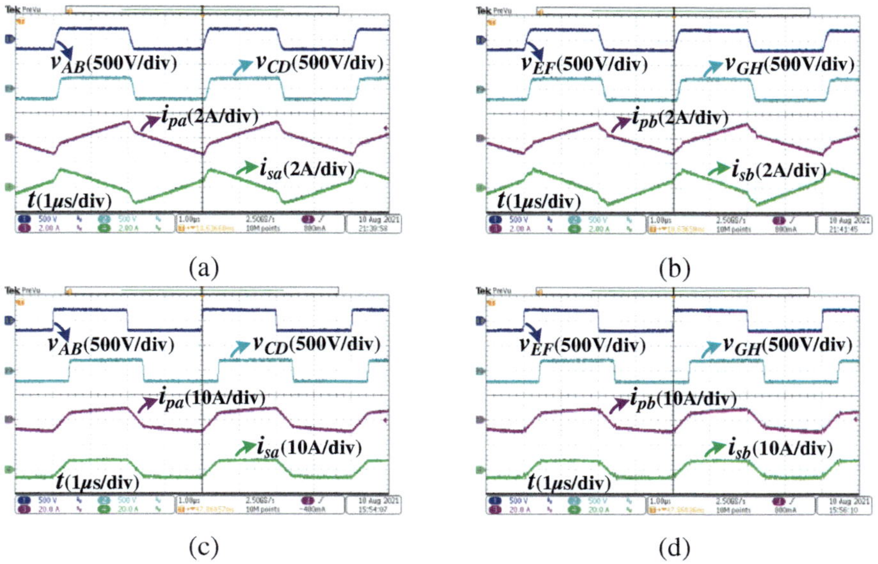

Fig. 6.15 Steady-state waveforms of the converter when $V_2 = 400$. **a** The waveforms of Module A of the converter at light load. **b** The waveforms of Module B of the converter at light load. **c** The waveforms of Module A of the converter at heavy load. **d** The waveforms of Module B of the converter at heavy load

negative, which is helpful for S_{1a} to achieve ZVS. As Fig. 6.16b shows, when S_{3a} is turned on, the polarity of i_{sa} is positive, which is helpful for S_{3a} to achieve ZVS. When the converter operates at heavy load, all switches can also achieve ZVS, which can be seen in Fig. 6.16c and d. Because the operation condition of the two DAB modes is the same, ZVS for all switches of module B can also be achieved.

Figure 6.17 shows the ZVS waveforms of the converter when $V_2 = 400$ V. As seen in Fig. 6.17a and b, the drain-source voltage of S_{1a} and S_{3a} have been reduced to zero before the driving signals are coming. Therefore, S_{1a} and S_{3a} can be achieved ZVS at light load. As Fig. 6.17c and d show, when the converter operates at heavy load, ZVS can also be achieved for S_{1a} and S_{3a}.

Figure 6.18 shows the ZVS waveforms of the converter when $V_2 = 420$ V. As Fig. 6.18a and b show, the polarity of i_{pa} and i_{sa} is helpful for S_{1a} and S_{3a} to achieve ZVS at light load. When the converter operated at heavy load, whose waveforms can be seen in Fig. 6.18a and b, S_{1a} and S_{3a} can achieve ZVS.

Figure 6.19 plots the efficiency comparisons with different solutions of the ISOP DAB converter. As seen, with the proposed solution, the efficiency is improved compared to the conventional individual inductor solution. Especially, the efficiency is significantly improved at light load. When the converter operates when the power is around 700~1100 W, the efficiency of the conventional solution is a little higher than that of the proposed method. It can be observed that the proposed solution has more advantages when the converter operates at light load or heavy load.

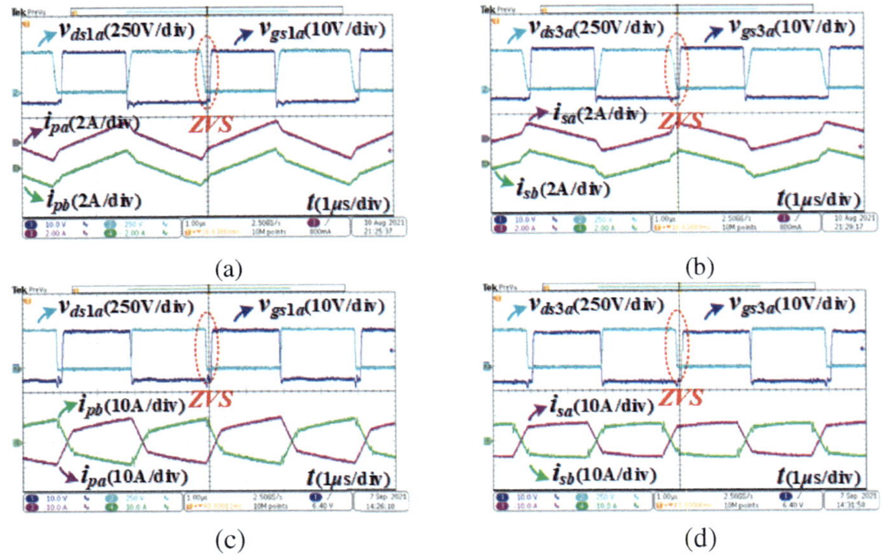

Fig. 6.16 ZVS waveforms of the converter when $V_2 = 380$. **a** Primary side ZVS waveforms at light load. **b** Secondary side ZVS waveforms at light load. **c** Primary side ZVS waveforms at heavy load. **d** Secondary side ZVS waveforms at heavy load

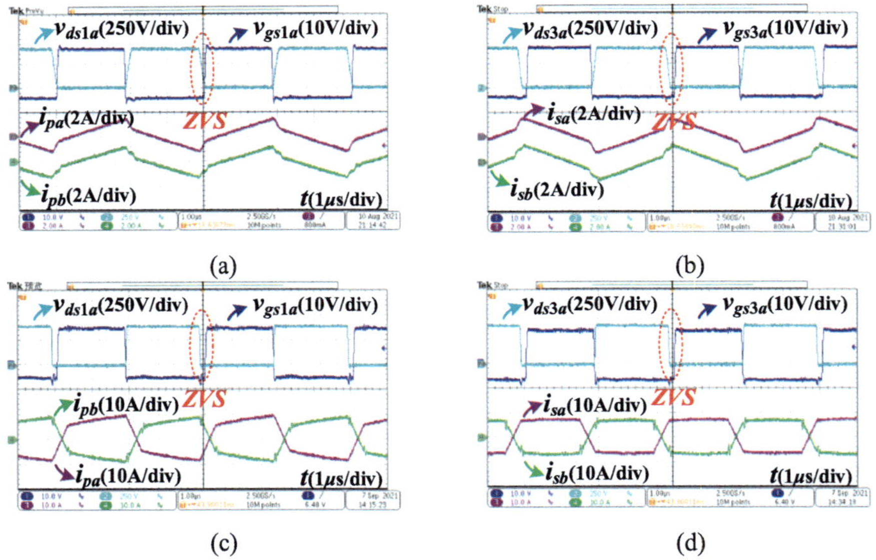

Fig. 6.17 ZVS waveforms of the converter when $V_2 = 400$. **a** Primary side ZVS waveforms at light load. **b** Secondary side ZVS waveforms at light load. **c** Primary side ZVS waveforms at heavy load. **d** Secondary side ZVS waveforms at heavy load

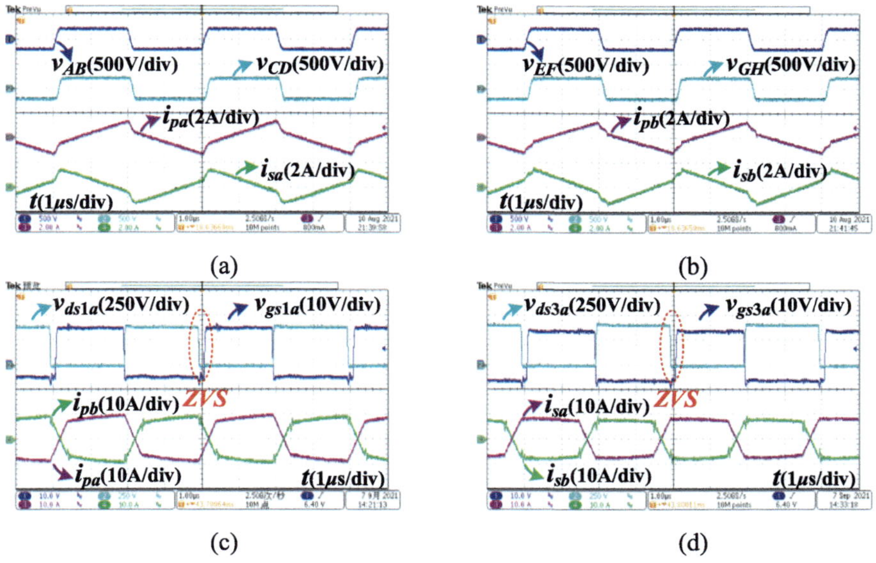

Fig. 6.18 ZVS waveforms of the converter when $V_2 = 420$. **a** Primary side ZVS waveforms at light load. **b** Secondary side ZVS waveforms at light load. **c** Primary side ZVS waveforms at heavy load. **d** Secondary side ZVS waveforms at heavy load

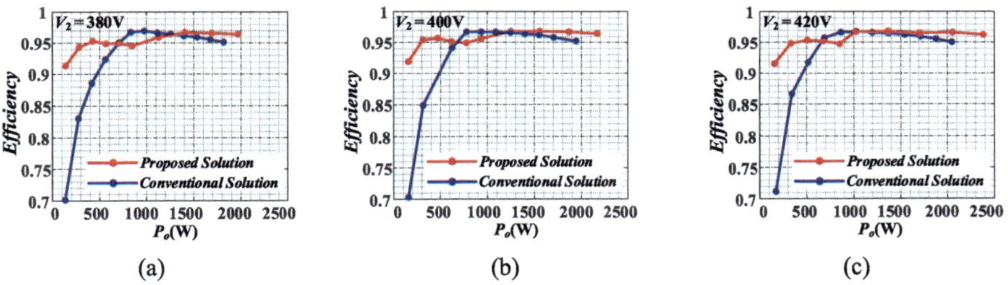

Fig. 6.19 Efficiency comparisons with different solutions of ISOP DAB converter. **a** $V_2 = 380$V. **b** $V_2 = 400$V. **b** $V_2 = 420$V

6.6 Conclusion

This chapter proposes an ISOP DAB converter based on the modulated coupled inductor for the DC transformer application, and a corresponding hybrid SPS control is also proposed. To optimize the RMS current, the modulated coupled inductor is utilized for the converter. Meanwhile, a corresponding control boundary is introduced to modulate the coupled inductor and control the equivalent leakage inductance, leading to optimization for the RMS current. To achieve full load range ZVS, the design process is studied for the parameters of the transformers, including the turns ratio and the magnetizing inductance. From the experimental results of a 2000 W prototype, it can be verified that the full load range ZVS can be achieved for all switches. Besides, by comparing with the conventional individual leakage inductors solution, the efficiency of the proposed method is significantly improved at light load and heavy load. Meanwhile, when the ISOP system has more DAB modules, the proposed ISOP DAB converter with the modulated coupled inductor can be regarded as a sub-module. Then, each sub-module can be designed and controlled with the proposed method.

References

1. Duong, T.-Q., & Choi, S.-J. (2022). Deadbeat control with bivariate online parameter identification for SPS-modulated DAB converters. *IEEE Access, 10*, 54079–54090.
2. Jafari, A., Nikoo, M. S., Karakaya, F., & Matioli, E. (2020). Enhanced DAB for efficiency preservation using adjustable-tap high-frequency transformer. *IEEE Transactions on Power Electronics, 35*(7), 6673–6677.
3. Chen, X., Xu, G., Han, H., Sun, Y., Liu, Y., & Su, M. (2021). Modulated coupled inductor for input-serial-output-parallel dual-active-bridge converter. *IEEE Transactions on Industrial Electronics, 69*(6), 6450–6455.
4. Shi, H., Wen, H., & Hu, Y. (2020). Deadband effect and accurate ZVS boundaries of GaN-based dual-active-bridge converters with multiple-phase-shift control. *IEEE Transactions on Power Electronics, 35*(9), 9886–9903.
5. Xu, G., Tang, J., Zhang, L., Xiong, W., Sun, Y., & Su, M. (2022). A hybrid extended phase shift modulation strategy for DAB converter with DC blocking capacitor to extend ZVS range and reduce RMS current. *IEEE Journal of Emerging Selected Topics on Power Electronics, 10*(5), 6192–6207.
6. Hu, S. C., & De Doncker, R. W. (2022). Natural boundary transition and inherent dynamic control of a hybrid-mode-modulated dual-active-bridge converter. *IEEE Transactions on Power Electronics, 37*(4), 3865–3877.

Chapter 7
Magnetizing Current Injection Based Push–Pull Dual Active Bridge Converter with Optimized Control

Abstract With fewer switches, the push–pull circuits are suited for low-voltage and high-current applications, such as the distributed generation system. In this chapter, a modified push–pull dual active bridge (PPDAB) dc-dc converter and its optimized control are proposed to achieve full load range zero voltage switching (ZVS). Considering the switch junction capacitors, the added current is injected by the magnetizing inductance to achieve full load range ZVS. Based on the RMS value of transformer current, the optimized control strategy is proposed to enhance the system efficiency with the precondition of achieving full load range ZVS of all switches. The working principles, ZVS region analysis, and design of key parameters are discussed in detail. Finally, the effectiveness of the proposed method for the PPDAB converter is verified by the results of a laboratory prototype.

Keywords Push-pull dual active bridge (PPDAB) dc-dc converter · Magnetizing inductance · Switch junction capacitor · Optimized control · Zero voltage switching (ZVS)

7.1 Topology and Operation of Modified PPDAB Converter Under EPS Contorl

To expand the soft switching region of the push–pull DAB converter, a pulse width modulation plus phase shift (PPS) control strategy was applied in (Li et al. in IEEE Trans Industrial Electronics 64:8514–8524, 2017). But, the full load range ZVS is still impossible in (Li et al. in IEEE Trans Industrial Electronics 64:8514–8524, 2017). A push–pull half-bridge converter with single phase shift (SPS) control can achieve ZVS of all switches at rated power (Zhang et al. in IEEE Trans Industrial Electronics 59:2761–2771, 2012). And this converter sup-presses voltage spikes by a simple structure. But, as with conventional DAB converters, this topology will lose ZVS at light load and under wide voltage range. Besides, the control degree of freedom is limited due to the half-bridge structure on the secondary side. Furthermore, since ZVS realization requires the switch junction capacitor to fully release energy

G. Xu and X. Chen, *Topology Deduction and Control for Dual-Active-Bridge Converters*, https://doi.org/10.1007/978-981-97-4856-3_7

by the leakage current, the switch junction capacitor has a significant impact on the ZVS range (Everts in IEEE Trans Power Electronics 32:7561–7576,2017). However, in some existing references for the push–pull DAB converter, the ZVS analysis is based only on the condition of current polarity and ignores the effect of the junction capacitor (Li et al. in IEEE Trans Industrial Electronics 64:8514–8524, 2017, Zhang et al. in IEEE Trans Industrial Electronics 59:2761–2771, 2012). In recent years, widening the ZVS range of the push–pull DAB converter by using magnetizing inductance has not been mentioned, which is an effective method to improve soft-switching performance for DAB converters (Riedel et al. in IEEE Trans Industrial Electronics 65:9438–9448, 2018). Yet topology change leads to differences in ZVS analysis when considering the junction capacitor and magnetizing inductance.

In order to avoid the problem of narrow ZVS range in low-voltage and high-current applications, this Chapter modifies the push–pull dual active bridge (PPDAB) dc-dc converter in (Zhang et al. in IEEE Trans Industrial Electronics 59:2761–2771, 2012) by replacing the half-bridge circuit in secondary side with a full-bridge circuit to add control degree of freedom and specifically introduces a suitable magnetizing inductance (L_m) to help realize the whole load range ZVS under EPS control, as shown in Fig. 7.1. The advantages of the proposed solution are as follows: (1) Through deep analysis of the ZVS performance with considering the switch junction capacitor, it is found that the two ZVS regions will be separated due to the existence of switch junction capacitors, which cannot be solved by the control strategy. (2) Besides, an optimized control law that considers the tradeoff between the minimum RMS current and calculation burden is proposed to achieve low current stress, under the precondition of full load range ZVS in a wide voltage range.

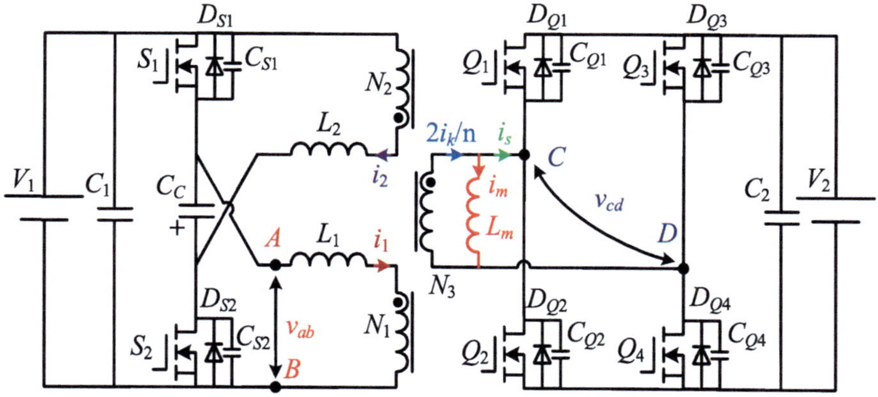

Fig. 7.1 Topology of the PPDAB converter

7.1.1 Topology and Typical Waveforms for Modified PPDAB Converter with Magnetizing Current Injection

The modified PPDAB converter is shown in Fig. 7.1. The push–pull side is made up of two switches, a clamping capacitor, and two leakage inductors. The high-voltage side is a full-bridge structure with four switches. And the magnetizing inductance is specially addressed, which could change the transformer current, thus providing the possibility to improve the ZVS performance. Since the theory of the backward power transfer is similar to that of the forward power transfer, this Chapter will focus on the analysis of the forward operation.

Just like other DAB converters, EPS control which modulates the phase shift angle and the duty cycle of the full bridge can be adopted to the PPDAB converter with magnetizing current injection. The typical waveforms are shown in Fig. 7.2. D_φ is the phase shift ratio between v_{ab} and v_{cd}, and D_y is the duty cycle of v_{cd}. V_1 is the input voltage; V_2 is the output voltage; V_{Cc} is the voltage across the clamp capacitor and V_{Cc} is equal to V_1 in the steady state. i_1 and i_2 are the currents of the leakage inductors, i_m is the magnetizing current, and i_s is the secondary side current. The transformer turns ratio is defined as $N_3/N_1 = N_3/N_2 = n$. The primary side equivalent junction capacitances of C_{S1} and C_{S2} are C_{oss1}, and C_{oss2} is the value of the secondary side junction capacitance. C_{oss1} and C_{oss2} are non-linear parameters related to the voltage across the switch. To facilitate analysis, a suitable capacitance value is selected as C_{oss} according to the switching datasheet. As shown, the converter has three working modes. When the magnetizing current is induced, the switching instance current will be changed for the transformer current.

7.1.2 Operation Principle Analysis of EPS Control Utilizing Magnetizing Current

As shown in Fig. 7.2, the operating condition of the PPDAB converter can be divided into three cases: $0 < D_\varphi < (1\text{-}D_y)/2$, $(1\text{-}D_y)/2 < D_\varphi < (1 + D_y)/2$, $(1 + D_y)/2 < D_\varphi < 1$. Moreover, the description of every stage can be analogously inferred and will not be discussed here. Since the reactive current is relatively small when D_φ is less than 0.5, only case 1 and the part of D_φ less than 0.5 in case 2 are used in practical application.

Based on the analysis, it is clear that there is a DC circuit on the low-voltage side of the PPDAB converter. This means that the leakage currents i_1 and i_2 both contain a DC component, and the amplitude of the DC current is equal to $P/(2V_1)$. The primary side current i_k is defined to be equal to i_1 without the DC component. Since current i_1 and current i_2 respectively superimpose the opposite AC component on the same DC component, the AC component can be eliminated by summing currents i_1 and i_2. In addition, the current $2i_k/n$ reflected on the secondary side is composed of the secondary side current i_s and the magnetizing current i_m. Thus, the relationships

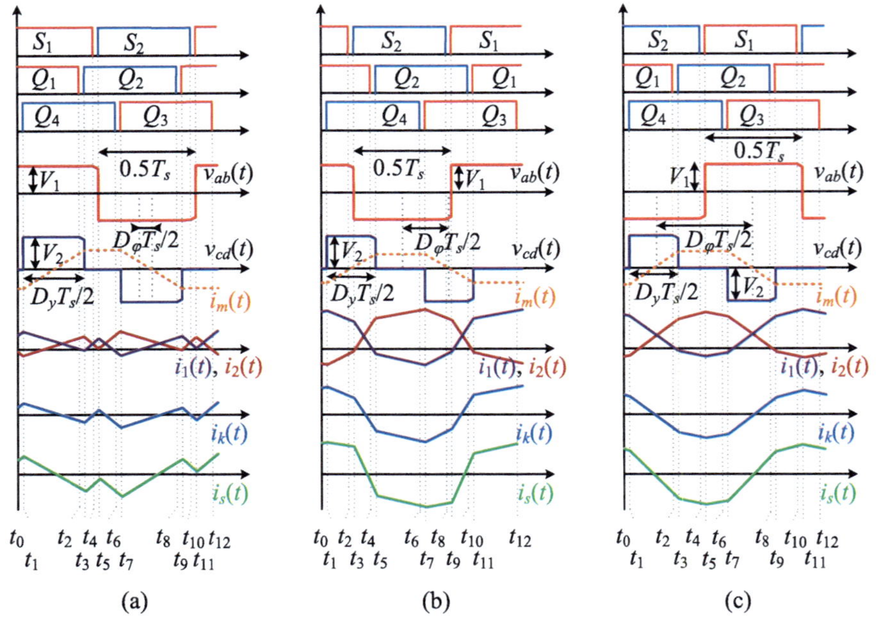

Fig. 7.2 Forward operation waveforms under EPS control. **a** Case 1: $0 < D_\varphi < (1\text{-}D_y)/2$. **b** Case 2: $(1\text{-}D_y)/2 < D_\varphi < (1 + D_y)/2$. **c** Case 3: $(1 + D_y)/2 < D_\varphi < 1$

among i_1, i_2, i_k, i_m and i_s can be given by

$$\begin{cases} i_k(t) = i_1(t) - P/(2V_1) \\ i_1(t) + i_2(t) = P/V_1 \\ 2i_k(t)/n = i_s(t) + i_m(t) \end{cases} \tag{7.1}$$

To analyze the ZVS conditions, it is important to deduce the expressions of the current i_k. According to the principle that the average value of i_k in each switching cycle must be equal to zero, the positive half-cycle expression of i_k in case 1 can be expressed as

$$i_k(t)=\begin{cases} \dfrac{V_1\left(2D_\varphi - D_y + kD_y\right)}{4Lf_s} + \dfrac{V_1(1-k)}{L}(t-t_0),\ t \in [t_0, t_2] \\ \dfrac{V_1\left(2D_\varphi + D_y - kD_y\right)}{4Lf_s} + \dfrac{V_1}{L}(t-t_2), \quad t \in (t_2, t_4] \\ \dfrac{V_1\left(1 - kD_y\right)}{4Lf_s} - \dfrac{V_1}{L}(t-t_4), \quad t \in (t_4, t_6] \end{cases} \tag{7.2}$$

where $k = V_2/(nV_1)$ is the voltage conversion ratio.

In the same way, the positive half-cycle expression of i_m in case 1 can be obtained as

$$i_m(t)=\begin{cases} -\dfrac{nV_1kD_y}{4L_mf_s}+\dfrac{nV_1k}{L_m}(t-t_0), & t\in[t_0,t_2] \\ \dfrac{nV_1kD_y}{4L_mf_s}, & t\in(t_2,t_6] \end{cases} \tag{7.3}$$

Based on (7.1)–(7.3), the average output power expressions can be derived. The transmission power is only related to D_φ and D_y. Therefore, for subsequent analysis, the average output power is normalized by $P_{base} = V_1{}^2/(4Lf_s)$. The normalized values of P_{case1_pu} and P_{case2_pu} are calculated as

$$P_{case1\_pu} = \frac{1}{P_{base}}\left(V_2\frac{2}{T_s}\int_0^{T_s/2} i_s(t)dt\right) = 4kD_yD_\varphi \tag{7.4}$$

$$P_{case2\_pu} = \frac{1}{P_{base}}\left(V_2\frac{2}{T_s}\int_0^{T_s/2} i_s(t)dt\right) = k\left[1-(1-2D_\varphi)^2-(1-D_y)^2\right] \tag{7.5}$$

7.2 Magnetizing Current Injection Analysis for the Modified PPDAB Converter

7.2.1 *ZVS Region Analysis Considering the Switch Junction Capacitors and the Magnetizing Inductance*

A minimum current is required on the inductor to complete the transition of the voltage across the junction capacitor from $2V_1$ (or V_2) to 0, and the minimum value of the current i_k is defined to be I_{k_min}. Furthermore, since the resonance process is very short and the magnetizing current does not change abruptly, the magnetizing current i_m can be considered to remain steady during the junction capacitor discharging. Because the analysis of ZVS is focused in this Chapter, only the equivalent ZVS circuits of Stages 1, 3, and 5 are respectively illustrated in Fig. 7.3. In these equivalent ZVS circuits, the magnetizing inductance is regarded as a constant current source with a current value of I_{m_toff}, and the instant when the corresponding switch is turned off is defined to be t_{off}.

Take Stage 1 in case 1 as an example for analysis. At t_0, the energy stored in the inductors starts to discharge C_{Q4} and charge C_{Q3}, as illustrated in Fig. 7.3a. The voltage across junction capacitor of Q_4 plus the voltage across junction capacitor of Q_3 must be equal to V_2. The dynamic equations of the secondary side current and junction capacitor voltages are expressed as

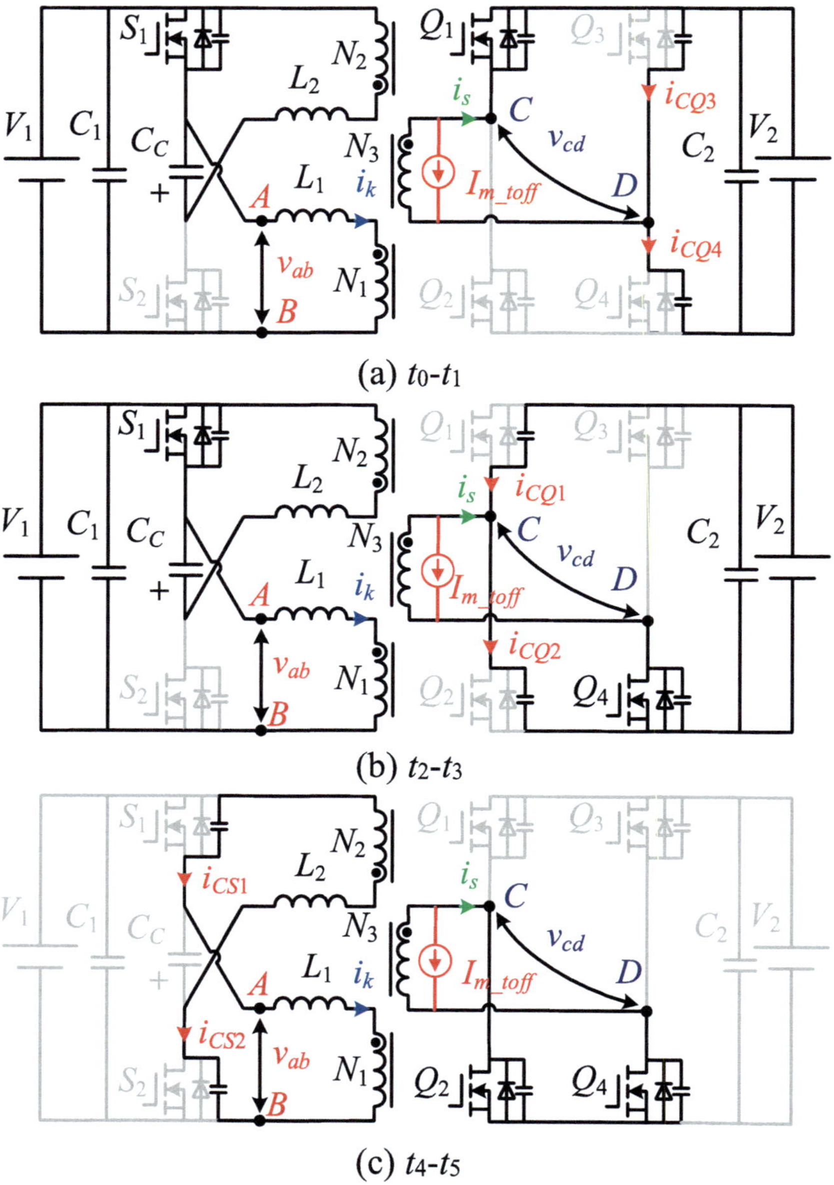

Fig. 7.3 Working stages of Case 1. **a** Stage 1: t_0-t_1. **b** Stage 2: t_2-t_3. **c** Stage 3: t_4-t_5

$$\begin{cases} i_s(t) = C_{oss2}\dfrac{du_{CQ3}(t)}{dt} - C_{oss2}\dfrac{du_{CQ4}(t)}{dt} \\ V_1 - L\dfrac{di_k(t)}{dt} = \dfrac{V_2 - u_{CQ4}(t)}{n} \end{cases} \tag{7.6}$$

And the initial conditions when C_{Q4} starts to discharge can be given by

$$\begin{cases} i_s(0) = \dfrac{2I_{k\_\min 1}}{n} - I_{m\_toff} \\ u_{CQ4}(0) = V_2 \end{cases} \tag{7.7}$$

Using (7.6) and (7.7), the drain-source voltage u_{CQ4} can be obtained as

$$\begin{aligned} u_{CQ4}(t) = & \left(\frac{n}{2}I_{m\_toff} - I_{k\_\min 1}\right)\sqrt{\frac{L}{C_{oss2}}}\sin\left(t\sqrt{\frac{1}{n^2C_{oss2}L}}\right) \\ & + V_1 n\cos\left(t\sqrt{\frac{1}{n^2C_{oss2}L}}\right) + V_2 - V_1 n \end{aligned} \tag{7.8}$$

In order to realize the switch Q_4 is turned on with ZVS, the junction capacitor C_{Q4} needs to be fully discharged. So, $I_{k\_\min1}$ obtained by (7.9) can reduce u_{CQ4} in (7.8) just to zero.

$$(V_2 - V_1 n)^2 = \left[\left(\frac{nI_{m\_toff}}{2} - I_{k\_\min 1}\right)\sqrt{\frac{L}{C_{oss2}}}\right]^2 + (V_1 n)^2 \tag{7.9}$$

Therefore, the ZVS current limit for Q_4 can be derived as

$$\begin{cases} i_k(t_0) \geq I_{k\_\min 1} = knV_1\sqrt{\dfrac{C_{oss2}}{L}\left(1 - \dfrac{2}{k}\right)} + \dfrac{n}{2}I_{m\_toff} \\ \qquad = knV_1\sqrt{\dfrac{C_{oss2}}{L}\left(1 - \dfrac{2}{k}\right)} - \dfrac{n^2V_1kD_y}{8L_mf}, \ (k > 2) \\ i_k(t_0) \geq \dfrac{n}{2}I_{m\_toff} = -\dfrac{n^2V_1kD_y}{8L_mf}, \qquad (1 < k \leq 2) \end{cases} \tag{7.10}$$

Likewise, the ZVS limits for the other stages can also be gained, and all the ZVS restrictions are listed in Table 7.1, where $a = 1 + (n^2L)/(2L_m)$, $t_1 = f_s(LC_{oss1})^{0.5}$, $T_{Q1} = 4knf_s[LC_{oss2}(2/k\text{-}1)]^{0.5}$, and $T_{Q3} = 4knf_s[LC_{oss2}(1\text{–}2/k)]^{0.5}$.

According to the ZVS restrictions in Table 7.1, the operating area where the converter can achieve ZVS for all switches in case 1 as well as in case 2 can be obtained. The condition of the ZVS region I can be expressed as follows

Table 7.1 ZVS Restrictions of the PPDAB Converter with Junction Capacitor and Magnetizing Inductance

Case	1: $0 < D_\varphi < (1-D_Y)/2$	2: $(1-D_y)/2 < D_\varphi < 0.5$
$S_1\&S_2$	$D_y \le \frac{1}{k}$	$D_\varphi \ge \frac{k-1+8\tau_1\sqrt{k}}{2k}$
$Q_1\&Q_2$	$\begin{cases} D_y \ge \dfrac{2D_\varphi + T_{Q1}}{ak-1}, (1 < k \le 2) \\ D_y \ge \dfrac{2D_\varphi}{ak-1}, \quad (k > 2) \end{cases}$	$D_y \ge \frac{2(1-D_\varphi)}{ak+1}$
$Q_3\&Q_4$	$\begin{cases} D_y \ge \dfrac{-2D_\varphi}{ak-1}, \quad (1 < k \le 2) \\ D_y \ge \dfrac{-2D_\varphi + T_{Q3}}{ak-1}, (k > 2) \end{cases}$	

$$\begin{cases} 0 < D_\varphi < \dfrac{1-D_y}{2} \\ D_y \le \dfrac{1}{k} \\ D_y \ge \dfrac{2D_\varphi + T_{Q1}}{ak-1}, \qquad (1 < k \le 2) \\ D_y \ge \dfrac{2D_\varphi}{ak-1}, D_y \ge \dfrac{-2D_\varphi + T_{Q3}}{ak-1}, \quad (k > 2) \end{cases} \tag{7.11}$$

The condition of the ZVS region II can be expressed as follows

$$\begin{cases} \dfrac{1-D_y}{2} < D_\varphi < \dfrac{1}{2} \\ D_\varphi \ge \dfrac{k-1+8\tau_1\sqrt{k}}{2k} \\ D_y \ge \dfrac{2(1-D_\varphi)}{ak+1} \end{cases} \tag{7.12}$$

7.2.2 Disconnection in the ZVS Region Caused by the Switch Junction Capacitors Without Magnetizing Inductance

Without the magnetizing inductance design, the original magnetizing inductance of the transformer is much larger than the theoretical value of the magnetizing inductance to achieve the whole load range ZVS. So, to simplify the analysis, the effect of magnetizing inductance L_m on the ZVS range can be disregarded, and $a = 1 + (n^2L)/(2L_m)$ as mentioned before is approximately equal to one. Substituting 1 for a into Table 7.1 and base on (7.4) and (7.5), the ZVS boundary lines considering the resonance process without magnetizing current can be drawn in Fig. 7.4. Figure 7.4a shows that the ZVS region cannot cover the power range over a wide voltage range. As shown in Fig. 7.4b, the soft switching ranges of case 1 and case 2 do not have any one intersection point when $k = 1.5$, so the soft switching loses at the dividing line between case 1 and case 2. Figure 7.5 shows the power transfer curves in the ZVS region. There is only one intersection point between the two ZVS regions ignoring junction capacitors ($C_{oss1} = C_{oss2} = 0$), as shown in Fig. 7.5a. However, when considering the resonance process, the two ZVS regions no longer have any intersecting parts, as demonstrated in Fig. 7.5b. Thus, without magnetizing current injection under EPS control, the converter cannot achieve whole load range ZVS.

The ZVS regions are plotted in Fig. 7.6 when the magnetizing inductance L_m is equal to 3.35 mH, which is about more than ten times the needed value to achieve the

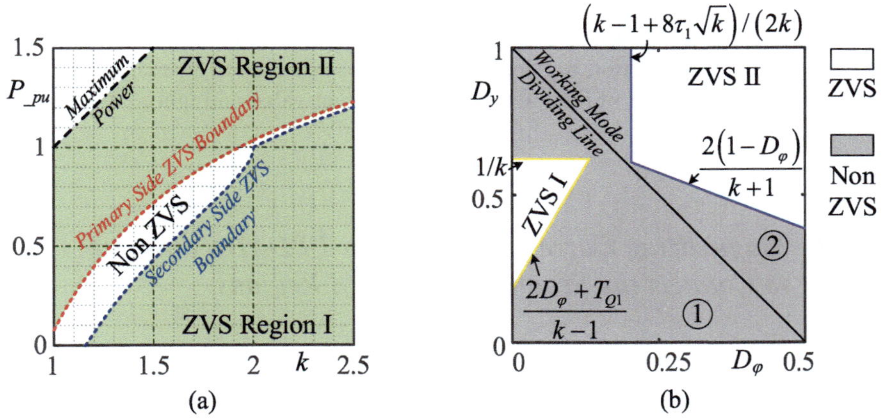

Fig. 7.4 ZVS region considering switch junction capacitors ($L_m = \infty$). **a** ZVS region versus $P_{\_pu}$ and k. **b** ZVS region versus D_y and D_φ with $k = 1.5$

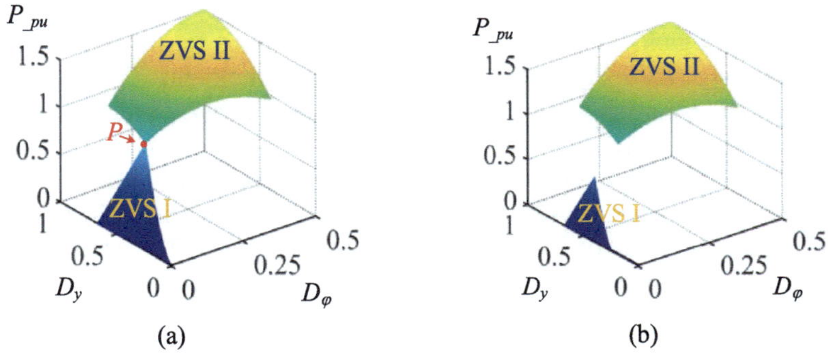

Fig. 7.5 Power transfer curves with $k = 1.5$ and $L_m = \infty$. **a** Ignoring switch junction capacitors. **b** Considering switch junction capacitors

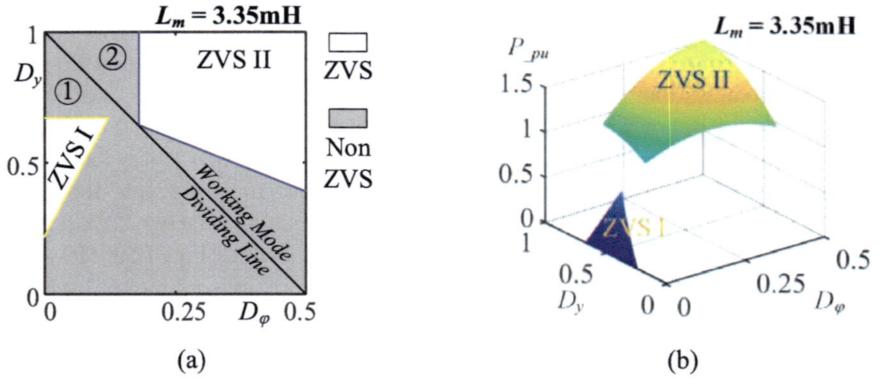

Fig. 7.6 **a** ZVS region when L_m is equal to 3.35 mH. **b** Power transfer curves when L_m is equal to 3.35 mH

whole load range ZVS. As seen, there is still the drawback that the two ZVS regions do not have an intersection region. Therefore, it is reasonable to treat the original magnetizing inductance of the transformer as a larger value and ignore its effect on the ZVS range.

7.2.3 Magnetizing Current Injection for PPDAB Converter to Connect the Separated Two ZVS Regions

For realizing ZVS in the whole load range, the magnetizing current is injected by designing the magnetizing inductance in this Chapter. Base on Table 7.1, the extended ZVS range with the magnetizing current injection is plotted in Fig. 7.7, and it is found that the two separated regions can be connected. The failure of ZVS at some loads can be solved by properly configuring the magnetizing inductance, and the full load range ZVS can be realized for different voltage conversion ratios as illustrated in Fig. 7.7a. As seen in Fig. 7.7a, the Non-ZVS region in Fig. 7.4a is replaced by the extended ZVS region, which provides the theoretical basis for the implementation of full load range ZVS. And taking $k = 1.5$ as an example, the ZVS regions with different D_y and D_φ are shown in Fig. 7.7b. The two expanded ZVS regions are linked by points C and E.

Compared with Figs. 7.4 and 7.7, the PPDAB converter with the magnetizing current injection can compensate for the previous inability to achieve soft switching around the dividing line of ZVS region I and ZVS region II. This also means that it is possible to design a seamless power control law to optimize the efficiency performance with full load range ZVS.

Figure 7.8 illustrates the ZVS regions with injected magnetizing current for $1 < k \leq 2$ and $k > 2$, respectively. As shown, the two ZVS regions have intersected parts, and points C, E and N respectively are the intersection points of the two ZVS

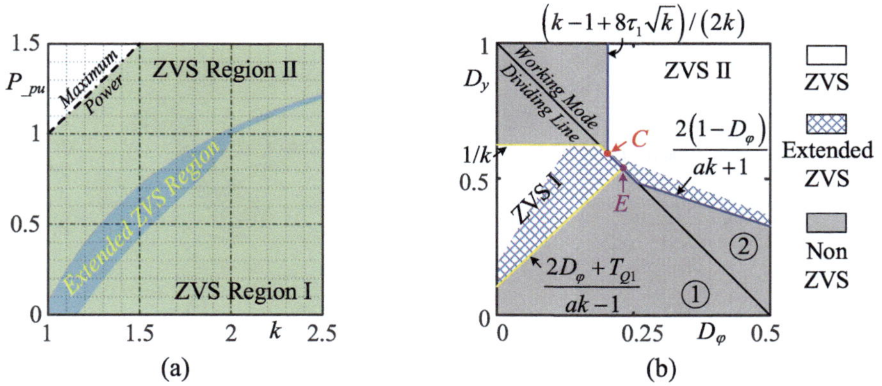

Fig. 7.7 ZVS region when L_m is selected to be a suitable value. **a** ZVS region versus $P_{\_pu}$ and k. **b** ZVS region versus D_y and D_φ with $k = 1.5$

regions. If the intersection point N is chosen as the endpoint of the linear control law, although the ZVS constraint of $k > 2$ can be satisfied, there must be a part of the control law go across the Non-ZVS region when $1 < k \leq 2$. Thus, ZVS for all switches is more difficult to achieve in the $1 < k \leq 2$ case. In order to achieve soft switching over the full load range, the PPDAB converter is analyzed and designed for $1 < k \leq 2$. The coordinates of intersection points C and E are calculated as

$$C: \begin{cases} D_{\varphi\_C} = \left(k - 1 + 8\tau_1\sqrt{k}\right)/(2k) \\ D_{y\_C} = \left(1 - 8\tau_1\sqrt{k}\right)/k \end{cases} \tag{7.13}$$

$$E: \begin{cases} D_{\varphi\_E} = (ak - 1 - T_{Q1})/(2ak) \\ D_{y\_E} = (1 + T_{Q1})/(ak) \end{cases} \tag{7.14}$$

7.3 Power Control Laws to Obtain Low RMS Current and Easy Engineering Implementation Based on Magnetizing Current Injection

7.3.1 Obtain a Power Control Law with Minimum Transformer Winding Current

From 7.2, a reasonable set of D_y and D_φ allows the PPDAB converter to operate in the region where all switches are able to achieve soft switching within whole load range. Provided that ZVS conditions can be satisfied, the RMS current is considered

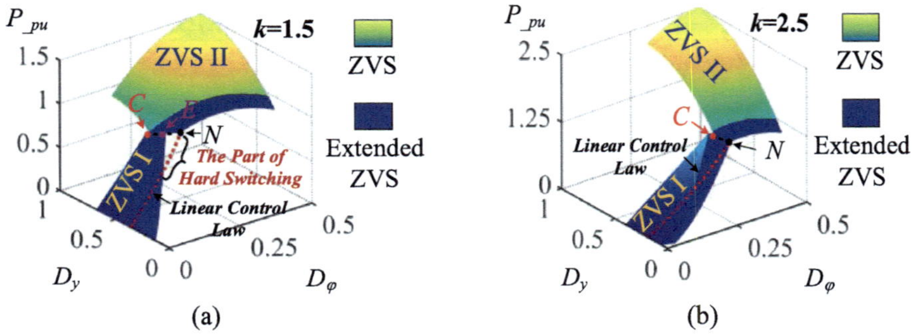

Fig. 7.8 The power transfer curves and the intersection points when L_m is selected to be a suitable value. **a** $1 < k \leq 2$. **b** $k > 2$

the optimization target to design an operation strategy. By reducing the RMS current I_{rms} flowing through the transformer winding, the losses in the converter can be decreased. From (7.1)–(7.3), the expression of the normalized value of I_{rms} in case 1 can be obtained as

$$\begin{aligned} I_{rms\_pu} &= \frac{I_{rms}}{I_{base}} = \frac{4knLf_s}{V_1}\sqrt{\frac{2}{T_s}\int_{t_0}^{t_6}\left(\frac{2}{n}i_k(t) - i_m(t)\right)^2 dt} \\ &= \frac{2\sqrt{3}}{3}k\sqrt{akD_y\left[akD_y(3-2D_y) + 12D_\varphi^2 + D_y^2 - 3\right] + 1} \end{aligned} \tag{7.15}$$

where $I_{base} = V_1/(4knLf_s)$.

In order to calculate D_y that minimizes I_{rms_pu} at the same power, it is necessary to obtain the relationship among I_{rms_pu}, $P_{\_pu}$, and D_y. For D_j in (7.15) is replaced by $P_{case1\_pu}/(4kD_y)$, and the relationship can be calculated as

$$I_{rms\_pu} = \frac{2\sqrt{3}}{3}k\sqrt{akD_y\left[akD_y(3-2D_y) + \frac{3(P_{case1\_pu})^2}{4k^2D_y^2} + D_y^2 - 3\right] + 1} \tag{7.16}$$

Taking the partial derivative of (7.16) with respect to D_y, an equation can be obtained. Let the numerator of this equation be equal to zero, and then substitute $P_{case1\_pu} = 4kD_yD_\varphi$ into this numerator, thus obtaining the equation as

$$(1 - 2ak)D_y^2 + 2akD_y - 4D_\varphi^2 - 1 = 0 \tag{7.17}$$

Solve (7.17) to get the stagnation points of D_y.

$$D_y = \frac{ak - \sqrt{-8akD_\varphi^2 + 4D_\varphi^2 + (ak-1)^2}}{2ak - 1} \tag{7.18}$$

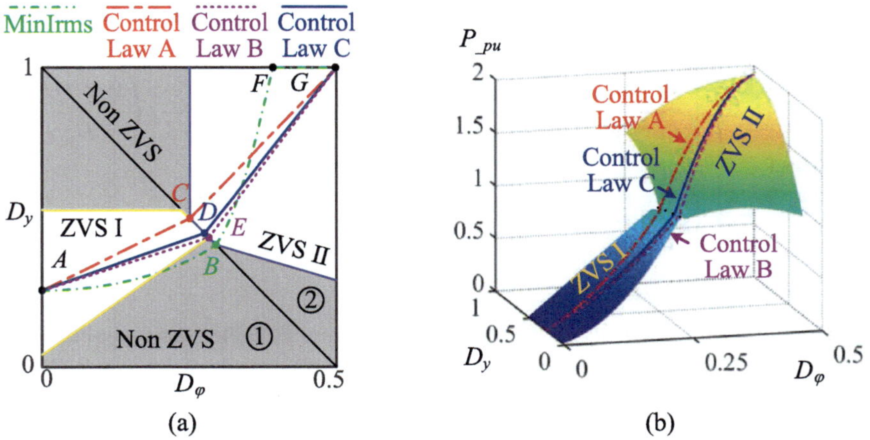

Fig. 7.9 **a** The four different power control laws. **b** Power transfer curves of the three linear control laws

D_y determined by (7.18) is the minimum point of (7.15), which is the optimal value to satisfy the conditions. Similarly, the minimum RMS current value in case 2 can be solved for as

$$D_y = \begin{cases} 1 - ak(1 - 2D_\varphi) + \sqrt{\begin{array}{c}(1 - 2D_\varphi)^2 + \\ [1 - ak(1 - 2D_\varphi)]^2\end{array}}, & \left(\begin{array}{c}\frac{ak-1}{2ak} \le D_\varphi < \\ \frac{1+\sqrt{k^2a^2-1}-ak}{2}\end{array}\right) \\ 1, & \left(\frac{1+\sqrt{k^2a^2-1}-ak}{2} \le D_\varphi \le \frac{1}{2}\right) \end{cases} \tag{7.19}$$

From (7.18) and (7.19), it is shown that D_y corresponding to the minimum RMS current is a nonlinear function of D_φ, leading to complex control algorithm. And the curve (*A*-*B*-*F*-*G*) in Fig. 7.9a is the control law MinIrms obtained by (7.18) and (7.19).

7.3.2 Optimal Power Control Law with Full Load Range ZVS and Easy Engineering Implementation

As shown, a partial curve of the control law MinIrms is located in the Non-ZVS region, which means that ZVS of some switches is not implemented in this part. To simplify the control and ensure that the power transfer capability can be fully utilized at the same time, this Chapter proposes to control D_y linearly according to D_φ, that is, to linearize the control laws in (7.18) and (7.19). The intersection points *C* and *E* are respectively selected as the middle points of the linear control laws, and their starting and ending points are chosen as points *A* and *G*. The dashed line (*A*-*C*-*G*) in

Fig. 7.9 is the power control law A, and the expression of which can be expressed by

$$D_{yLawA} = \begin{cases} \left[\dfrac{4ak^2 - 4k}{\left(k + 8\tau_1\sqrt{k} - 1\right)(2ak - 1)} - 2\right]D_\varphi + \dfrac{1}{2ak - 1}, & \begin{pmatrix} 0 \le D_\varphi \\ < D_{\varphi\_C} \end{pmatrix} \\ \left(\dfrac{k}{8\tau_1\sqrt{k} - 1} + 1\right)(1 - 2D_\varphi) + 1, & \begin{pmatrix} D_{\varphi\_C} \le D_\varphi \\ \le 0.5 \end{pmatrix} \end{cases} \tag{7.20}$$

The dotted line (*A-E–G*) is the power control law B, which only applies in the range of $1 < k \le 2$, because the intersection point E exists only at $1 < k \le 2$. And its corresponding expression can be expressed as

$$D_{yLawB} = \begin{cases} \dfrac{2\left[4kn\tau_2\sqrt{2/k - 1}(1 - 2ak) - ak + 1\right]}{\left(4kn\tau_2\sqrt{2/k - 1} - ak + 1\right)(2ak - 1)}D_\varphi + \dfrac{1}{2ak - 1}, & \begin{pmatrix} 0 \le D_\varphi \\ < D_{\varphi\_E} \end{pmatrix} \\ \left(1 - \dfrac{ak}{1 + 4kn\tau_2\sqrt{2/k - 1}}\right)(1 - 2D_\varphi) + 1, & \begin{pmatrix} D_{\varphi\_E} \le D_\varphi \\ \le 0.5 \end{pmatrix} \end{cases} \tag{7.21}$$

Since the intersection points C and E contain non-integer powers of the voltage conversion ratio k, there are square root operations of k in both control laws A and B, which will seriously slow down the operation speed of the controller in engineering implementation. For decreasing the complexity of the implementation, the functions are better to contain only integer powers of k. The square root terms containing k in the intersection point E are replaced by Taylor Formula, resulting in the point D, whose point coordinate expression is

$$D : \begin{cases} D_{\varphi\_D} = (ak - 1 - 4kn\tau_2 K)/(2ak) \\ D_{y\_D} = (1 + 4kn\tau_2 K)/(ak) \end{cases} \tag{7.22}$$

where $t_2 = f_s(LC_{oss2})^{0.5}$ and $K = 1/(2k^3)\text{-}2/k^2 + 7/(2\,k)\text{-}1$, and K is always bigger than $(2/k\text{-}1)^{0.5}$ when $k > 1$. So, the position of point D is always on the upper left of the position of point E.

Using point D as the middle point of the power control law C, the solid line (*A-D-G*) can be obtained as shown in Fig. 7.9. Its corresponding expression can be expressed as

$$D_{yLawC} = \begin{cases} \frac{2[4kn\tau_2K(1-2ak)-ak+1]}{(4kn\tau_2K-ak+1)(2ak-1)}D_\varphi + \frac{1}{2ak-1}, & \begin{pmatrix} 0 \le D_\varphi \\ < D_{\varphi\_D} \end{pmatrix} \\ \left(1-\frac{ak}{1+4kn\tau_2K}\right)(1-2D_\varphi)+1, & \begin{pmatrix} D_{\varphi\_D} \le D_\varphi \\ \le 0.5 \end{pmatrix} \end{cases} \quad (7.23)$$

The control law C in (7.23) is a linear monotonic function and it can be rewritten as a function of D_φ versus D_y, so it is the same whether D_y or D_φ is determined first in the control process. However, in the literature about the control strategy of the DAB converter, the phase shift ratio D_φ is generally generated from the output of the voltage control loop first, and then other control variables are derived based on this. This process is also used in this Chapter.

In order to have the curve of control law C all in the ZVS region, the position of point D needs to be located below point C. The ZVS condition in the whole load range for all switches can be obtained as

$$\left(\frac{4kn\tau_2K+1}{ak} \le \frac{1-8\tau_1\sqrt{k}}{k}\right) \cap \left(\frac{k+8\tau_1\sqrt{k}-1}{2k} \le \frac{ak-4kn\tau_2K-1}{2ak}\right) \quad (7.24)$$

Simplifying the soft switching condition in (7.24), the following equation is obtained.

$$\left(\frac{4kn\tau_2K+1}{1-8\tau_1\sqrt{k}} \le a\right) \cap \left(8\tau_1\sqrt{k}-1 \le \frac{-4kn\tau_2K-1}{a}\right) \quad (7.25)$$

Using L_m and L to eliminate a and transforming the equation, the following equation can be derived.

$$\left(\frac{4kn\tau_2K+1}{1-8\tau_1\sqrt{k}} \le 1+\frac{n^2L}{2L_m}\right) \cap \left(\frac{-4kn\tau_2K-1}{8\tau_1\sqrt{k}-1} \le 1+\frac{n^2L}{2L_m}\right) \quad (7.26)$$

Hence the range of the magnetizing inductance is derived as

$$L_m \le \frac{n^2\left(1-8\tau_1\sqrt{k}\right)}{2\left(8\tau_1\sqrt{k}+4kn\tau_2K\right)}L \quad (7.27)$$

where $K = 1/(2k^3)-2/k^2 + 7/(2\,k)-1$.

The power transfer curves of the three different control laws A, B, and C are drawn in Fig. 7.9b according to (7.4), (7.5), (7.20), (7.21) and (7.23). As can be seen, these power transfer curves not only achieve ZVS in the range of whole load, but also allow each power value to be obtained from a corresponding set of D_y and D_φ. Among these three linear control laws, control law C is easily accomplished by a digital signal processor in practical engineering implementation.

7.3.3 Comparison of Control Laws

To give a further comparison of the linear control laws, the RMS current values of these three are control laws analyzed in detail for the given voltage conversion ratio and output power. The RMS current of the control law MinIrms is defined as I_{rmsMin_pu}, and the RMS current of the linear control law is defined as I_{rmsX_pu}. Then the error between them can be expressed as

$$\Delta E_X = \frac{I_{rmsX\_pu} - I_{rmsMin\_pu}}{I_{rmsMin\_pu}} \times 100\% \tag{7.28}$$

where X is control laws A, B or C. So, ΔE_A, ΔE_B and ΔE_C represent the errors of control laws A, B and C, respectively. As seen, ΔE_X can reflect the error between the RMS current of the linear control law I_{rmsX_pu} and the minimum RMS current I_{rmsMin_pu}. The larger ΔE_X is, the higher it represents I_{rmsX_pu} than I_{rmsMin_pu}.

At different voltage conversion ratios and output powers, Fig. 7.10 shows deviations between I_{rmsX_pu} and I_{rmsMin_pu}. In addition, these differences become larger and then smaller as $P_{\_pu}$ increases. Therefore, the maximum error $\Delta E_{\max}$ is solvable and varies as the voltage transfer ratio k changes. As indicated in Fig. 7.10a, $\Delta E_{A\max}$ reaches 26.4%, and I_{rmsA_pu} is larger than the minimum RMS current at light load under the same operating conditions. According to Fig. 7.10b and c, only when $1 < k < 1.06$, $\Delta E_{B\max}$ and $\Delta E_{C\max}$ will reach about 7.5%. Both the two linear control laws have a minimal RMS difference over most of the operating range, which means that their RMS currents are slightly larger than the minimum current RMS at any k.

Compared to the control law MinIrms, the control law C slightly increases the RMS current but achieves the full load range ZVS as well as being easier to implement in a digital signal controller. Compared with control law A, the RMS current under control law C is smaller, especially at light loads. Meanwhile, compared with control law B, the calculation of D_y in control law C is simpler to implement in engineering. Hence, the control law C, which is named proposed control law below, is chosen as the control strategy of the PPDAB converter with magnetizing current injection, which can improve the system efficiency with the precondition of achieving soft switching for all switches at whole load range.

7.3.4 Optimized Control Strategy Implementation

Figure 7.11 illustrates the control block of the proposed control law. Based on the error between V_2 and V_{2_ref}, the PI regulator generates D_φ required for the pulse width modulation (PWM) signal to ensure a constant output voltage. The external phase shift ratio D_φ is only used in the range of −0.5 to 0.5 in practice. According to the proposed control law, the internal phase shift ratio D_y is calculated with D_φ and k by (7.21), where k can be gained from V_1 and V_2. In addition, D_y is restricted to

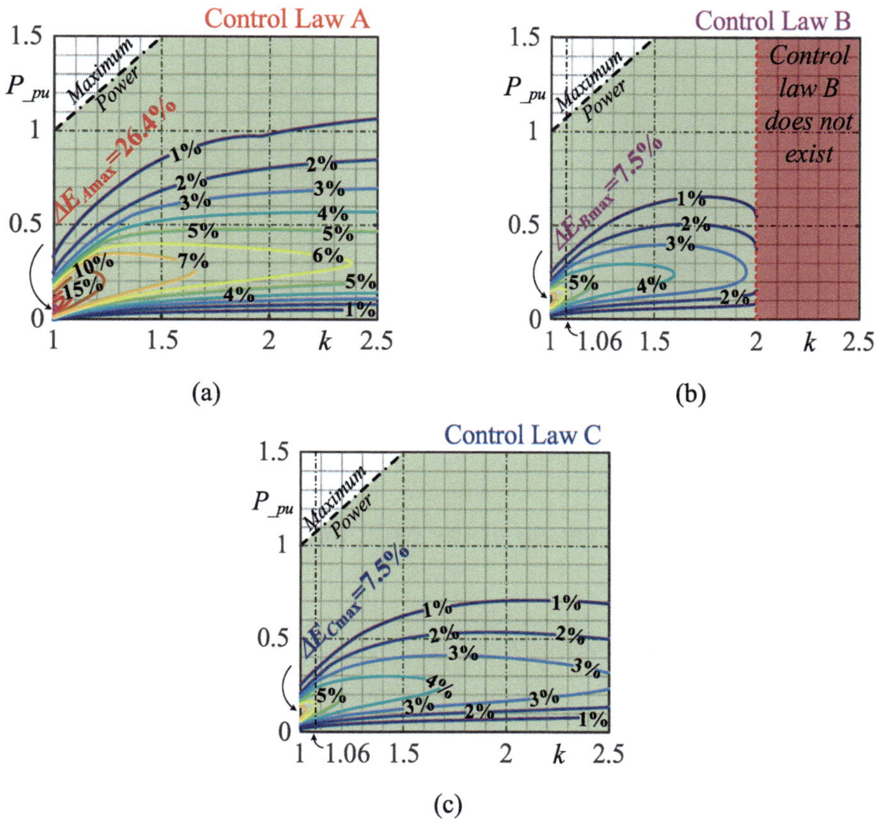

Fig. 7.10 The ΔE_X value for the three linear control laws. **a** Contour line plots of ΔE_A. **b** Contour line plots of ΔE_B. **c** Contour line plots of ΔE_C

the range of 0 to 1. Finally, the PWM generator outputs the PWM signals required for each switch based on the input D_φ and D_y.

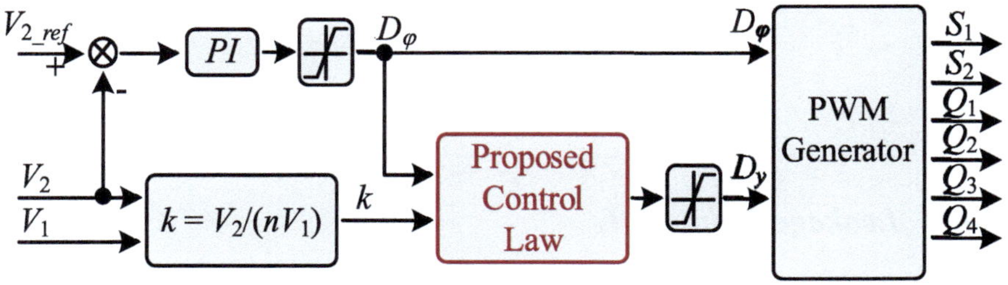

Fig. 7.11 Implementation of the proposed control law

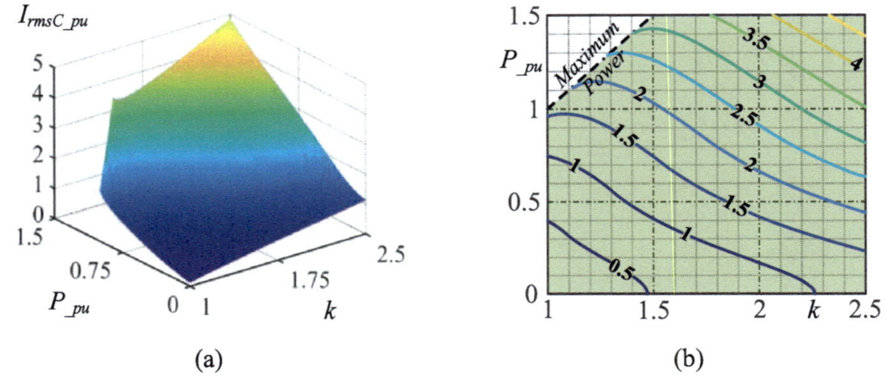

Fig. 7.12 The I_{rmsC_pu} curve for different output powers and voltage conversion ratios. **a** Three-dimensional graph. **b** Contour line plots

7.4 Design Consideration of the Modified PPDAB Converter

7.4.1 Voltage Conversion Ratio k

The relationships among k, $P_{\_pu}$ and I_{rmsC_pu} under the proposed control law are illustrated in Fig. 7.12. It is observed that the RMS value of transformer current increases as the voltage conversion ratio k increases at the same output power $P_{\_pu}$. Therefore, it is preferred to design k as small as possible in full load ZVS region.

7.4.2 Turns Ratio N

The turns ratio is determined by the DC bus voltage and the voltage conversion ratio. It can be expressed as

$$n = V_2/(k_{\min} V_{1\max}) \tag{7.29}$$

7.4.3 Leakage Inductor L

The leakage inductance determines the maximum value of the power transfer capability of the PPDAB converter. From (7.5), the maximum power is kP_{base}, and $P_{base} = V_1^2/(4Lf_s)$. The minimum value of kP_{base} must be larger than the rated power P_r. Therefore, the range of L can be deduced as

$$L \leq \left(kV_1^2\right)_{\min}/(4P_r f_s) \tag{7.30}$$

7.4.4 *Magnetizing Inductance L_m*

The change of the magnetizing inductance will change the soft switching condition in (7.24) and then change the ZVS region. Besides, it will also affect the RMS current in the transformer winding. Therefore, the magnetizing inductance should be optimally designed to make a tradeoff between the ZVS condition and overall efficiency. According to the derivation process of the constraints on the magnetizing inductance in 7.3.2, L_m should be satisfied as

$$L_m \leq \frac{n^2\left(1 - 8\tau_1\sqrt{k}\right)}{2\left(8\tau_1\sqrt{k} + 4kn\tau_2 K\right)} L \tag{7.31}$$

Indeed, taking into account other parasitic parameters, the actual magnetizing inductance of the transformer should be slightly smaller than the calculated value of (7.31). With the core selected and the number of winding turns determined, the air gap of the transformer is polished to ensure that L_m meets the theoretical constraint. In addition, according to (7.30) and (7.31), there is a relationship between L_m and P_r: the higher the rated power, the smaller the required magnetizing inductance.

7.5 Experimental Results

Figure 7.13 illustrates the laboratory prototype of the PPDAB converter. Based on the design considerations, the specifications of the laboratory prototype are listed in Table 7.2. The required value of the magnetizing inductance is obtained by grinding the air gap in the core middle column of the transformer. Due to the small leakage inductance of the transformer under the sandwich winding structure, the required leakage value is considered to be provided by the external leakage inductor only, as shown in the Fig. 7.13. In addition, the primary side switches and the secondary side switches are selected as IPP051N15N5 and C3M0120090D, respectively. And, it is worth noting that two MOSFETs are connected in parallel as a primary side switch to reduce conduction losses and benefit system thermal design. By consulting the drain-source voltage versus capacitance C_{oss} curves in the switching datasheets, the charge-equivalent junction capacitances of the primary and secondary switches under rated operation can be calculated. Noteworthy, the sandwich winding structure is adopted to reduce the equivalent leakage inductance of the transformer, and the leakage paraments of the transformer windings can be ignored compared to the

Fig. 7.13 The laboratory prototype of the PPDAB converter

required leakage inductance. Figure 7.14 illustrates the sandwich structure of the transformer, which can be divided into 3 layers: the first- and third-layer windings are both the secondary winding N_3, and the second-layer winding consists of the primary side windings N_1 and N_2 together.

Table 7.2 Specifications of the Laboratory Converter

Parameters	Values
Input voltage (V_1)	18–36 V
Output voltage (V_2)	270 V
Switching frequency (f_s)	100 kHz
Output rated power (P_r)	500 W
Voltage conversion ratio (k)	1.072–2.144
Turns ratio (n)	7
Leakage inductors (L)	$L_1 = L_2 = 2.31$ mH
Magnetizing inductance (L_m)	196mH
Primary side switches (S_1 and S_2)	IPP051N15N5 (2 parallel)
Secondary side switches (Q_1 - Q_4)	C3M0120090D
Junction capacitances (C_{oss1} and C_{oss2})	$C_{oss1} = 2.8$ nF, $C_{oss2} = 65$ pF

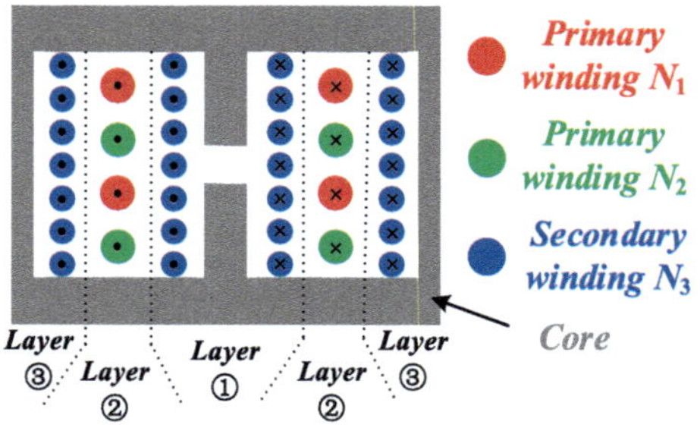

Fig. 7.14 The sandwich winding structure of the transformer

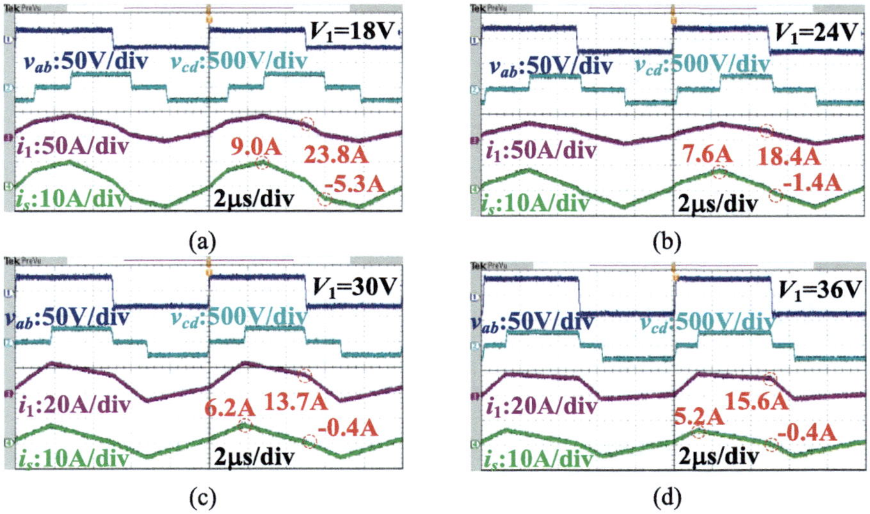

Fig. 7.15 Experimental waveforms when $P = 500$ W. **a** $V_1 = 18$ V. **b** $V_1 = 24$ V. **c** $V_1 = 30$ V. **d** $V_1 = 36$ V

Figures 7.15, 7.16 7.17 illustrate the experimental steady state waveforms of the PPDAB converter working under the proposed control law for different operating conditions. The average value of i_1 which contains the DC component is not equal to zero in a switching period, as shown in these figures. Also, the experimental waveform of i_1 agrees with the theoretical analysis. At the same load, the inter-bridge phase shift ratio D_y increases as the amplitude of v_{ab} increases (meaning that the voltage conversion ratio k becomes lower), while the peak of i_s becomes smaller. Moreover, the differences between the current i_1 waveform and the current i_s waveform prove that the injection of magnetizing current affects the secondary current waveform.

In addition, the current limits derived from the previous theoretical analysis are all listed in Table 7.3, where $I_{DC} = P/(2V_1)$. By measuring the instantaneous current values when switches are turned off and comparing them with the minimum current conditions in Table III, it is observed that the current waveforms in Figs. 7.15, 7.16 7.17 all meet the required limits. Thus, all switches can be turned on with ZVS at light, half and full loads after injecting the magnetizing current, which is in agreement with the previous analysis that all switches of the modified PPDAB converter can operate under soft switching over the total load range under the proposed control law.

As seen, for the modified PPDAB converter, ZVS of all switches are achieved with magnetizing current injection. Complementary drive signals are used for two switches on the same bridge arm, so one of them can achieve ZVS if the other one realizes ZVS. Therefore, S_1, Q_1 and Q_3 are taken as examples to show soft switching waveforms. According to the ZVS experimental waveforms in Fig. 7.18, the drain-source voltage v_{dsS1} reaches zero before the gate drive signal v_{gsS1} changes to be positive. Therefore, ZVS of the switch S_1 can be realized within whole load

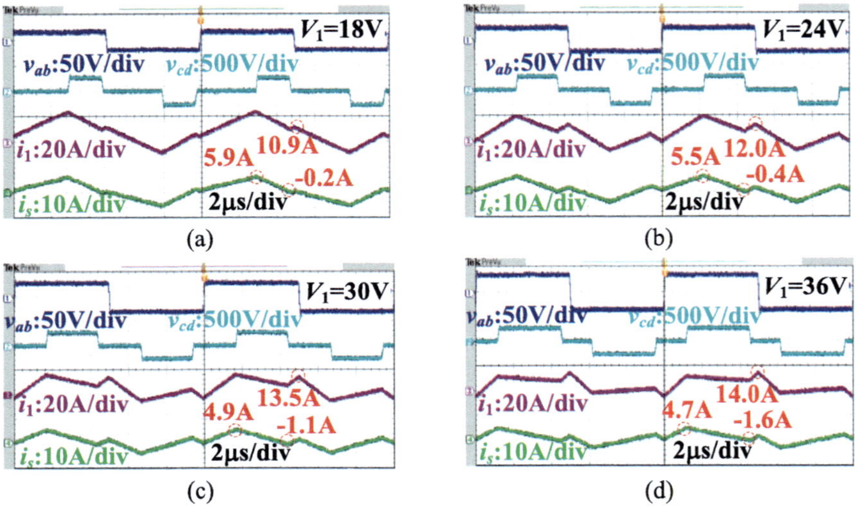

Fig. 7.16 Experimental waveforms when $P = 250$ W. **a** $V_1 = 18$ V. **b** $V_1 = 24$ V. **c** $V_1 = 30$ V. **d** $V_1 = 36$ V

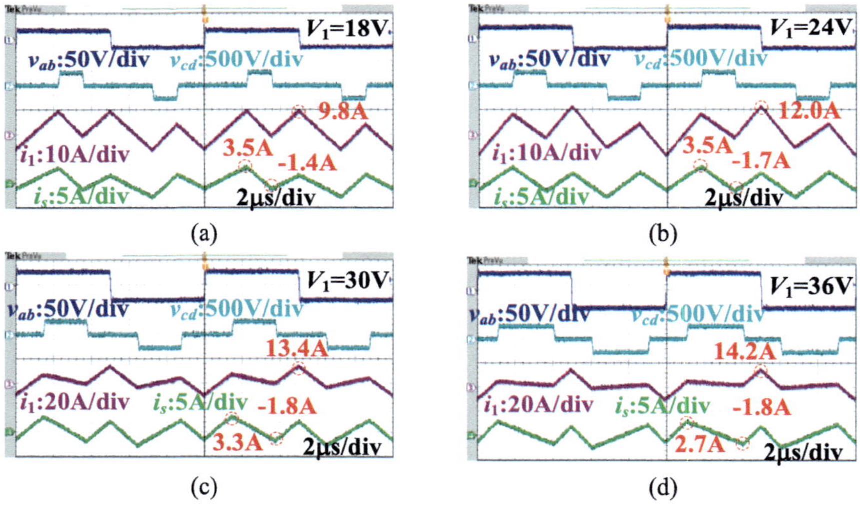

Fig. 7.17 Experimental waveforms when $P = 50$ W. **a** $V_1 = 18$ V. **b** $V_1 = 24$ V. **c** $V_1 = 30$ V. **d** $V_1 = 36$ V

range. In Fig. 7.19, the rising edge of v_{gsQ1} appears only after the junction capacitor C_{Q1} has finished discharging. These demonstrate the achievement of ZVS for Q_1 under different loads. Figure 7.20 illustrates the ZVS waveforms of the switch Q_3 at different loads. Before v_{gsQ3} changes from zero to positive, the voltage across the

Table 7.3 ZVS Current Limits

V_1	18 V		24 V		30 V		36 V	
Case	1	2	1	2	1	2	1	2
S_1&S_2(A)	$\geq I_{DC}$	$\geq I_{DC} + 1.8$	$\geq I_{DC}$	$\geq I_{DC} + 2.1$	$\geq I_{DC}$	$\geq I_{DC} + 2.4$	$\geq I_{DC}$	$\geq I_{DC} + 2.6$
Q_1&Q_2(A)	≤ 0	≤ 0	$\leq$ -0.2	≤ 0	$\leq$ -0.3	≤ 0	$\leq$ -0.4	≤ 0
Q_3&Q_4(A)	≥ 0.1	≥ 0	≥ 0	≥ 0	≥ 0	≥ 0	≥ 0	≥ 0

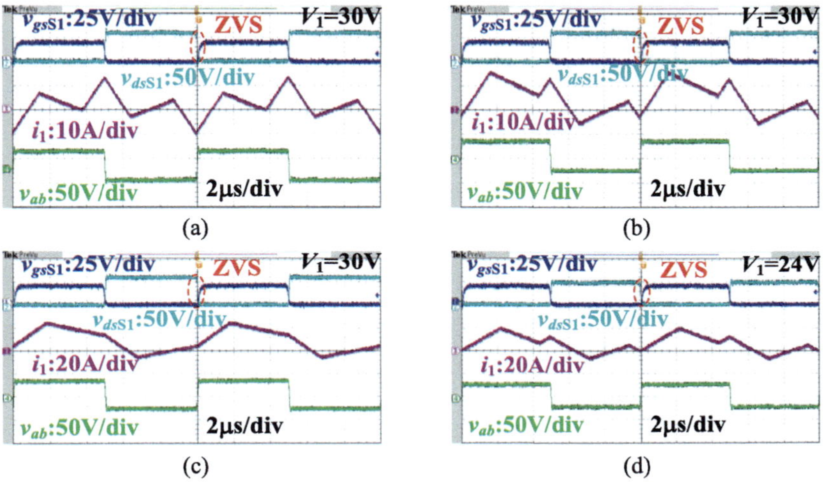

Fig. 7.18 ZVS operation of S_1. **a** $P = 50$ W. **b** $P = 250$ W. **c** $P = 500$ W. **d** $V_1 = 24$ V and $P = 250$ W

junction capacitor of Q_3 has been reduced to zero. Thus, it is confirmed that full load range ZVS of Q_3 can be achieved with the proposed control law.

Figure 7.21 respectively illustrates the experimental steady state waveforms of the reverse power transfer mode with different operating conditions including input voltages and loads. Unlike the forward power transfer mode, the phase shift rate D_φ between the voltages v_{AB} and v_{CD} in the reverse operation mode is negative. And the phase shift rate decreases as the load increases to transmit more power. As shown in Fig. 7.21, the DC component of the current i_1 is replaced with a negative value from the positive value in the former forward operation mode.

The soft switching waveforms of the modified PPDAB converter in the reverse operation mode are illustrated in Fig. 7.22. v_{gsS1}, v_{gsQ1} and v_{gsQ3} are the gate drive signals of switches S_1, Q_1 and Q_3, respectively. And, v_{dsS1}, v_{dsQ1} and v_{dsQ3} are drain-source voltages of switches S_1, Q_1 and Q_3, respectively. As seen, the drain-source voltage reaches zero before the gate drive signal becomes positive. Therefore, all switches can achieve ZVS in the reverse operation by the proposed method.

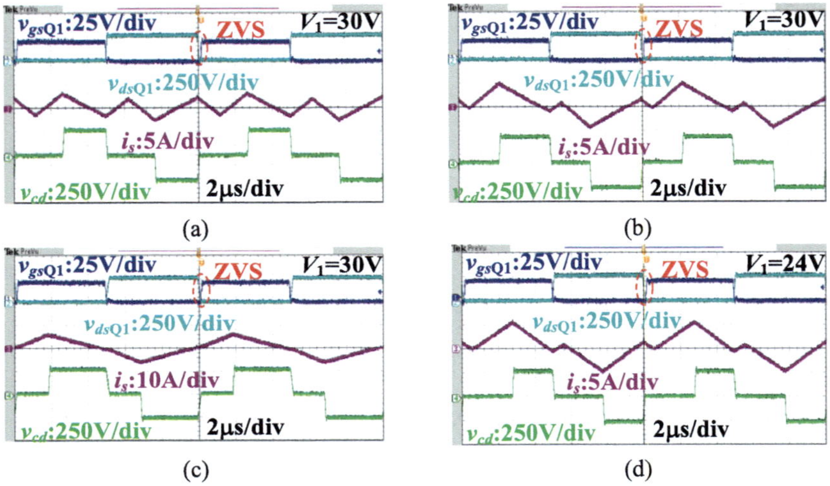

Fig. 7.19 ZVS operation of Q_1. **a** $P = 50$ W. **b** $P = 250$ W. **c** $P = 500$ W. **d** $V_1 = 24$ V and $P = 250$ W

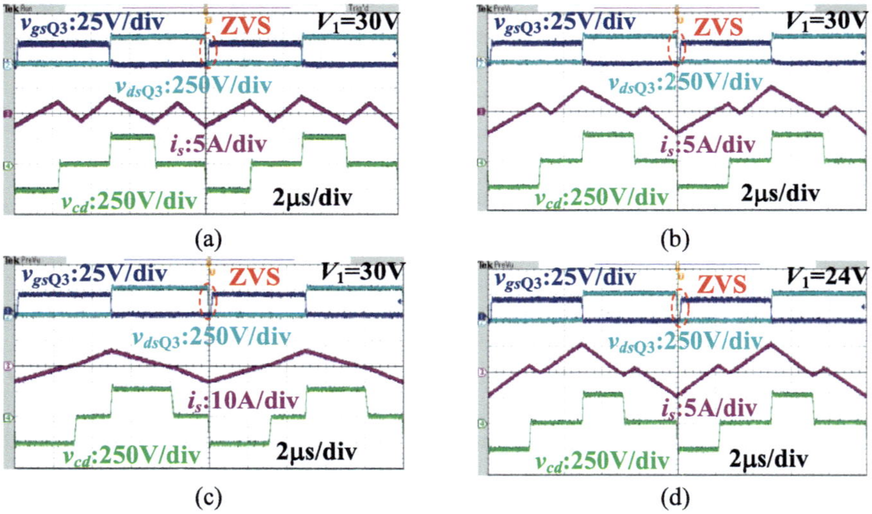

Fig. 7.20 ZVS operation of Q_3. **a** $P = 50$ W. **b** $P = 250$ W. **c** $P = 500$ W. **d** $V_1 = 24$ V and $P = 250$ W

Figure 7.23 shows the transition waveforms of the converter with the proposed control law. When the output power suddenly changes, V_2 almost remains constant under the regulation of the PI module and i_s changes as the power changes. The transient response results show that i_s can change smoothly under the proposed control when the load has a sept change.

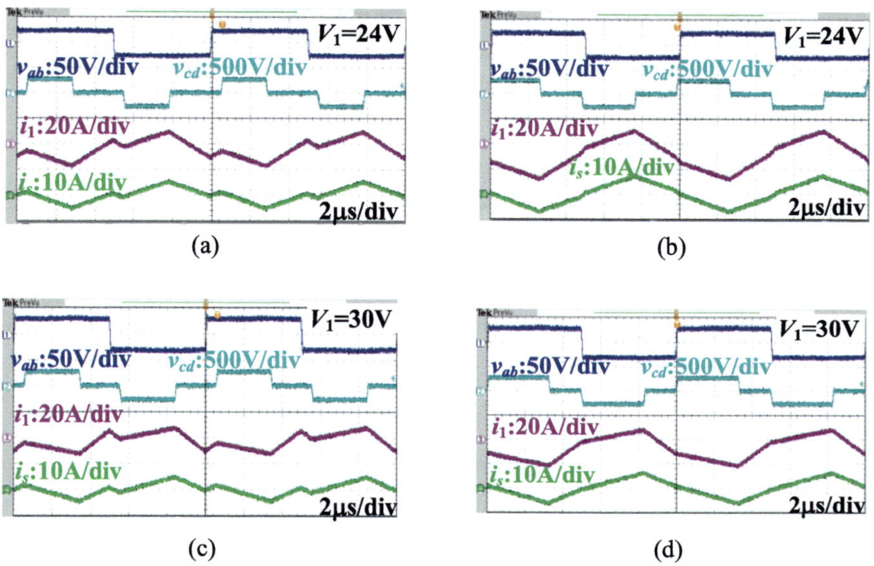

Fig. 7.21 Experimental waveforms in the reverse operation mode. **a** $V_{in} = 24$ V and $P = 250$ W. **b** $V_{in} = 24$ V and $P = 500$ W. **c** $V_{in} = 30$ V and $P = 250$ W. **d** $V_{in} = 30$ V and $P = 500$ W

For different input voltages, the efficiency comparisons between the proposed method and the traditional EPS control strategy (the magnetizing inductance of this method is not optimal designed) are shown in Fig. 7.24. When the output power is between 200 and 500 W, within the ranges that ZVS is lost under the traditional method, the injection of magnetizing current can solve this problem, so the system efficiency is improved. For the operating conditions where all switches can be realized ZVS, the introduction of the magnetizing inductance slightly increases the additional loss, leading to a reduction in efficiency, as seen in Fig. 7.24a. However, the efficiency performance is improved by reducing the switching losses with the optimized method except for these conditions. On the other hand, the proposed control law improves the maximum efficiency of the converter to 97.6%.

A comparison of the circuit components of the proposed push–pull DAB converter and the conventional DAB converter is listed in Table 7.4. Although additional passive components are introduced, including a filter capacitor, a leakage inductor, and a transformer winding, the push–pull DAB converter reduces the two active power switches on the low-voltage side compared to the full-bridge DAB converter.

Figure 7.25 illustrates the loss breakdown comparisons between the push–pull DAB converter and the conventional DAB converter for different loads and input voltages. For the fairness of the comparison, the low-voltage side switch of the conventional DAB converter was reselected, and its leakage inductor and transformer were redesigned to meet the transfer power requirement. As shown in Fig. 7.25a, the total loss of a push–pull DAB converter is lower than that of a conventional DAB converter at full load due to the advantage of the few switches on the low-voltage side,

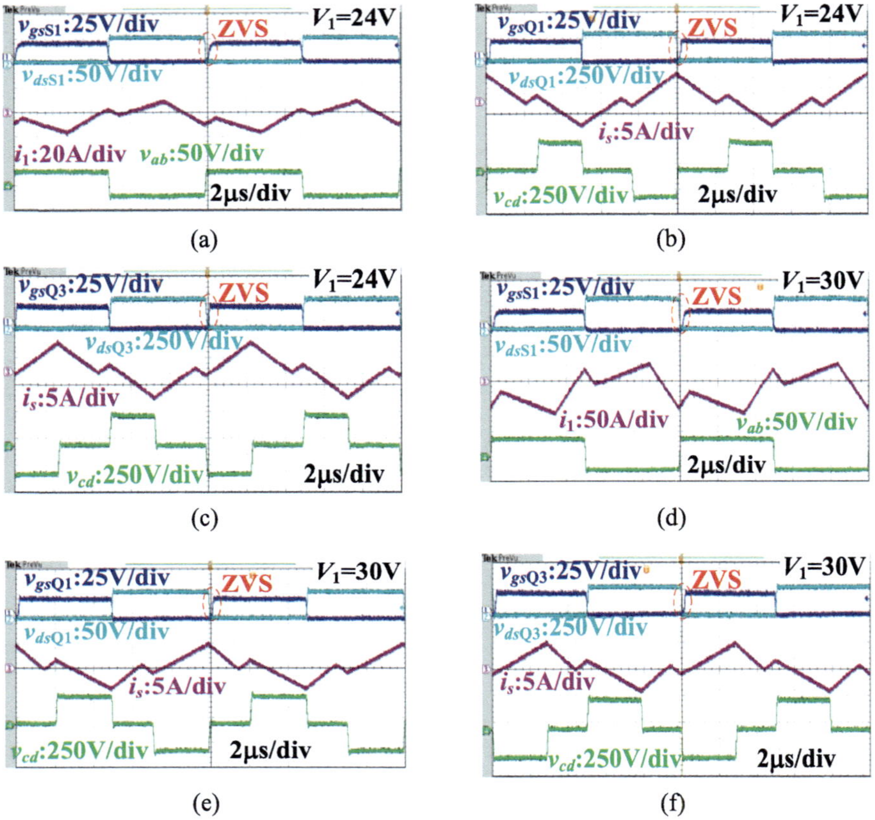

Fig. 7.22 Soft switching waveforms in the reverse operation mode. **a** and **b** and **c** $V_{in} = 24$ V, $P = 250$ W. **d** and **e** and **f** $V_{in} = 30$ V, $P = 250$ W

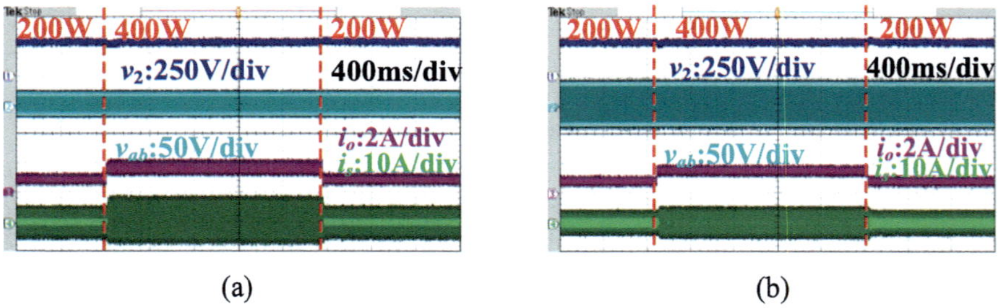

Fig. 7.23 Transition waveforms for the load change. **a** $V_1 = 18$ V. **b** $V_1 = 36$ V

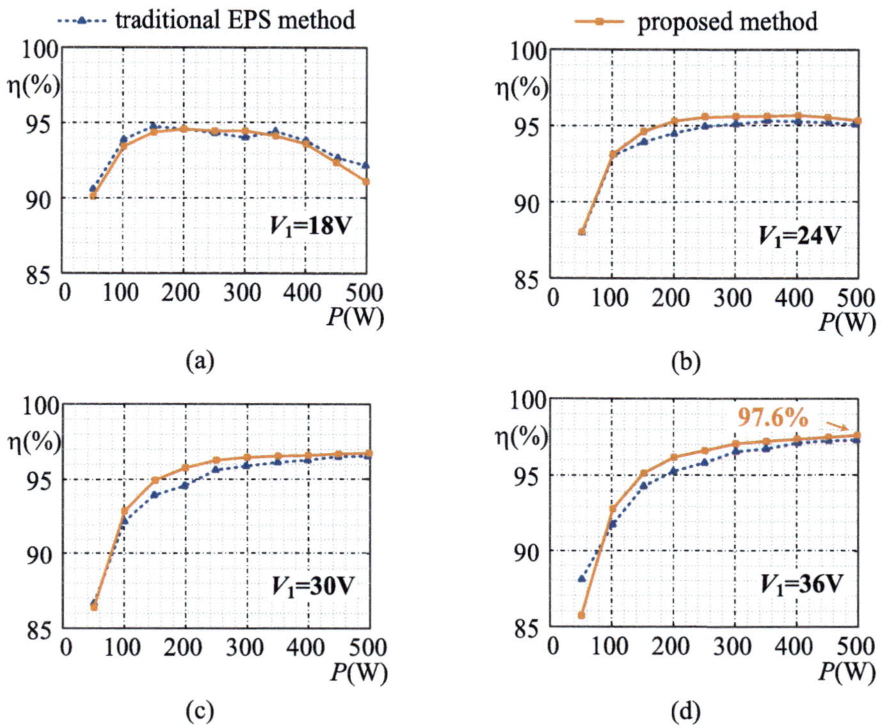

Fig. 7.24 Measured efficiency comparison between proposed method and traditional method. **a** $V_1 = 18$ V. **b** $V_1 = 24$ V. **c** $V_1 = 30$ V. **d** $V_1 = 36$ V

Table 7.4 Comparison of the Circuit Components between the Proposed Push–pull DAB Converter and the Conversion DAB Converter

Number of circuit components	Proposed push–pull DAB converter	Conventional DAB converter
Active power switch	6	8
Transformer	A three-winding transformer	A double-winding transformer
Leakage inductor	2	1
Filter capacitor	3	2

which reduces the switch conduction loss and switch turn-off loss. So, the push–pull DAB converter has a slight efficiency advantage over the conventional DAB converter in low-voltage and high-current fields. Although the total loss is slightly higher than that of conventional DAB converters at light loads, the push–pull DAB converter can reduce the number of driver circuits.

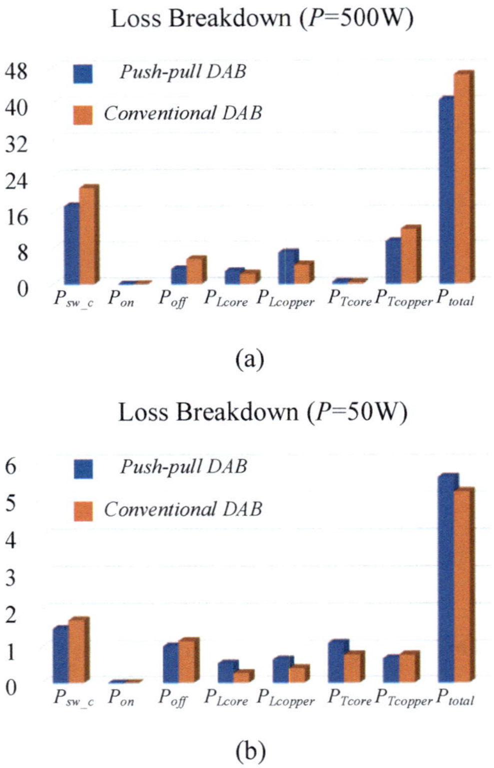

Fig. 7.25 Loss breakdown comparisons **a** $V_1 = 18$ V and $P = 500$ W. **b** $V_1 = 36$ V and $P = 50$ W

7.6 Conclusion

To overcome this shortcoming, this Chapter modifies the PPDAB converter and proposes an EPS control strategy based on magnetizing current injection. Firstly, a modified PPDAB converter is obtained by replacing the half-bridge circuit on the secondary side with a full-bridge circuit. And, the current and power characteristics of the modified PPDAB converter under EPS control are analyzed. Then, the impacts of switch junction capacitors on the ZVS range of the converter are analyzed. The two ZVS regions will be disconnected, instead of just intersecting at a particular point with conventional EPS control. In order to create a connected ZVS region within full load range, the magnetizing inductance is introduced to inject the auxiliary current to help achieve ZVS. Based on current expressions and power transfer characteristics as well as ZVS current limiting conditions, a power control law is proposed with the RMS current as the optimization target, which not only achieves ZVS for all switches within the whole load range, but also reduces calculation burden of the digital signal processor. Finally, an experimental laboratory prototype is built to verify the validity of the above theoretical analysis. Therefore, the proposed PPDAB converter with optimized control is suitable for the distributed generation system.

References

1. Li, S., Xiangli, K., & Smedley, K. M. (2017). A Control map for a bidirectional PWM plus phase-shift-modulated push-pull DC-DC converter. *IEEE Trans Industrial Electronics, 64*(11), 8514–8524.
2. Zhang, Z., Thomsen, O. C., & Andersen, M. A. E. (2012). Optimal design of a push-pull-forward half-bridge (PPFHB) bidirectional DC-DC converter with variable input voltage. *IEEE Trans Industrial Electronics, 59*(7), 2761–2771.
3. Everts, J. (2017). Closed-form solution for efficient ZVS modulation of DAB converters. *IEEE Trans Power Electronics, 32*(10), 7561–7576.
4. Riedel, J., Holmes, D. G., McGrath, B. P., & Teixeira, C. (2018). Maintaining continuous ZVS operation of a dual active bridge by reduced coupling transformers. *IEEE Trans Industrial Electronics, 65*(12), 9438–9448.

Part IV
Dual-Active-Bridge Converter with Additional Auxiliary High-Frequency Networks

Chapter 8
Light-Load Efficiency Enhancement of High-Frequency Dual-Active-Bridge Converter Under SPS Control

Abstract Dual-active-bridge (DAB) converter is an attractive topology to be used as a DC transformer (DCT) in the DC power distribution system. Because the voltage gain for DC transformer is relatively fixed, single-phase-shift (SPS) control is popularly adopted for DAB converter to achieve simple power control. Aiming at increasing the light load efficiency of high-frequency DAB converter under SPS control, this chapter proposes a concept to achieve full load range zero-voltage-switching (ZVS) with minimum RMS current in certain voltage variation range. In the proposed method, the added magnetizing inductance creates the possibility to achieve ZVS for all power switches. And, the voltage gain of the converter is specially controlled by turns ratio, which is no longer the conventional unity voltage gain condition. The key parameters optimization, including the turns ratio, magnetizing inductance and leakage inductance of the transformer, are introduced based on ZVS and current stress analysis. A 250 kHz, 400/48 V, 1000 W GaN-based DAB prototype is built, which shows performance improvements at light load.

Keywords Dual-active-bridge · Single-phase-shift (SPS) control · Full load range ZVS

8.1 Basic Concept of the Proposed Method

A half-bridge DAB converter is used to illustrate the basic concept of the proposed method as shown in Fig. 8.1. And the magnetizing inductor of the transformer is specially addressed.

To achieve ZVS, the energy stored in the inductive components should be enough to charge/discharge the switch junction capacitors when the switches are turned off. Fig. 8.2 shows the operating waveforms of DAB converter under SPS control. i_{min_p} and i_{min_s} are the threshold current for primary side switches and secondary side switches to achieve ZVS. i_{p_p} at t_4 and i_{p_s} at t_2 are the turn off current of two sides switches. When the turn off current i_{p_p} and i_{p_s} are larger than the threshold current i_{min_p} and i_{min_s} respectively, the switches can achieve ZVS.

G. Xu and X. Chen, *Topology Deduction and Control for Dual-Active-Bridge Converters*, https://doi.org/10.1007/978-981-97-4856-3_8

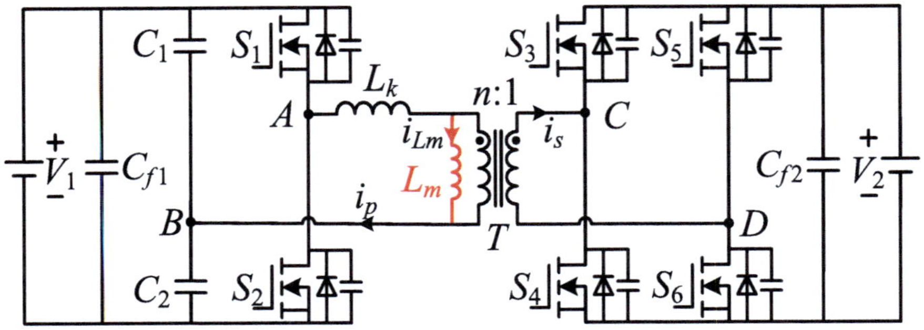

Fig. 8.1 The topology of DAB converter

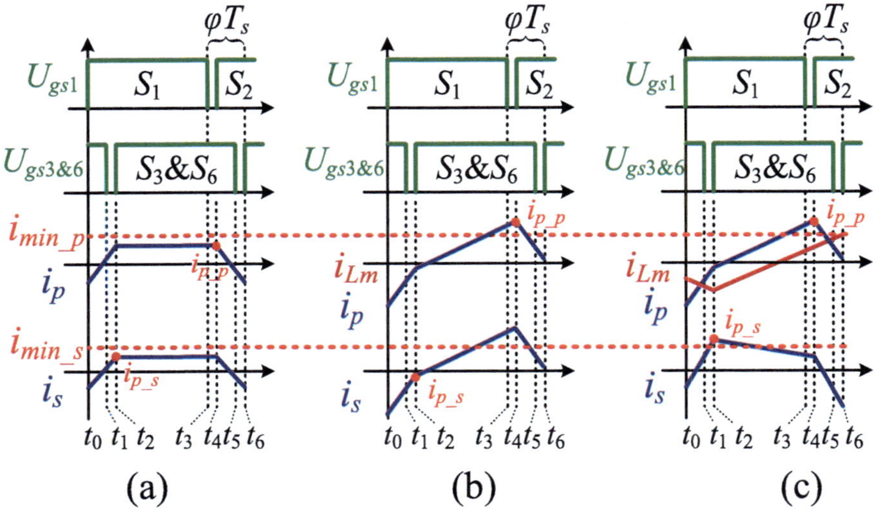

Fig. 8.2 The operation waveform of DAB converter under SPS control at light load. **a** Unity voltage gain ($n = V_1/2V_2$). **b** Voltage mismatching ($n < V_1/2V_2$). **c** Proposed method ($n < V_1/2V_2$, and considering magnetizing inductor)

Fig. 8.2a shows the light-load waveforms under SPS control for the conventional DAB converter which is worked under unity voltage gain ($n = V_1 /2V_2$). As seen, the turn off current i_{p_p} at t_4 and i_{p_s} at t_2 are all small than the threshold current i_{min_p} and i_{min_s}, which means both two sides switches cannot achieve ZVS. This can explain the reason of hard switching for SPS control at light load.

In order to achieve ZVS at light load and make i_{p_p} to be larger than the threshold current i_{min_p}, the turns ratio n can be designed to be smaller than $V_1/2V_2$ ($n < V_1 / 2V_2$). As seen in Fig. 8.2b, when $n < V_1 /2V_2$, the slope of i_p can be increased, and the turn off current i_{p_p} at t_4 will be increased to be larger than the threshold current i_{min_p}, which will help primary side switches to be achieved ZVS at light load.

However, the existing slope in secondary current will lead to hard switching for secondary side switches. As Fig. 8.2b shows, at t_2, the turn off current i_{p_s} of the secondary side switches becomes negative, which will cause hard switching for secondary side switches. To achieve ZVS for secondary side switches, as Fig. 8.2c shows, the magnetizing inductance is proposed to be added into consideration, which is rarely studied in previous works. The polarity of i_{p_s} will be changed through injecting the magnetizing current, and ZVS can be realized for the secondary side switches.

As analyzed above, the utilizing of the magnetizing inductance can provide a possibility to achieve full load ZVS for DAB converter under SPS control, and through designing the mismatching point and magnetizing inductance, DAB converter can achieve ZVS for all switches at light load.

8.2 ZVS Analysis with Magnetizing Inductance

8.2.1 ZVS Condition Analysis Considering Magnetizing Inductance and Switch Junction Capacitors

To achieve ZVS for all the switches, the energy in the inductive components must be high enough to charge/discharge the switch junction capacitors sufficiently. For the primary side, i_p will approach its maximum value when S_1 is turned off, and L_k then charges/discharges the capacitors.

Figure 3a shows the equivalent ZVS circuit considering junction capacitors after S_1 is turned off. As illustrated, i_p charges C_2 and C_{s1}, and discharges C_1 and C_{s2}. The total energy includes three parts: the energy stored in L_k, energy in all the capacitors and the energy flowing into V_2. During the charging/discharging process, the sum of all energy in the junction capacitors is a constant value, while the energy of the inductor is decreased and fed back to secondary side. These energy relations can be expressed as

$$E_L(t_3) = E_L(t_4) + E_{feedback} = \frac{1}{2} L_k i_{p\_p}^2 \tag{8.1}$$

The relationship of leakage inductor current and capacitor voltage can be expressed as

$$i_p(t) = C_{s1}\frac{du_{s1}(t)}{dt} - C_{s2}\frac{du_{s2}(t)}{dt} = 2C_{s1}\frac{du_{s1}(t)}{dt} \tag{8.2}$$

where $u_{s1}(t)$ and $u_{s2}(t)$ are the voltages of switch junction capacitors for S_1 and S_2, respectively.

When ZVS resonant process is finished, the energy stored in L_k feeds back to V_2, which is calculated as

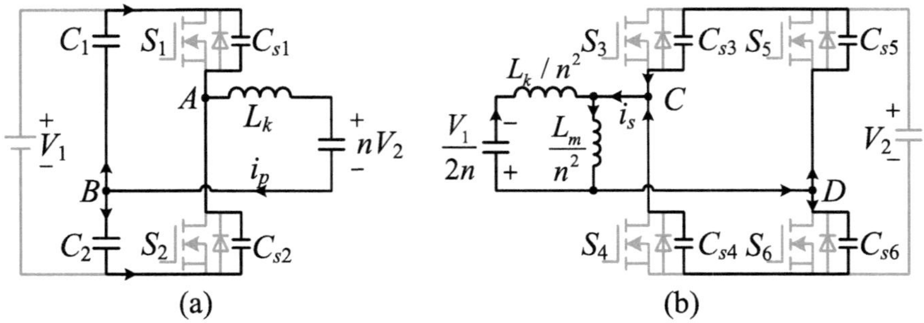

Fig. 8.3 Equivalent ZVS circuit. **a** The equivalent ZVS circuit after S_1 is turned off. **b** Equivalent ZVS circuit after S_3 and S_6 are turned off

$$E_{feedback} = \int_{t_3}^{t_4} nV_2 \cdot i_p(t)dt \tag{8.3}$$

The switch junction capacitors will be charges/discharges sufficiently, in another word, the remaining energy of the inductor is greater than zero after the resonant process. Therefore, the range of i_{p_p} for ZVS realization of primary side switches can be obtained as (8.4) according to (8.1), (8.2) and (8.3).

$$i_{p\_p}(\varphi) \geq \sqrt{\frac{4nV_1V_2C_{s1}}{L_k}} \tag{8.4}$$

where φ is the phase-shift ratio of the two bridges, L_k is the leakage inductance.

For ZVS realization of the secondary side switches, the polarity of i_{p_s} needs to be considered. Figure 3b illustrates the equivalent ZVS circuit considering magnetizing inductor and the switch junction capacitors after S_3 and S_6 are turned off. As illustrated, the direction of i_s is changed through canceling i_p by injecting the magnetizing current. Therefore, the sufficient condition for secondary switches to achieve ZVS is

$$i_{p\_s}(\varphi) > 0 \tag{8.5}$$

8.2.2 *ZVS Range Extension with Magnetizing Inductance*

The power transfer of the proposed converter is expressed as

$$P(\varphi) = \frac{nV_1V_2\varphi(1 - 2\varphi)}{2L_kf} \tag{8.6}$$

where f is the switching frequency. To normalize the calculation, the base power can be defined when the phase-shift-ratio φ equals to 0.25 and the voltages are matched ($V_1 = 2nV_2$), which is expressed as

$$P_{base} = \frac{V_1^2}{32L_k f} \tag{8.7}$$

Then the normalized power by φ is shown as

$$P^*(\varphi) = \frac{P(\varphi)}{P_{base}} = 8G\varphi(1 - 2\varphi) \tag{8.8}$$

where G is the ratio of the voltages at two ports, and $G = 2nV_2/V_1$. The base current is defined as

$$I_{base} = \frac{P_{base}}{V_1} = \frac{V_1}{32L_k f} \tag{8.9}$$

According to the above analysis, i_{p_p} and i_{p_s} are the major concerns for the switches to realize ZVS. The normalized i_{p_p} can be calculated as

$$i^*_{p\_p}(\varphi) = \frac{i_{p\_p}(\varphi)}{I_{base}} = 16G\varphi - 4G + 4 \tag{8.10}$$

And the normalized i_{p_s} can be obtained in a similar way.

$$i^*_{p\_s}(\varphi) = \frac{i_{p\_s}(\varphi)}{nI_{base}} = 16\varphi + 4G(1 + k_L) - 4 \tag{8.11}$$

where k_L is the ratio between the leakage inductance and the magnetizing inductance: $k_L = L_k/L_m$.

According to (8.10) and (8.11), Fig. 8.4 plots the ZVS range versus normalized power P^* and G, with different k_L. As Fig. 8.4a shows, Region B cannot achieve ZVS for secondary side switches, Region C cannot achieve ZVS for primary side switches, and Region A cannot achieve ZVS for all the switches. So, at light load, none ZVS region will always occurs no matter how to design the gain G. This indicates that it is impossible to achieve ZVS at light load for all the switches under SPS control.

In order to achieve ZVS at light load for both side switches, the magnetizing current is considered, thus k_L is no longer equal to zero. As Fig. 8.4b shows, ZVS range will be expanded when k_L increases from 0 to 0.3, and full load range ZVS can be achieved when G is in the range between point D to F. Namely, with the proper design of G and k_L, the DAB converter can be achieved full load range ZVS within variable gain range ΔG shown in Fig. 8.4b.

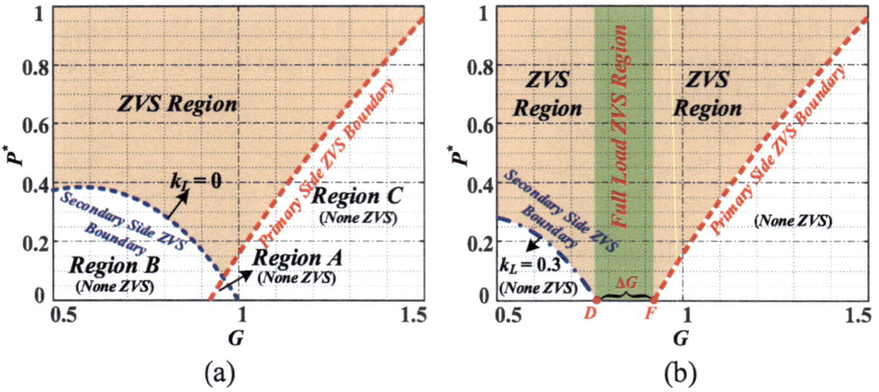

Fig. 8.4 ZVS range comparison. **a** ZVS range versus P^* and G with $k_L = 0$. **b** ZVS range versus P^* and G with $k_L = 0.3$

8.3 Design Consideration to Achieve Full Load Range ZVS with Minimum RMS Current

As the analysis above, to achieve ZVS for all switches in full load range, G and k_L should be properly designed. Meanwhile, the RMS value of i_p and i_s need to be optimized to decrease the conduction loss. Since the working principle is symmetrical under different power direction, only the case of forward power transfer mode is analyzed.

8.3.1 *Optimized Conversion Gain G Design to Achieve ZVS for Primary Side Switches with Minimum Primary RMS Current*

To achieve ZVS for the primary side switches, G should be properly designed to ensure that i_{p_p} is high enough to charge/discharge junction capacitors, and also to obtain a relatively lower RMS current. In addition, L_k and G should also guarantee the maximum power transfer. The normalized RMS current value of of i_p is expressed as

$$I^*_{p\_rms}(\varphi) = \frac{I_{p\_rms}(\varphi)}{I_{base}} = 4\sqrt{\frac{G^2 - 64G\varphi^3 + 48G\varphi^2 - 2G + 1}{3}} \tag{8.12}$$

Fig. 8.5a plots the RMS value contour line versus G and P^*, and the ZVS range is marked according to (8.4). As seen, in the full load ZVS region, for a given output power, the RMS current is decreased as G increases. So, to optimize RMS current, G should be designed as large as possible in full load ZVS region. Fig. 8.5b illustrates the

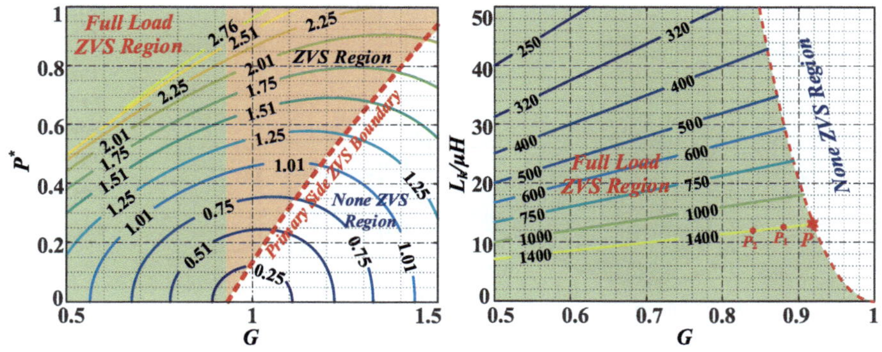

(a) (b)

Fig. 8.5 L_k and G optimization ($f = 250$ kHz). **a** The RMS value contour line versus G and P^*. **b** The maximum transferred power contour line versus L_k and G

maximum transferred power contour line versus L_k and G. To satisfy the maximum transferred power, L_k should be designed according to the power curves as shown in Fig. 8.5b. Meanwhile, G should be designed as large as possible with light load ZVS. In this chapter, the rated power of the prototype is set as 1000 W. And the maximum power is designed to be larger than the rated power about 40% to leave some power margins. Then, the recommended design is $G = 0.92$, $L_k = 13\mu H$, which is shown as point P in Fig. 8.5b.

8.3.2 Magnetizing Current Injection Optimization for ZVS of Secondary Side Switches

For ZVS realization of secondary side switches, the magnetizing current should be properly designed to inject reactive energy for secondary switches. Meanwhile, the RMS value of i_s needs to be optimized to decrease the related loss. The normalized RMS current value of i_s is expressed as

$$I^*_{s\_rms}(\varphi) = \frac{I_{s\_rms}(\varphi)}{nI_{base}} = 4\sqrt{\frac{[16(3-4\varphi)\varphi^2 + (1+k_L)G - 2](1+k_L)G + 1}{3}} \tag{8.13}$$

Fig. 8.6 plots $I^*_{s\_rms}$ versus P^* and k_L, and the ZVS range is marked according to (8.5). As Fig. 8.6a shows, to achieve full load range ZVS for secondary side switches, k_L should be designed at full load ZVS region. As Fig. 8.6b shows, in full load ZVS region, with a certain P^*, $I^*_{s\_rms}$ is decreased as k_L decreases. To optimize RMS current, k_L should be designed as small as possible in full load ZVS region.

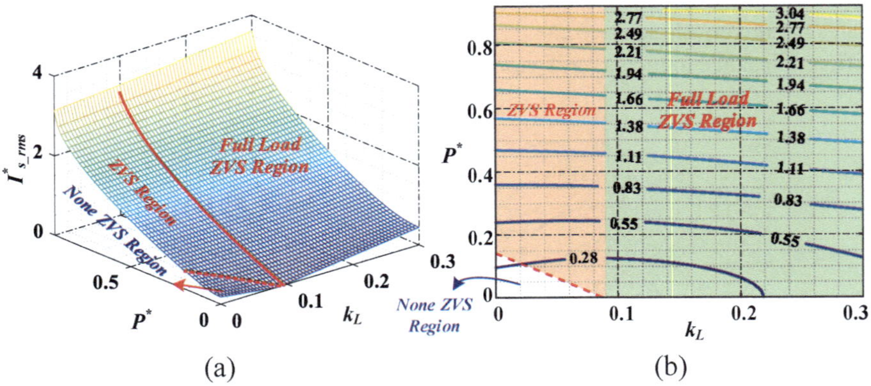

Fig. 8.6 $I^*_{s\_rms}$ versus P^* and k_L ($G = 0.92$). **a** 3D plots. **b** Contour line plots

Therefore, k_L can be set as 0.1, which can achieve both full load range ZVS and relatively small RMS current.

8.4 Experimental Results and Comparison

A 1000 W prototype of a GaN-based DAB converter is built to verify the effectiveness of the proposed solution, which is shown in Fig. 8.7. The detailed specifications are designed according to Figs. 8.5 and 8.6, which are shown in Table 8.1. Meanwhile, the parameters for conventional DAB design are also included.

Fig. 8.8 illustrates the ZVS waveforms of the primary side switch and the secondary side switch at no load condition with the proposed method. v_{gs1} and v_{gs3} are the driving signals of S_1 and S_3, respectively. v_{ds1} and v_{ds3} are the voltage across S_1 and S_3, respectively. i_p and i_s are the primary side current and secondary side current of the transformer, respectively. As Fig. 8.8a shows, when S_1 is turned on, i_p is negative, which discharges the junction capacitor of S_1, and ZVS is achieved. As

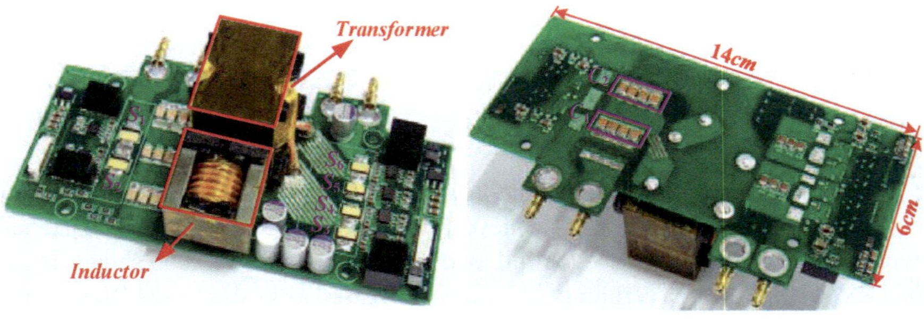

Fig. 8.7 The experimental prototype

Table 8.1 Parameters of the prototype

Item	Detail (Proposed)	Detail (Conventional)
Rated Power (P)	1000 W	
Primary side Voltage (V_1)	400 V	
Secondary side Voltage (V_2)	48 V	
Leakage Inductance (L_k)	13μH	
Magnetizing Inductance (L_m)	130μH ($k_L = 0.1$)	Ignored ($k_L \approx 0$)
Turns Ratio (n)	23:6 ($G = 0.92$)	25:6 ($G = 1$)
Switching Frequency (f)	250 kHz	
Half bridge capacitance (C_1, C_2)	4μF	
Primary side switches (S_1, S_2)	GS66508T	
Secondary side switches ($S_3 \sim S_6$)	GS61008T	

Fig. 8.8b shows, when S_3 is turned on, i_s is positive, which discharges the junction capacitor of S_3, and ZVS is achieved. Note that, the Chanel M is the math utility of the scope, and the calculation formula is ($i_p - i_s/n$), which represents the magnetizing current. As seen, the amplitude of i_{Lm} is slightly higher than i_p, which can change the polarity of i_s, and achieve ZVS for secondary side switches.

Figs. 8.8c and d illustrate the ZVS waveforms of the primary side switch and the secondary side switch at no load condition with the conventional method. As seen, both the primary side and secondary side switches are suffering hard switching, because i_p and i_s are so small that the energy stored in the inductor cannot charge/discharge the junction capacitor of switches sufficiently.

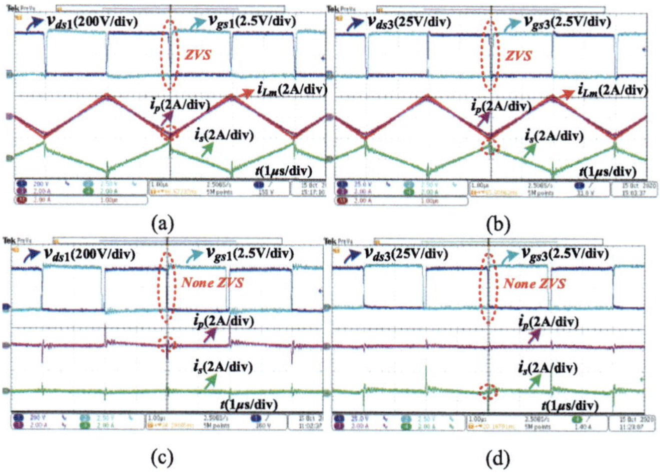

Fig. 8.8 ZVS waveforms at no load. **a** Primary side switch with proposed method. **b** Secondary side switch with proposed method. **c** Primary side switch with conventional method. **d** Secondary side switch with conventional method

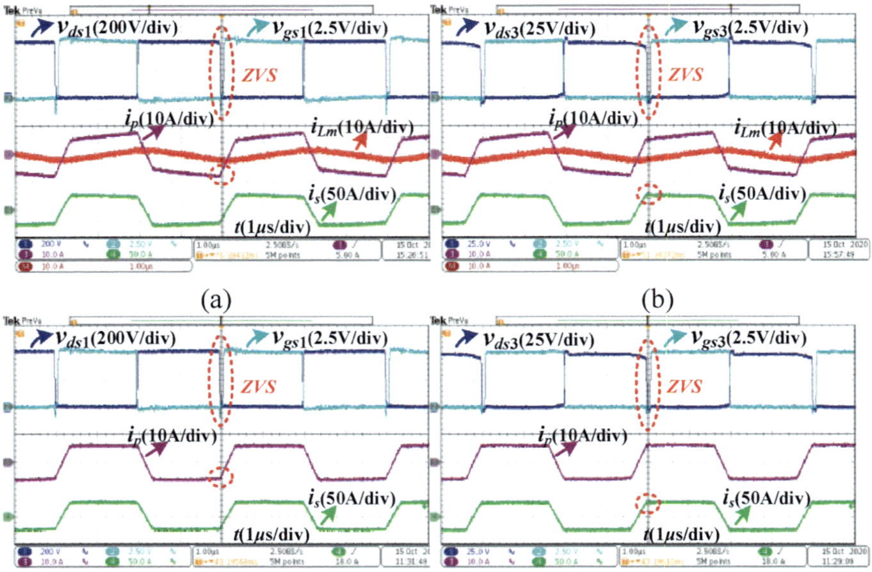

(a) (b) (c) (d)

Fig. 8.9 ZVS waveforms when $P = 1000$ W. **a** Primary side switch with proposed method. **b** Secondary side switch with proposed method. **c** Primary side switch with conventional method. **d** Secondary side switch with conventional method

Fig. 8.9 shows the ZVS waveforms with two methods when the output power is 1000 W. As seen, when the converter works at heavy load condition, both two methods can achieve ZVS sufficiently for all switches.

Fig. 8.10a shows the measured RMS current of i_p with different G, and the corresponding power points P_1 and P_2 are denoted in Fig. 8.5. As for the conventional DAB converter, G is equal to 1, and the RMS current is the smallest. However, when $G = 1$, full load rang ZVS cannot be achieved. When G is decreased, full load range ZVS can be achieved. But, smaller G will lead to higher RMS current. So, G should be as large as possible if full load range ZVS is achieved. Then the optimal point P ($G = 0.92$) can be selected for obtaining both full load range ZVS and relatively small RMS current. As for the transformer secondary side RMS current I_{s_rms} shown in Fig. 8.10b, its value is almost the same with different G.

The loss breakdown comparisons of the converter between the proposed method and the conventional method at different loads are shown in Fig. 8.11. As Fig. 8.11a shows, when the converter works at light load (100 W), the total loss of the conventional method is higher than that of the proposed method about 7 W, which is mainly caused by hard switching loss, leading to low efficiency at light load. As Fig. 8.11b shows, when the converter works at heavy load condition (1000 W), the power loss with the proposed method is higher than that of the conventional method about 5 W, which is mainly caused by the increased conduction and copper loss. The loss

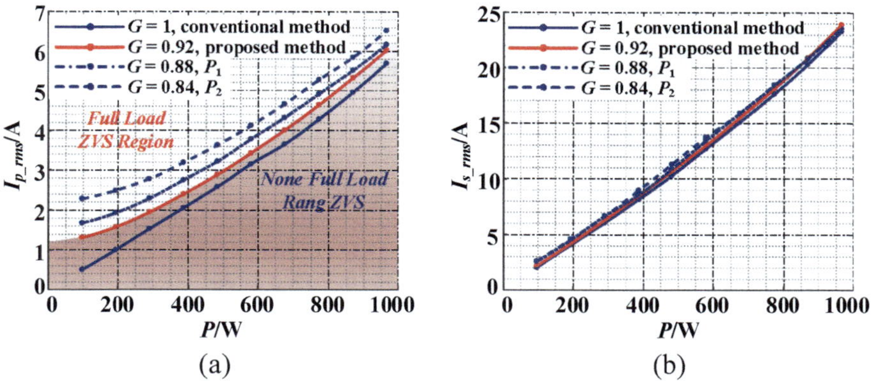

Fig. 8.10 The measured RMS current of i_p and i_s with different G. **a** The measured I_{p_rms}. **b** The measured I_{s_rms}

difference between the two methods accounts for only 0.5% of the total power at 1000 W.

Fig. 8.12 plots the efficiency of the proposed method and the conventional DAB converter under SPS control. As seen, when working at light load, the proposed method has better efficiency than the conventional DAB converter, which is mainly caused by losing ZVS for the conventional method. And at heavy load, the efficiency of the proposed method is slightly lower than that of the conventional DAB converter because the injected magnetizing current could slightly increase the conduction loss.

There are also some works on the improvement of the light load efficiency. Table 8.2 illustrates the ZVS range and the light load efficiency at different working points with different methods. As seen, only the proposed method can achieve full load range ZVS with SPS control scheme. Besides, with the proposed method, the efficiency at 10% and 20% of the rated power is higher than that of other works.

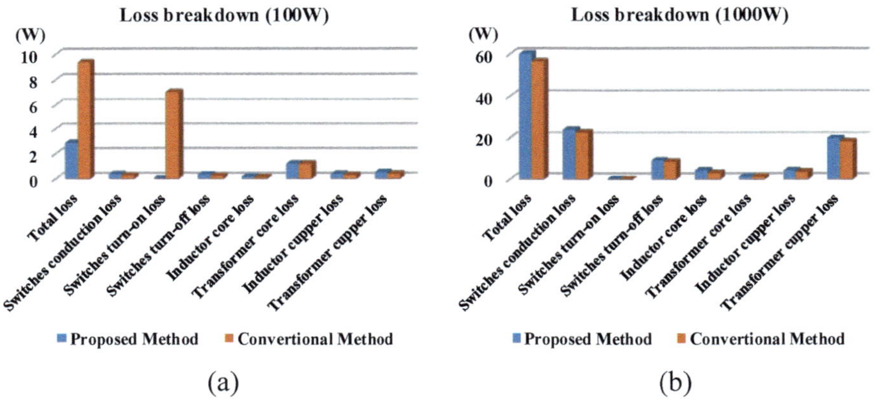

Fig. 8.11 Loss breakdown comparisons. **a** At light load (100 W) **b** At heavy load (1000 W)

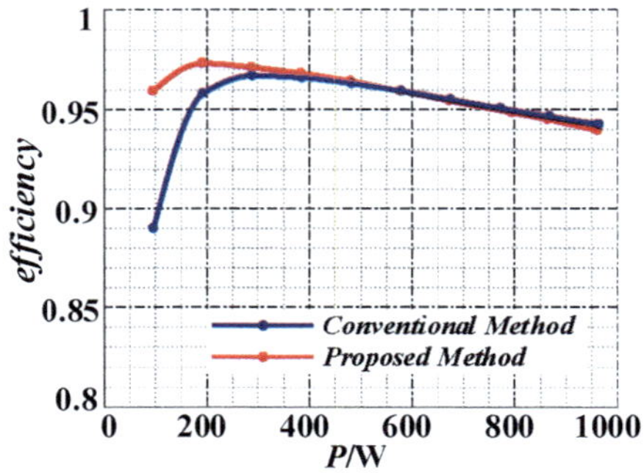

Fig. 8.12 Efficiency comparison between the proposed method and the conventional method

Table 8.2 Comparison with other works

Item	Full load range ZVS	Light load efficiency	
		10% rated power	20% rated power
Proposed Method	Yes	96%	97.3%
Conventional Method	No	89%	96%
Method in [1]	No	< 94%	96%
Method in [2]	No	96%	94%

8.5 Conclusions

This chapter proposed a method to enhance the light-load efficiency for DAB converter. With the consideration of magnetizing inductance, full load range ZVS can be achieved under SPS control. The voltages gain point and the magnetizing inductance are specially designed to achieve full load range ZVS for all switches considering switch junction capacitors, and the efficiency at light load is improved. ZVS model analysis, ZVS range and design consideration are presented in this chapter. The experimental results from a 250 kHz, 400/48 V, 1000 W prototype have verified that the performance of the DAB converter with proposed method is enhanced. Besides, by comparing the proposed method with other recent works, it shows the benefit of the proposed method. Meanwhile, the basic concept of the proposed method can also be applied into other types DAB converter for achieving better features, such as full bridge DAB converter, push–pull DAB converter, the three phase DAB converter, etc.

References

1. Jafari, A., Nikoo, M. S., Karakaya, F., & Matioli, E. (2020). Enhanced DAB for efficiency preservation using adjustable-tap high-frequency transformer. *IEEE Transactions on Power Electronics, 35*(7), 6673–6677.
2. Liu, P., & Duan, S. (2021). A ZVS range enhancement strategy for the DAB converter by using blocking capacitors. *IEEE Journal of Emerging and Selected Topics in Power Electronics, 9*(2), 1389–1398.

Chapter 9
Optimized EPS Control to Achieve Full Load Range ZVS with Seamless Transition for Dual Active Bridge Converters

Abstract Dual active bridge (DAB) converters are widely studied for applications requiring isolation, bidirectional power flow and high efficiency. To solve the problem of limited zero voltage switching (ZVS) range and achieve seamless transition for extended-phase-shift (EPS) control, the magnetizing current is utilized and an optimized control law is proposed in this chapter. With the injection of magnetizing current, the two ZVS regions under EPS control will be creatively intersected with a large region, instead of disconnected or just intersected at a particular point as in previous studies. The ZVS model considering switch junction capacitors, ZVS region analysis with magnetizing current injection and power transfer characteristics are analyzed. Based on these analyses, the optimized EPS control is proposed to achieve full load range ZVS, low root-mean-square (RMS) current and maximum transmission power capability. A 250 kHz, 500W GaN-based DAB prototype is built to verify the effectiveness of the proposed solution.

Keywords Dual-active-bridge (DAB) · Extended phase-shift (EPS) · Zero-voltage switching (ZVS)

9.1 Topology and Conventional EPS Control

The topology of DAB dc-dc converter is depicted in Fig. 9.1. Both of the input and output sides are active full bridges, and they are linked by a high frequency transformer. The active bridge in V_1 side generates a high frequency voltage u_{AB}, and the other one generates voltage u_{CD}. Bidirectional power transfer can be controlled by manipulating the phase shift ratio between the two high frequency voltages and their duty cycles.

Figure 9.2 shows the operation principle of EPS control. D_φ is the duty cycle corresponding to the phase-shift angle between u_{AB} and u_{CD}, and D_y represents the duty cycle of u_{AB}. V_1 and V_2 stand for the voltages of high side and low side respectively. T_s is the switching period. i_k and i_s are the current of leakage inductance

G. Xu and X. Chen, *Topology Deduction and Control for Dual-Active-Bridge Converters*, https://doi.org/10.1007/978-981-97-4856-3_9

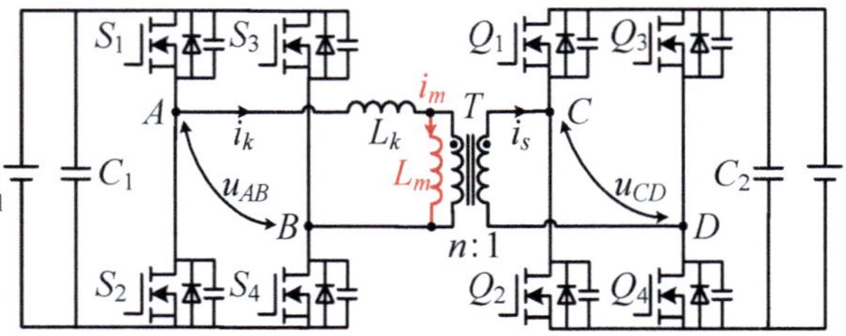

Fig. 9.1 Topology of the DAB dc-dc converter

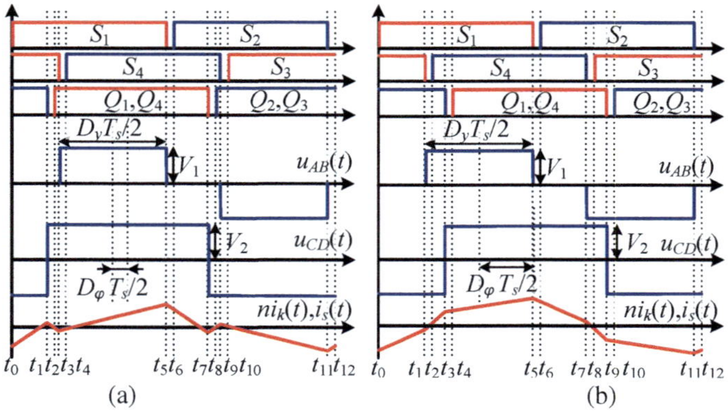

Fig. 9.2 Working waveforms under EPS control. **a** Case 1: $0 \leq D_\varphi \leq (1\text{-}D_y)/2$. **b** Case 2: $(1\text{-}D_y)/2 \leq D_\varphi \leq 0.5$

L_k and transformer current in V_2 side respectively. n:1 represents the transformer turns ratio.

According to the difference in the position of the rising edges of u_{AB} and u_{CD}, the forward working condition can be divided into three cases: $0 < D_\varphi < (1\text{-}D_y)/2$, $(1\text{-}D_y)/2 < D_\varphi < (1 + D_y)/2$, $(1 + D_y)/2 < D_\varphi < 1$. Since serious circulating energy will be generated when D_φ is larger than 0.5 [1], only two cases are used in real application. The operation waveforms of these two cases are shown in Fig. 9.2a and b respectively.

9.2 ZVS Analysis Considering Junction Capacitors

9.2.1 *Instant Current Expressions for Leakage Inductance*

ZVS conditions are related to the current values when power switches are turned OFF; hence it is necessary to calculate the expressions of i_k based on Fig. 9.2. Due to symmetrical characteristic of the PWM, the current values and conditions of ZVS are

Table 9.1 Current Values in Case 1

Stage	Value	Time
$t_0 \sim t_1$	$i_k(t) = i_k(t_0) + (t - t_0)nV_2/L_k$, $i_k(t_1) = (k - D_y)nV_2T_s/(4kL_k)$	$(1 - D_y + 2D_\varphi)T_s/4$
$t_2 \sim t_3$	$i_k(t) = i_k(t_2) - (t - t_2)nV_2/L_k$ $i_k(t_3) = (kD_y - D_y + 2kD_\varphi)nV_2T_s/(4kL_k)$	$(1 - D_y - 2D_\varphi)T_s/4$
$t_4 \sim t_5$	$i_k(t) = i_k(t_4) + (t - t_4)(1 - k)nV_2/kL_k$ $i_k(t_5) = (D_y - kD_y + 2kD_\varphi)nV_2T_s/(4kL_k)$	$D_yT_s/2$

Table 9.2 Current Values in Case 2

Stage	Value	Time
$t_0 \sim t_1$	$i_k(t) = i_k(t_0) + (t - t_0)nV_2/L_k$ $i_k(t_1) = [2k - (k + 1)D_y - 2kD_\varphi]nV_2T_s/(4kL_k)$	$(1 - D_y)T_s/2$
$t_2 \sim t_3$	$i_k(t) = i_k(t_2) + (t - t_2)(1 + k)nV_2/(kL_k)$ $i_k(t_3) = (k - 1 + 2D_\varphi)nV_2T_s/(4kL_k)$	$(D_y + 2D_\varphi - 1)T_s/4$
$t_4 \sim t_5$	$i_k(t) = i_k(t_4) + nV_2(1 - k)(t - t_4)/(kL_k)$ $i_k(t_5) = (D_y - kD_y + 2kD_\varphi)nV_2T_s/(4kL_k)$	$(1 + D_y - 2D_\varphi)T_s/4$

analyzed only within one half cycle from t_0 to t_6, as denoted in Fig. 9.2. The average value of i_k over one switching period must be equal to zero at steady state, which satisfies $i_k(t_0) = -i_k(t_6)$. According to this, through integrating the voltage crossing the inductance, the current values at the instants when switches are turned OFF can be summarized as in Tables 9.1 and 9.2 for case 1 and case 2 respectively, where $k = nV_2/V_1$ is the voltage conversion ratio.

9.2.2 *ZVS Restrictions Considering Switch Junction Capacitors*

Generally, to achieve ZVS, the energy stored in the inductance should be enough to charge/discharge the switch junction capacitors. As a result, the minimum values of i_k when the power switches are turned OFF, defined to be I_{min}, need to be guaranteed to provide enough energy for ZVS. Meanwhile, the instants when i_k reaches zero are defined to be t_m.

The equivalent ZVS circuits for case 1 and case 2 are illustrated in Figs. 9.3 and 9.4 respectively. Since the impedance for the junction capacitor is much larger compared with other parasitic impedances during the charging process, and the minimum charging current highly depends on the junction capacitors rather than other parasitic parameters, only junction capacitors are considered in ZVS analysis.

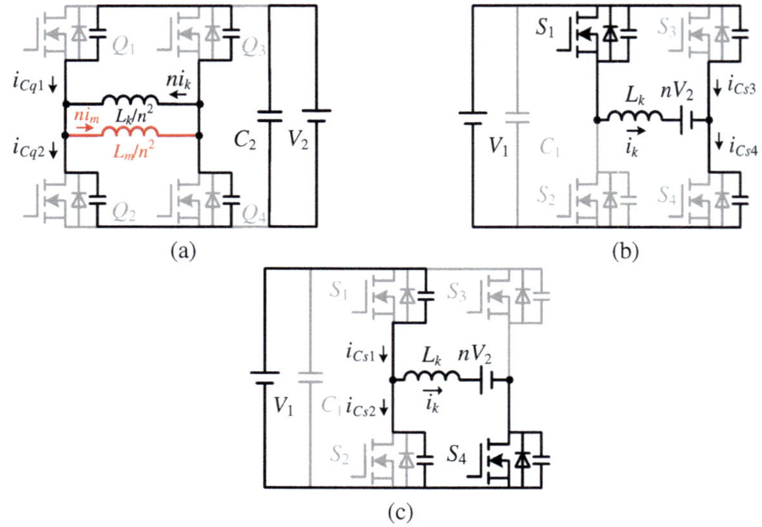

Fig. 9.3 Equivalent ZVS circuits of case 1. **a** t_1-t_2. **b** t_3-t_4. **c** t_5-t_6

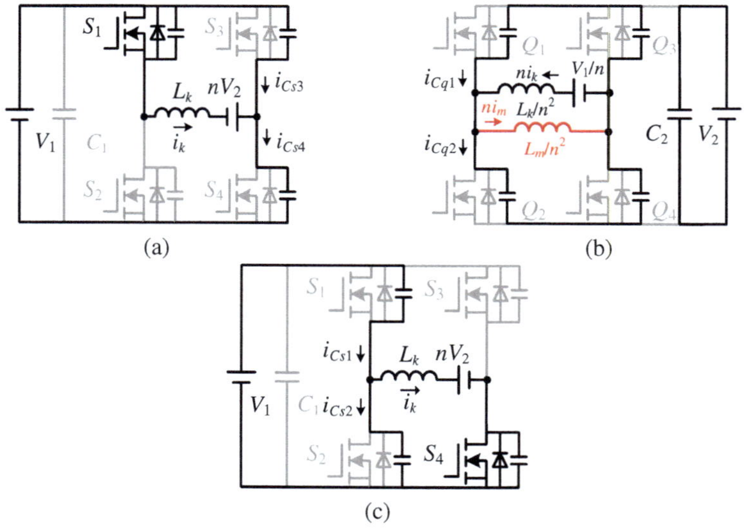

Fig. 9.4 Equivalent ZVS circuits of case 2. **a** t_1-t_2. **b** t_3-t_4. **c** t_5-t_6

Take case 1 for instance. At t_1, Q_2 and Q_3 are turned OFF. During time t_1 to t_2, i_s discharges the junction capacitors of Q_1 and Q_4 while charges the junction capacitors of Q_2 and Q_3, as shown in Fig. 9.3a. The junction capacitors on V_2 side resonate with the leakage inductance equivalent to V_2 side. In this process, there is no energy loss, and the ZVS condition for Q_1 and Q_4 can be obtained as

$$i_s(t_1) = ni_k(t_1) > 0 \tag{9.1}$$

At t_3, S_3 is turned OFF. i_k starts to discharge the junction capacitor of S_4 and charge the junction capacitor of S_3. In this process, the energy of L_k is transmitted to the voltage sources V_1 and nV_2 as shown in Fig. 9.3b. Hence, at $t = t_3$, $i_k(t_3)_{min} = I_{min1}$, $u_{Cs3} = 0$, $u_{Cs4} = V_1$, and at $t = t_{m1}$, $i_k(t_{m1})_{min} = 0$, $u_{Cs3} = V_1$, $u_{Cs4} = 0$.

During the stage from t_3 to t_{m1}, the leakage current can be expressed as

$$i_k(t) = C_s\frac{du_{Cs4}}{dt} - C_s\frac{du_{Cs3}}{dt} = -2C_s\frac{du_{Cs3}}{dt} \tag{9.2}$$

where C_s represents the junction capacitor on the V_1 side, $C_s = C_{s1} = C_{s2} = C_{s3} = C_{s4}$.

According to the energy balancing principle, the following relationship holds

$$E_{t_3} = E_{t_{m1}} + E_{loss} + E_{delivered1} \tag{9.3}$$

Assuming that the energy of circuit is lossless, it can be deduced as

$$L_k I_{min1}^2/2 = E_{delivered1} \tag{9.4}$$

$$\begin{aligned} E_{deliver1} &= \int_{t_3}^{t_{m1}} [nV_2 i_k(t) - V_1 i_{Cs4}(t)]dt \\ &= \int_{t_3}^{t_{m1}} [nV_2(-2C_s\frac{du_{Cs3}}{dt}) - V_1 C_s\frac{du_{Cs4}}{dt}]dt = C_s(\frac{nV_2}{k})^2(1-2k) \end{aligned} \tag{9.5}$$

Thus, the ZVS condition for S_4 can be derived as

$$\begin{cases} i_k(t_3) \le I_{min1} = -nV_2\sqrt{2C_s(1-2k)/L_k}/k, & 0 < k < 0.5 \\ i_k(t_3) \le I_{min1} = 0, & 0.5 \le k < 1 \end{cases} \tag{9.6}$$

S_1 is turned OFF at t_5. The junction capacitor of S_2 starts to be discharged by i_k, and its drain-source voltage begins to decrease as illustrated in Fig. 9.3c. In the same way, the leakage current expression and the relationship with energy can be listed as

$$i_k(t) = C_s\frac{du_{Cs1}}{dt} - C_s\frac{du_{Cs2}}{dt} = 2C_s\frac{du_{Cs1}}{dt} \tag{9.7}$$

$$\begin{aligned} E_{deliver2} &= \int_{t_5}^{t_{m2}} [nV_2 i_k(t) - V_1 i_{Cs1}(t)]dt \\ &= \int_{t_5}^{t_{m2}} [nV_2 \cdot 2C_s\frac{du_{Cs1}}{dt} - V_1 C_s\frac{du_{Cs1}}{dt}]dt = C_s(\frac{nV_2}{k})^2(2k-1) \end{aligned} \tag{9.8}$$

Therefore, to achieve ZVS for S_2, it should satisfy the condition that

Table 9.3 Constraints of ZVS for All Power Switches

Case	1: $0 \le D_\varphi \le (1\text{-}D_y)/2$	2: $(1\text{-}D_y)/2 \le D_\varphi \le 0.5$
Q_1-Q_4	$D_y \le k$	$D_\varphi \ge (1\text{-}k)/2$
S_3&S_4	$D_y \ge \dfrac{2kD_\varphi + a\sqrt{(1-2k)}}{1-k}, k \le 0.5$ $D_y \ge \dfrac{2kD_\varphi}{1-k}, \quad k > 0.5$	$D_y \ge \frac{2k(1-D_\varphi)+a\sqrt{(1+2k)}}{1+k}$
S_1&S_2	$D_y \ge \dfrac{a\sqrt{(2k-1)} - 2kD_\varphi}{1-k}, k > 0.5$ $D_y \ge \dfrac{-2kD_\varphi}{1-k}, \quad k \le 0.5$	$D_y \ge \dfrac{a\sqrt{(2k-1)} - 2kD_\varphi}{1-k}, k > 0.5$ $D_y \ge \dfrac{-2kD_\varphi}{1-k}, \quad k \le 0.5$

$$\begin{cases} i_k(t_5) \ge I_{min2} = 0, & 0 < k < 0.5 \\ i_k(t_5) \ge I_{min2} = nV_2\sqrt{2C_s(2k-1)/L_k}/k, & 0.5 \le k < 1 \end{cases} \tag{9.9}$$

Substituting the parameters in Table 9.1 into (9.1) (9.6) and (9.9), the constraints of ZVS can be summed up in Table 9.3. Likewise, the restrictions of ZVS for the case 2 can also be obtained, as listed in Table 9.3, where $a = (32L_kC_s)^{1/2}/T_s$.

9.2.3 *Discontinuous Power Transfer in ZVS Region Without Magnetizing Current*

The ZVS regions and Non ZVS regions, taking the junction capacitors into consideration, are illustrated in Fig. 9.5 based on Table 9.3. As seen in Fig. 9.5, ZVS region is divided into two parts, and these two parts are not connected under different voltage ratios. Therefore, the control can't be seamlessly changed if the converter needs to work in both two ZVS regions. To better illustrate this phenomenon, the transferred power can also be calculated as follows.

According to Fig. 9.2 and Table 9.1 as well as Table 9.2, the average power expressions of case 1 in (9.10) and case 2 in (9.11) can be obtained.

$$P_{\text{case1}} = V_1 \cdot \frac{2}{T_s}\int_0^{T_s/2} i_1(t)dt = V_1 \cdot \frac{2}{T_s}\int_{t_3}^{t_5} i_k(t)dt = \frac{(nV_2)^2 D_y D_\varphi}{2kL_k f_s} \tag{9.10}$$

$$\begin{aligned} P_{\text{case2}} &= V_1 \cdot \frac{2}{T_s}\int_0^{T_s/2} i_1(t)dt = V_1 \cdot \frac{2}{T_s}\int_{t_1}^{t_5} i_k(t)dt \\ &= (nV_2)^2[1-(1-2D_\varphi)^2-(1-D_y)^2]/(8kL_k f_s) \end{aligned} \tag{9.11}$$

Then, the plots for relationships of power, duty cycle, phase shift angle and ZVS regions are shown in Fig. 9.6. The output power is normalized by $P_{base} = (nV_2)^2/$

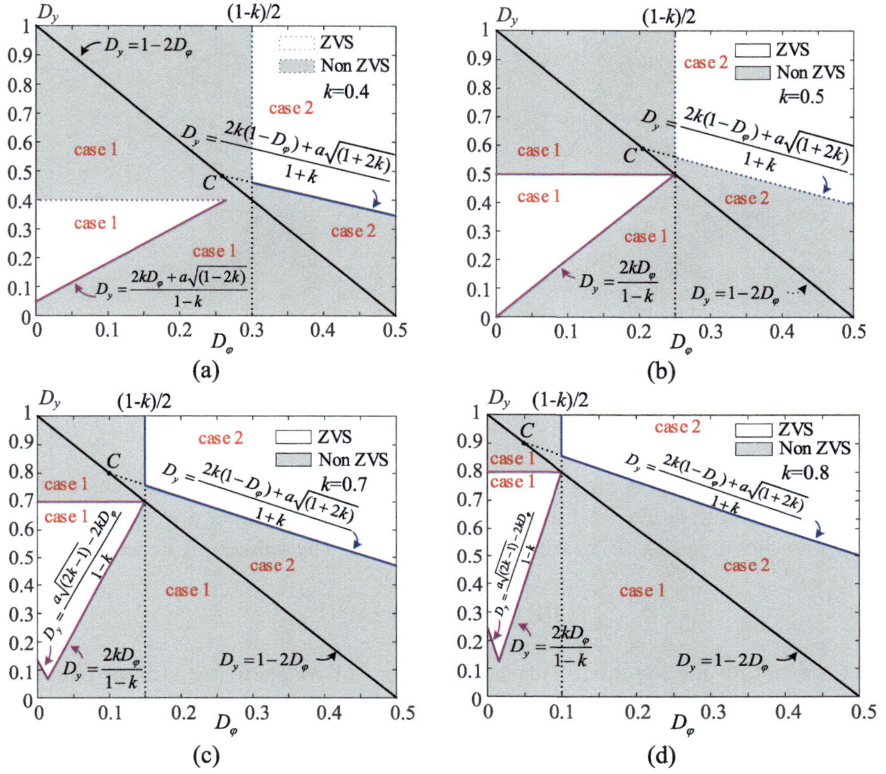

Fig. 9.5 ZVS regions when the switch junction capacitors are considered. **a** $k = 0.4$. **b** $k = 0.5$. **c** $k = 0.7$. **d** $k = 0.8$

$(8L_k f_s)$. As seen in Fig. 9.6a, when switch junction capacitors are ignored, the two ZVS regions only interacted at one particular point P. While the two ZVS regions are separated when the junction capacitors are considered, as demonstrated in Fig. 9.6b, which means that it is impossible to achieve full load range ZVS no matter how to control. To avoid this, the operation of DAB can only work at the lower ZVS region, however that will sacrifice the capability of the maximum power transfer.

9.2.4 *Magnetizing Current Injection to Achieve Entire Load Range ZVS*

In order to solve the problem of disconnected ZVS regions, the magnetizing inductance is utilized to extend ZVS range, which is usually ignored in previous studies. The waveforms with magnetizing current are shown in Fig. 9.7. As seen, the magnetizing inductance has an effect on the waveforms of i_s, whereas the waveforms of i_k

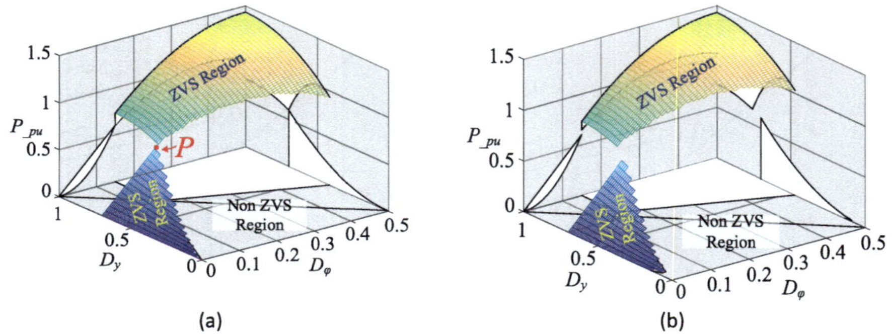

Fig. 9.6 Power transfer curves under different junction capacitors when $L_m = \infty$. **a** Ignoring switch junction capacitors. **b** With 65 pF junction capacitors

remain the same as shown in Fig. 9.2. The intervals utilizing the magnetizing current, the one from t_1 to t_2 in case 1 as shown in Fig. 9.7a and another one from t_3 to t_4 in case 2 as shown in Fig. 9.7b, are reanalyzed below. The value of i_s is determined by

$$i_s(t) = n[i_k(t) - i_m(t)] \tag{9.12}$$

Considering the symmetry of the waveform of magnetizing current over one switching cycle, the value of i_m at t_1 based on Fig. 9.7a can be denoted as

$$i_m(t_1) = -nV_2T_s/(4L_m) \tag{9.13}$$

Meanwhile, the expression of i_m at t_3 based on Fig. 9.7b is obtained in the same way, and it is the same as (9.13).

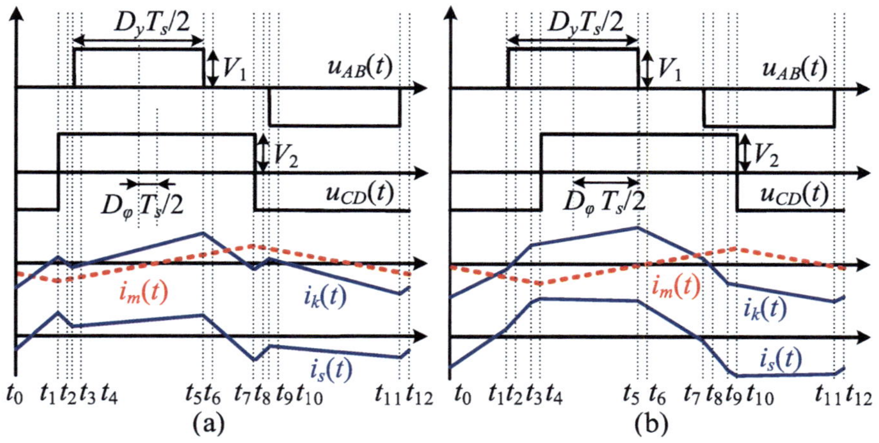

Fig. 9.7 Current waveforms considering L_m. **a** Case 1. **b** Case 2

According to Fig. 9.3a, the ZVS condition for switches of V_2 side in case 1 can be drawn as

$$i_s(t_1) = n[i_k(t_1) - i_m(t_1)] \geq 0 \tag{9.14}$$

Based on (9.13), (9.14) and Table 9.1, the constraint of ZVS can be deduced as

$$D_y \leq k(1 + L_k/L_m) \tag{9.15}$$

As illustrated in Fig. 9.4b, there is a voltage source V_1/n to provide energy. ZVS for switches Q_1-Q_4 in case 2 only needs to satisfy that

$$i_s(t_3) = n[i_k(t_3) - i_m(t_3)] \geq 0 \tag{9.16}$$

Based on (9.13), (9.16) and Table 9.2, the restriction of ZVS can be expressed as

$$D_\varphi \geq [1 - k(1 + L_k/L_m)]/2 \tag{9.17}$$

In terms of (9.15) (9.17) and Table 9.3, the regions when ZVS regions are connected utilizing the magnetizing inductance are illustrated in Fig. 9.8. After calculation, the coordinates of intersection point C and point D are

$$C: \begin{cases} D_{\varphi\_C} = (1 - k - a\sqrt{1+2k})/2 \\ D_{y\_C} = k + a\sqrt{1+2k} \end{cases} \tag{9.18}$$

$$D: \begin{cases} D_{\varphi\_D} = [1 - k(1 + L_k/L_m)]/2 \\ D_{y\_D} = k(1 + L_k/L_m) \end{cases} \tag{9.19}$$

The position of point D can be changed by changing the value of magnetizing inductance L_m. As seen from Fig. 9.8, to connect the ZVS regions, point D shall be at the upper left of point C. Therefore, the following relationship for L_m to connect

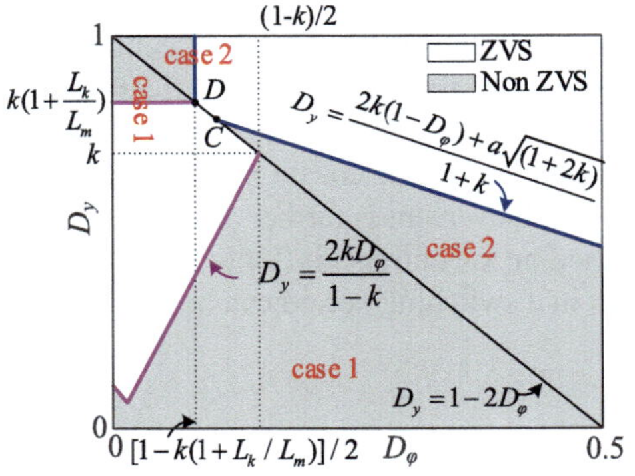

Fig. 9.8 ZVS regions when the magnetizing current is considered

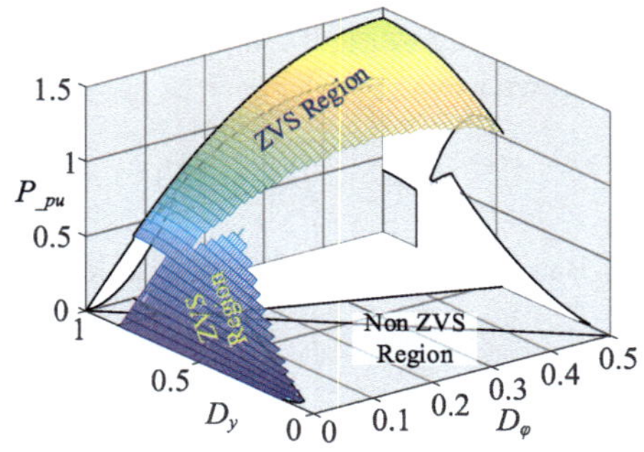

Fig. 9.9 Power transfer curve with magnetizing inductance

two ZVS regions are established.

$$\left(\frac{1-k(1+L_k/L_m)}{2} \le \frac{1-k-a\sqrt{1+2k}}{2}\right) \cap \left(k+a\sqrt{1+2k} \le k\left(1+\frac{L_k}{L_m}\right)\right) \tag{9.20}$$

From (9.20), the range of the magnetizing inductance can be calculated as

$$L_m \le kL_k/(a\sqrt{1+2k}) \tag{9.21}$$

The corresponding curve for the normalized power $P_{\_pu}$ is illustrated in Fig. 9.9. As shown, compared with Fig. 9.6, the two ZVS regions are intersected with a large region, instead of disconnected or just intersected at a particular point. This suggests the possibility to manipulate variables D_y and D_φ in order to operate the DAB dc-dc converter under soft switching for the whole operation range.

9.2.5 ZVS Seamless Power Control Law to Achieve ZVS with Low Rms Current and Maximum Power Transfer Capability

As analyzed above, reasonable selection of D_y and D_φ could enable ZVS of all switches in the full load range at different voltage ratios k. Since different combinations of D_y and D_φ will lead to different RMS currents I_{rms} of leakage inductance in a given power, an optimization of D_y and D_φ to achieve low I_{rms} at the precondition of ZVS achievement is further discussed.

Based on the currents in Table 9.1, the normalized value of leakage RMS current I_{rms_pu} in a switching period can be calculated as

$$
\begin{aligned}
I_{rms\_pu} &= \sqrt{2\int_{t_0}^{t_6} i_k^2(t)dt/T_s/I_{base}} \\
&= 2\sqrt{3(kD_y^3 - 2D_y^3 + 3D_y^2 + 12kD_yD_\varphi^2 - 3kD_y + k^2)/3}
\end{aligned}
\tag{9.22}
$$

where $I_{base} = nV_2/(8L_kf_s)$.

For a given power, taking the derivation of I_{rms_pu} versus D_y leads to

$$
\frac{\partial I_{rms\_pu}}{\partial D_y} = \frac{-(9k - 18D_y - 9kD_y^2 + 18D_y^2 + \frac{9k^3P_{\_pu}^2}{4D_y^2})}{3\sqrt{3(kD_y^3 - 2D_y^3 + 3D_y^2 + 12kD_y(\frac{kP_{\_pu}}{4D_y})^2 - 3kD_y + k^2)}}
\tag{9.23}
$$

By setting (9.23) equal to zero and replacing $P_{\_pu}$ with $4D_yD_\varphi/k$, an equation can be got as follow

$$
k - 2D_y - kD_y^2 + 2D_y^2 + 4kD_\varphi^2 = 0
\tag{9.24}
$$

Solving (9.24) leads to the solutions of D_y

$$
D_{y\_I_{min}1a} = (1 - \sqrt{(1-k)^2 - 4k(2-k)D_\varphi^2})/(2-k)
\tag{9.25}
$$

$$
D_{y\_I_{min}1b} = (1 + \sqrt{(1-k)^2 - 4k(2-k)D_\varphi^2})/(2-k)
\tag{9.26}
$$

When D_y is larger than D_{y_Imin1a}, $\partial I_{rms\_pu}/\partial D_y$ is larger than zero, and when D_y is less than D_{y_Imin1a}, $\partial I_{rms\_pu}/\partial D_y$ is less than zero, so (9.25) is the solution to obtain minimal RMS current. In a similar way, the minimum solution of I_{rms_pu} for case 2 can be obtained as

$$
D_{y\_I_{min}2} = \begin{cases} /k & \text{if } ((1-k)/2 \le D_\varphi \le (k - 1 + \sqrt{1-k^2})/(2k)) \\ 1, & \text{if } ((k - 1 + \sqrt{1-k^2})/(2k) \le D_\varphi \le 1/2) \end{cases}
\tag{9.27}
$$

As seen in (9.25) and (9.27), the expressions for D_{y_Imin} are nonlinear functions that lead to complex calculation. In addition, as seen from the ZVS region shown in Fig. 9.10, the curve for minimum RMS current which is from point A to B, then goes to E and ends with F (A-B-E–F) will go across the non ZVS region. To simplify the complex control, starting points and ending points of the curve can be taken to linearize the control law in (9.25) and (9.27). In addition, a new intersection point C, given as (9.18), can be chosen as the middle point to satisfy ZVS condition. As a result, three segments as the dotted lines (A-C-E–F) can be used. To further simplify the control under the precondition of realizing maximum power transfer capability,

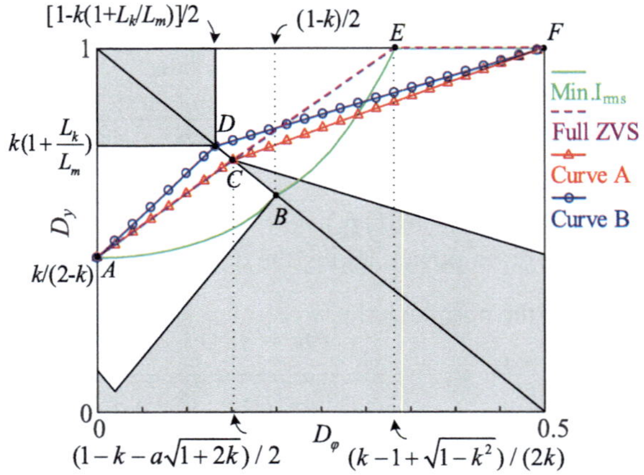

Fig. 9.10 Power control laws for different goals

the two segments in the upper ZVS region (*C-E* and *E–F*) can be linearized into one based on the starting point *C* and ending point *F*. Finally, curve A shown in Fig. 9.10 (the solid line with triangles, *A-C-F*) is obtained, and the corresponding expression is

$$D_{y\_A}=\begin{cases}\dfrac{-2(k+\frac{k}{k-2}+a\sqrt{2k+1})}{k+a\sqrt{2k+1}-1}D_{\varphi}+\dfrac{k}{2-k}, & \text{case1}\\[2ex] \dfrac{-2(k+a\sqrt{2k+1}-1)}{k+a\sqrt{2k+1}}D_{\varphi}+\dfrac{2k+2a\sqrt{2k+1}-1}{k+a\sqrt{2k+1}}, & \text{case2}\end{cases} \tag{9.28}$$

The coefficients of (9.28) have square-root functions, it is still complicated if digital controller such as DSP is used in engineering implementation. The middle point *C* can be modified to point *D* to make the calculation simple. As results, curve B shown in Fig. 9.10 (the solid line with circles, *A-D-F*) could also be a control candidate according to the actual situation. The expression of the solid line with circles is

$$D_{y\_B}=\begin{cases}\dfrac{2[L_k k+L_m k+(k-2)L_m/k]}{L_m-L_k k-L_m k}D_{\varphi}+\dfrac{k}{2-k}, & \text{case1}\\[2ex] \dfrac{2(L_m-L_k k-L_m k)}{(L_k+L_m)k}D_{\varphi}+\dfrac{2L_k k-L_m+2L_m k}{(L_k+L_m)k}, & \text{case2}\end{cases} \tag{9.29}$$

Based on (9.10), (9.11), (9.28) and (9.29), the two control laws are plotted on the power transfer surface, as shown in Fig. 9.11. Control law A is the curve for control in (9.28), and also, control law B represents (9.29). As seen, the two power transfer curves are monotonously increasing with the increasing of D_{φ} and D_y. No step change occurs in the curves, and the curves change seamlessly. Compared with (9.28), the calculation of D_y in (9.29) is simple and can be easily implemented by

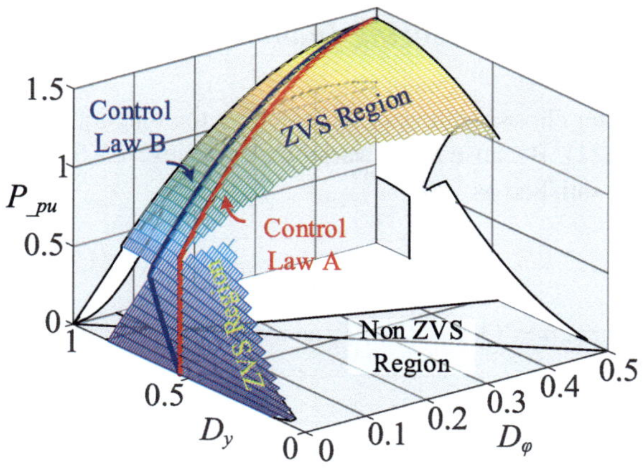

Fig. 9.11 The simplified control laws on the surface of the power transfer

digital controller. Therefore, (9.29) is selected to be the control law to achieve simple calculation, and full load range ZVS in this chapter.

9.3 Design Consideration

9.3.1 *Turns Ratio N*

The change of k will change the position of point C in Fig. 9.10. To ensure that point C is below line $D_y = 1$, k should be designed in a suitable range. Then, the relationship between the turns ratio and voltage gain can be expressed as

$$n = k_{\max}(V_1/V_2)_{\min} \tag{9.30}$$

9.3.2 *Leakage Inductance L_k*

Since the leakage inductance is the key component for power transmission in DAB converter, the choice of L_k can be determined by the maximum power. It can be derived from (9.11) that the maximum power is P_{base}/k. Besides, the minimum value of P_{base}/k should be larger than rated power P_r value, which is expressed as

$$P_{base}/k_{\max} = (nV_2)^2/(8k_{\max}L_k f_s) \geq P_r \tag{9.31}$$

Then the range of L_k can be deduced from (9.31).

9.3.3 Magnetizing Inductance L_m

After choosing the suitable devices, the magnetizing inductance can be designed by (9.21). Because k is a variable, to achieve ZVS in whole voltage range, L_m should be satisfied as

$$L_m \leq k_{\min} L_k / (a\sqrt{1 + 2k_{\min}}) \tag{9.32}$$

where $a = (32L_kC_s)^{1/2}/T_s$ as mentioned before.

9.3.4 Implementation of the Control Strategy

To achieve ZVS in full load range containing forward and reverse transmission, an absolute value of D_φ can be taken to the proposed control law. Based on (9.29), the control law can be expressed as

$$D_{y\_B1} = \frac{2[L_k k + L_m k + (k-2)L_m/k]}{L_m - L_k k - L_m k}|D_\varphi| + \frac{k}{2-k}, \; |D_\varphi| \leq |D_{\varphi\_C}| \tag{9.33}$$

$$D_{y\_B2} = \frac{2(L_m - L_k k - L_m k)}{(L_k + L_m)k}|D_\varphi| + \frac{2L_k k - L_m + 2L_m k}{(L_k + L_m)k}, \; |D_\varphi| > |D_{\varphi\_C}| \tag{9.34}$$

Figure 9.12 is the control block diagram to show how the unified control is implemented. Voltage V_2 is compared with the reference voltage V_{ref} to obtain the error which generates D_φ by a PI compensator. Since $D_\varphi = 0.5$ and $D_\varphi = -0.5$ correspond to the maximum power in forward and reverse power transfer respectively, D_φ is limited to the range [-0.5,0.5]. When $|D_\varphi| \leq |D_{\varphi\_C}|$, D_y is obtained based on (9.33). In contrast, D_y is obtained by (9.34) when $|D_\varphi| > |D_{\varphi\_C}|$. And D_y is limited to the range [0,1]. In addition, V_1 and V_2 are used to calculate the voltage conversion ratio k.

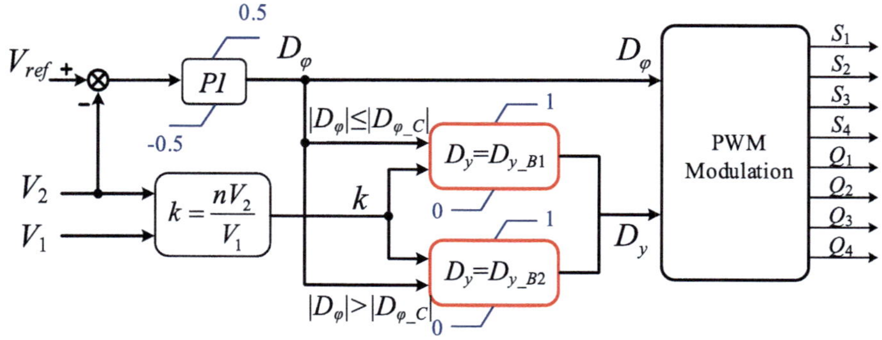

Fig. 9.12 Control block diagram with the proposed control law

9.4 Comparisons

Figure 9.13 shows leakage RMS currents comparisons among minimum I_{rms} control law and the two control laws after tradeoff between ZVS region and complexity. As seen, for different power transfer, the operating RMS currents under the three control methods are very close.

Figure 9.14 shows comparisons of RMS currents with different EPS control methods. The results show that RMS currents with the control laws proposed in [4] and [7] are significantly higher than the RMS current with minimum I_{rms} control law. In addition, the differences of RMS values among control law A, control B and minimum I_{rms} control law are very small, and the differences become smaller as the $P_{\_pu}$ increases, which implies that control law A and control law B can still obtain very low RMS current. Besides, the curves between control law A and control law B in Fig. 9.14 are almost coincident, which shows that the RMS currents are almost the same under the two control laws. Hence, control law B can also achieve small leakage RMS current.

A further comparison is made between proposed control law and minimum I_{rms} control law. The specifications used in comparison are: 1) $V_1 = 200$–400 V, $V_2 = 48$ V; 2) turns ratio n: 20/6; 3) rated power $P_r = 500$ W; 4) switching frequency $f_s = 250$ kHz; 5) GaN device GS66508T is chosen as the power switches of V_1 side and its junction capacitor is 65 pF, and GS61008T are the power switches of V_2 side with a 250 pF junction capacitor; 6) voltage conversion ratio $k = 0.4$–0.8. L_k is designed to be 32 μH by (31). Also, L_m is designed to be 142 μH to satisfy (32). Take case when $V_1 = 400$ V as an example, the comparisons of the calculated losses are shown in Fig. 9.15, including switches losses, inductor losses, transformer losses and total losses. As seen, since switches conduction loss, inductor and transformer copper losses are closely relative to RMS current, their loss curves versus power are similar, and the losses with proposed control law are a bit larger than that with minimum

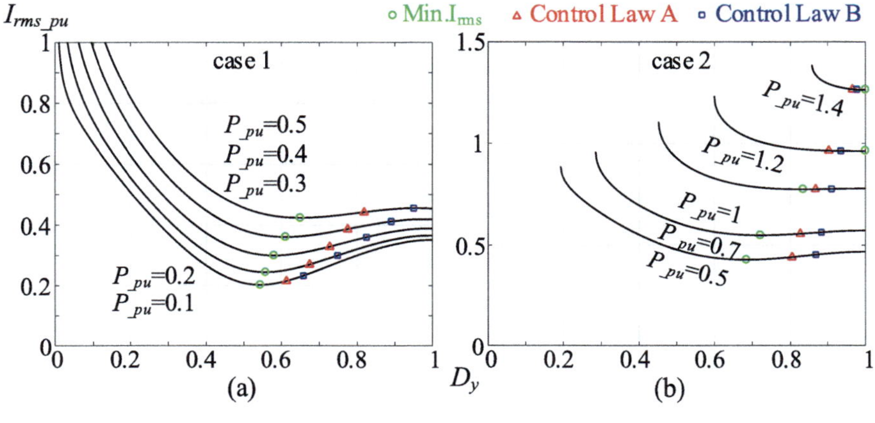

Fig. 9.13 Curves of leakage RMS current versus D_y for three different control laws in given power ($k = 0.6$). **a** Case 1. **b** Case 2

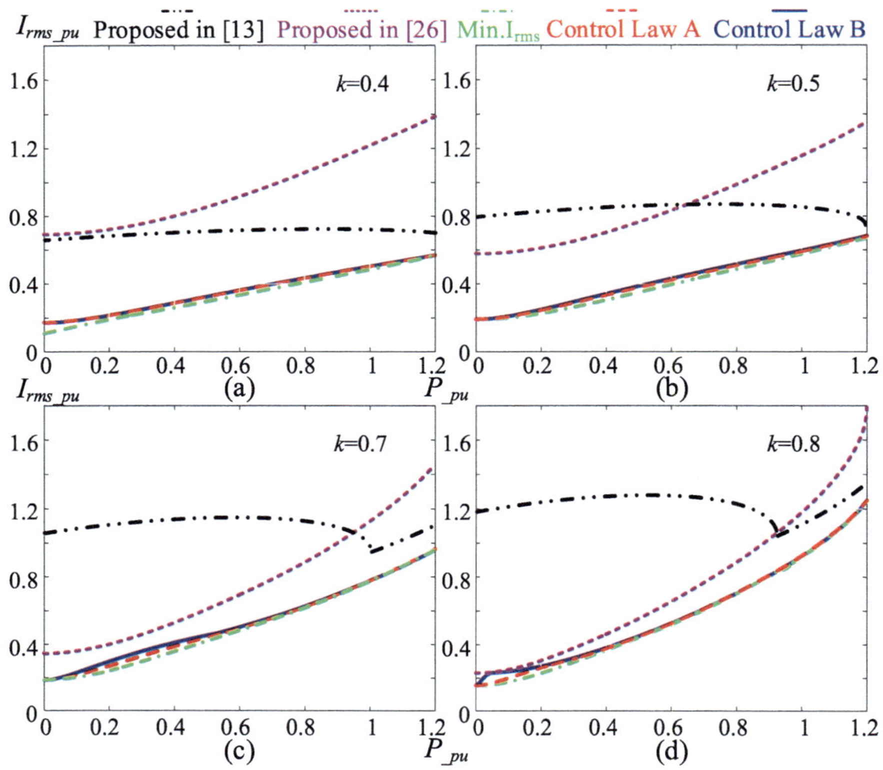

Fig. 9.14 Comparisons of RMS current with different EPS control methods. **a** $k = 0.4$. **b** $k = 0.5$. **c** $k = 0.7$. **d** $k = 0.8$

I_{rms} control law. Inductor iron loss with proposed control law is larger than that with minimum I_{rms} control law, too. For switches turn-off losses, due to the injection of magnetizing current, the turn-off currents in secondary side are increased at light load, $P_{sw.off.PC}$ is higher than $P_{sw.off.MIC}$. As the input power increases, input current becomes larger, and the effect of the magnetizing current decreases. When input power increases to 231 W, hard switching of S_3&S_4 causes switch turn-on loss. The transformer iron loss is not changed because the voltage-second during half switch period at the V_2 side of the transformer is constant. The turn-on process of power device has three possible modes, including full ZVS, partial ZVS and non ZVS. It is worth noting that when analyzing switches turn-on loss, the partial ZVS mode is not taken into account to simplify the calculation, which causes a sharp change as shown in Fig. 9.15c. As seen in Fig. 9.15c, the total loss of minimum I_{rms} control law is less than that of proposed control law in light load. With the addition of switch turn-on loss, the total loss of minimum I_{rms} control law becomes larger in heavy load. Therefore, due to the complex formula of minimum I_{rms} control law of (9.25) and (9.27) and the related issues caused by hard switching, the proposed control law is more preferred.

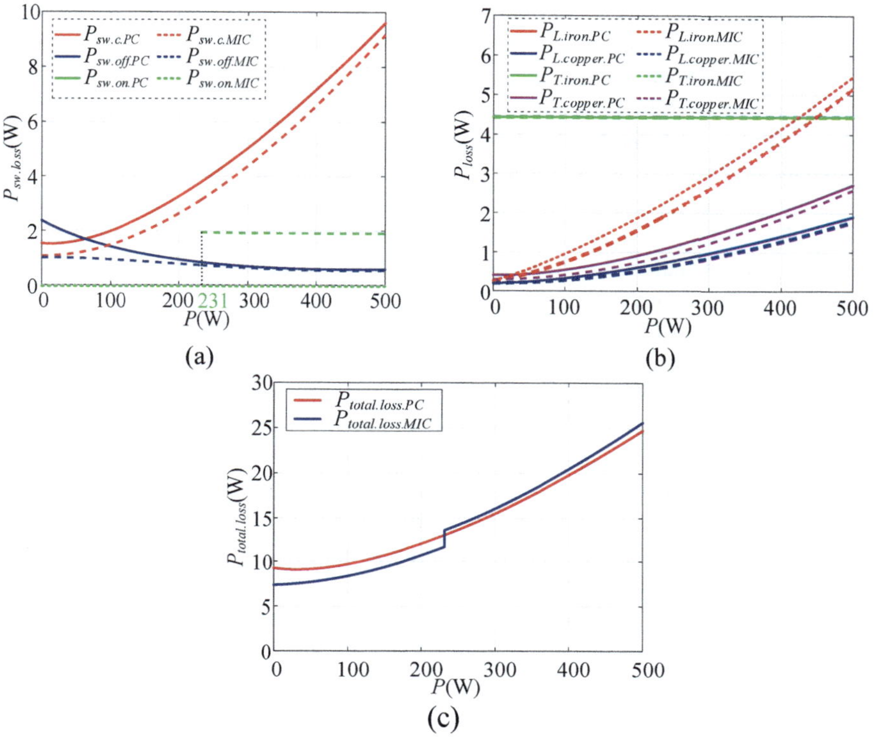

Fig. 9.15 Comparisons of losses. **a** Switches losses. **b** Inductor and transformer losses. **c** Total losses

9.5 Experimental Results

To validate the feasibility of the proposed solutions, a GaN device-based laboratory prototype with the same specifications used in theoretical analysis was built, as shown in Fig. 9.16.

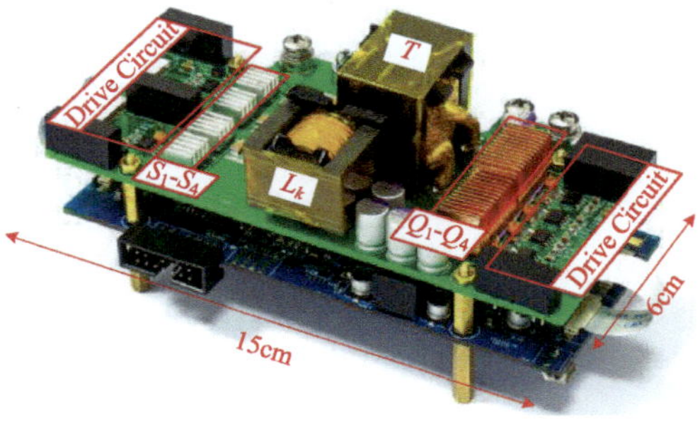

Fig. 9.16 The GaN device based experimental prototype

The steady-state waveforms under full load (500 W), half load (250 W) and light load (50 W) for different V_1 voltages including 200, 300, and 400 V are shown in Figs. 9.17, 9.18 and 9.19, respectively. As seen from these figures, under a same voltage conversion ratio k, the duty cycle of u_{AB} increases as the load increases, which agrees with the analysis. Furthermore, transformer secondary current i_s is different from leakage current i_k after considering with a magnetizing inductance.

Meanwhile, to directly judge whether ZVS is achieved or not, the current values at switch turn OFF instants are measured, and the corresponding calculated ZVS current limits as analyzed in Section III.B are listed in Table 9.4. In this table, I_{zvs_s} represents the current values of i_s at the rising edge of u_{CD}. If the condition for I_{zvs_s} is satisfied, ZVS for Q_1 to Q_4 is achieved. I_{zvs_1} and I_{zvs_2} stand for the current values of i_k at rising edge and falling edge of u_{AB} respectively. They respectively determine the achievement of ZVS for S_3&S_4 and S_1&S_2. By the way, the two operations for case 1 and case 2 shown in Fig. 9.2 can be identified by the waveforms of i_k.

Through making point-to-point comparisons of the denoted current values in the experiment waveforms and the values shown in Table 9.4, the conclusion can be made that the current values denoted from Figs. 9.17, 9.18, and 9.19 under different input voltages and loads all meet the conditions in Table 9.4, which demonstrates the realization of ZVS for all power switches.

One case when $V_1 = 300$ V is taken to show the specific soft switching waveforms under different loads, as displayed from Figs. 9.20, 9.21, and 9.22. u_{gs} is the driver signal, and u_{ds} is the drain-source voltage. Figure 9.20 shows driver signal and drain-source voltage of switch S_1. S_1 is turned on at peak current, and the current is significantly less than zero, so ZVS is achieved from light load to full load. Figure 9.21 illustrates ZVS occurrence of switch S_3. In all the loads, the current at the time of S_3

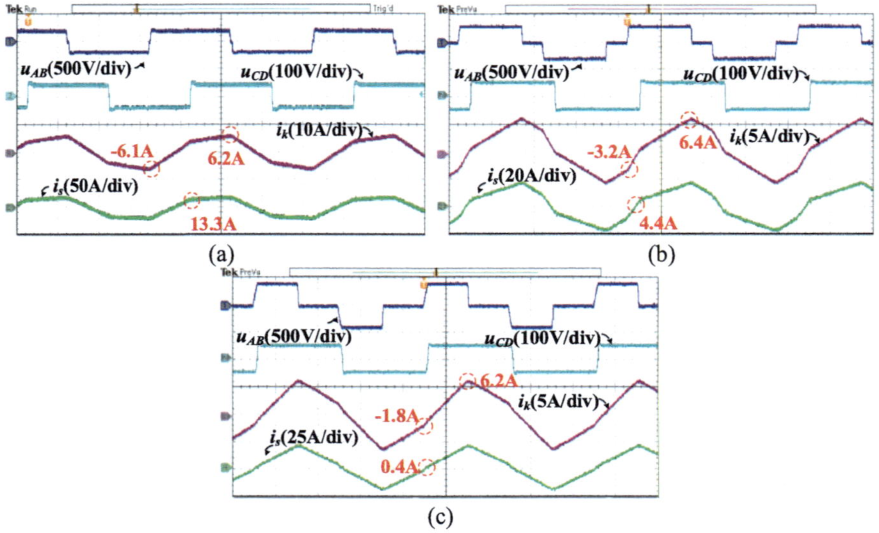

Fig. 9.17 Experimental waveforms under full load. **a** 200 V. **b** 300 V. **c** 400 V

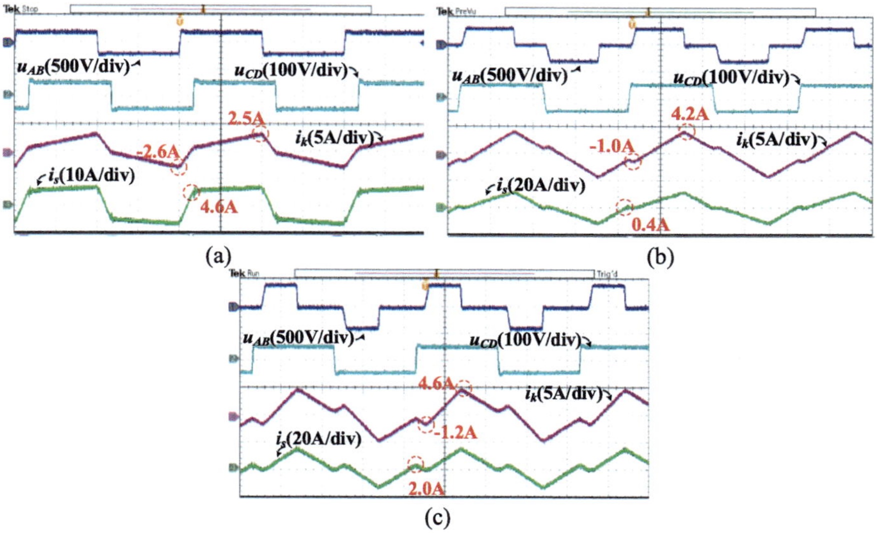

Fig. 9.18 Experimental waveforms under half load. **a** 200 V. **b** 300 V. **c** 400 V

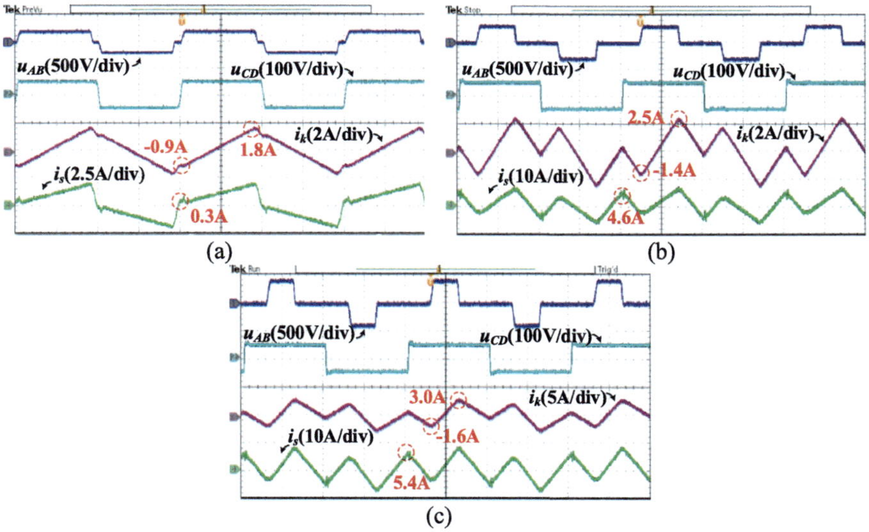

Fig. 9.19 Experimental waveforms under light load. **a** 200 V. **b** 300 V. **c** 400 V

turn-on is large, which demonstrates the ZVS achievement of S_3. Figure 9.22 shows driver signal and drain-source voltage of switch Q_2. As seen, before u_{gs} changes from zero to positive, u_{ds} already goes to zero, so ZVS of Q_2 can be obtained.

With the intention to show the dynamic performance of the experiment prototype, the dynamic waveforms, when output power is switched from 200 to 400 W and

Table 9.4 Calculated ZVS Current Limits for ZVS Achievement

V_1	200 V		300 V		400 V	
Case	1	2	1	2	1	2
I_{ZVS_S} (A)	≥ 0	≥ 0	≥ 0	≥ 0	≥ 0	≥ 0
I_{ZVS_1} (A)	≤ 0	$\leq$ -0.65	≤ 0	$\leq$ -0.87	$\leq$ -0.36	$\leq$ -1.08
I_{ZVS_2} (A)	$\geq$ 0.31	$\geq$ 0.31	$\geq$ 0.16	$\geq$ 0.16	≥ 0	≥ 0

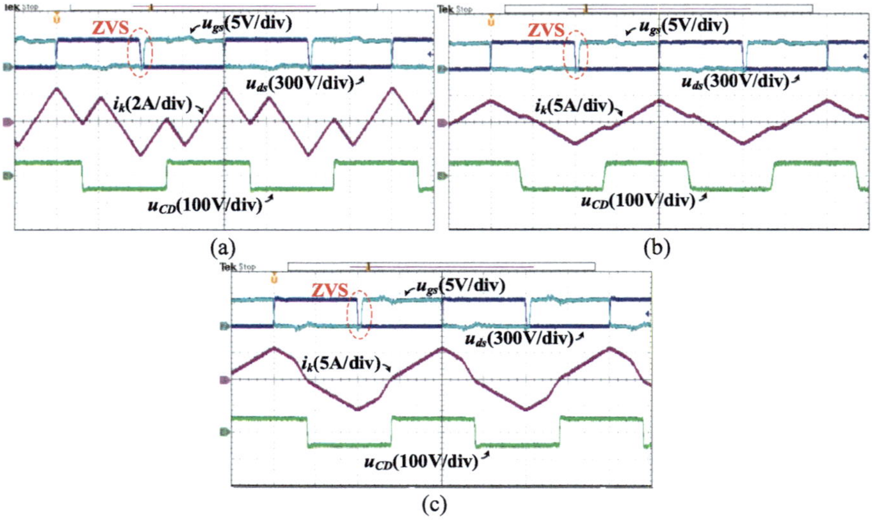

Fig. 9.20 Soft-switching waveforms for S_1&S_2. **a** light load. **b** half load. **c** full load

then switched back to 200 W, are shown in Fig. 9.22. As seen, the output voltage V_2 can keep constant when the load changes, and the leakage current is smoothly changed. Figure 9.23 indicates that the control is a seamless closed loop control. The experimental efficiency curves of the converter with the proposed control law for different V_1 side voltages are illustrated in Fig. 9.24. As shown, at light load, high efficiency will be obtained if the input voltage is low, mainly because of reduced reactive current. At heavy load, it has higher efficiency when the input voltage is 300 V. In addition, the maximum efficiency point is larger than 96% with 250 kHz operating frequency.

In addition, a series of comparisons among the proposed solution and recently reported solutions in the literature are made, and the results are listed in Table 9.5. Compared with the existing solutions, the proposed solution operates at 250 kHz switching frequency and only needs four operation modes, which is feature of high switching frequency and a medium control complexity without auxiliary components. Besides, the highest efficiency is 96.3% and the power density is 29 W/inch^3.

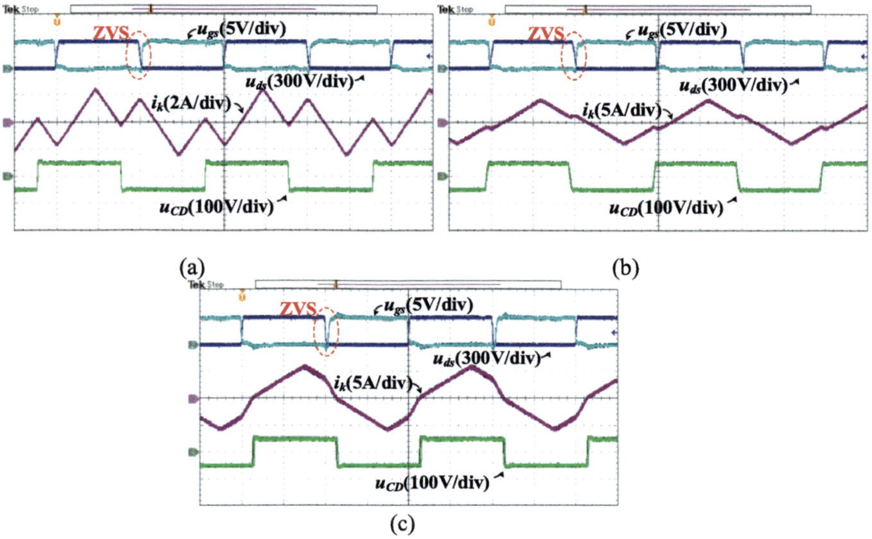

Fig. 9.21 Soft-switching waveforms for S_3&S_4. **a** light load. **b** half load. **c** full load

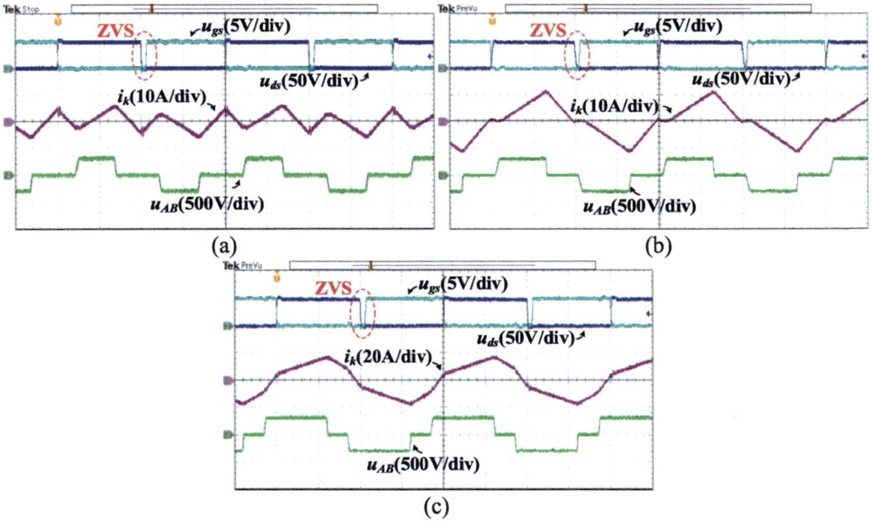

Fig. 9.22 Soft-switching waveforms for Q_1&Q_2. **a** light load. **b** half load. **c** full load

9.6 Conclusion and Future Work

In this chapter, an optimized EPS control is proposed for DAB dc-dc converters. The ZVS regions based on the ZVS working modes considering switch junction capacitors are analyzed, and it shows that the ZVS region is separated into two parts

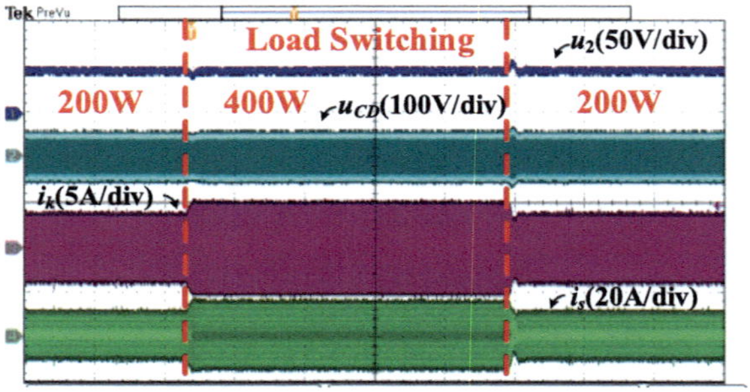

Fig. 9.23 Load switching waveforms

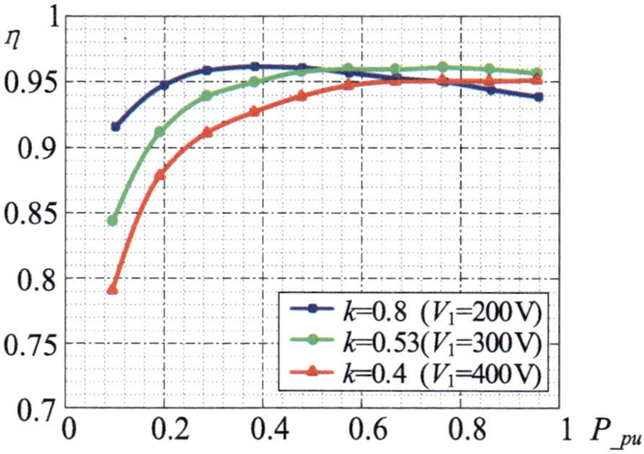

Fig. 9.24 Measured efficiency curves with different input voltages

with conventional control methods. To achieve ZVS for all power switches in full load range and obtain seamless transition, the magnetizing current is utilized, and a power control law is proposed. This optimized control law also has low RMS current and maximum power transmission capability. Finally, experimental results with a 250 kHz, 500 W laboratory prototype demonstrate that the proposed solutions are effective.

In addition to ZVS achievement, high power density is also an important demand in some engineering applications. GaN based devices are feature of low on-resistance, low junction capacitance and very small turn-off energy. They are not only beneficial to achieve ZVS and improve efficiency for DAB converter, but also helpful for power density improvement. To increase the power density, future modifications can be made to redesign the magnetics using plane transformer, and also to redesign 3D structure to make the prototype more compact. Also, the impact of turns ratio n on the close-loop system performance and the designing of the PI controller are worth of further study in the future.

Table 9.5 Comparisons of the Proposed Method and the Existing Method for DAB

Article	Addition Component	Frequency (kHz)	Power Density (W/inch^3)	Efficiency	Dynamic Behavior		Control Complexity	Modulation
					On/Off-line	Operation modes		
[2]	No	200	39	95.26%	online	2 modes	low	SPS
[3]	Two capacitors and a switch	100	5–15	97.5%	online	2 modes	low	SPS
[4]	No	20	—	91%	online	4 modes	medium	EPS
[5]	No	2.5	—	93%	offline	4 modes	high	TPS
[6]	No	50	4–6	95%	online	9 modes	high	TPS
This chapter	No	250	29	96.3%	online	4 modes	medium	EPS

References

1. Bai, H., & Mi, C. (2008). Eliminate reactive power and increase system efficiency of isolated bidirectional dual-active-bridge DC–DC converters using novel dual-phase-shift control. *IEEE Transactions on Power Electronics, 23*(6), 2905–2914.
2. Zhu, M., Shao, C., Wang, S., Hang, L., He, Y. & Fan, S. (2019). System design of dual active bridge (DAB) converter based on GaN HEMT device. 22nd International Conference on Electrical Machines and Systems (pp. 1–6). IEEE
3. Yaqoob, M., Loo, K., & Lai, Y. M. (2018). Fully soft-switched dual-active-bridge series-resonant converter with switched-impedance-based power control. *IEEE Transactions on Power Electronics, 33*(11), 9267–9281.
4. Shi, H., et al. (2018). Minimum-backflow-power scheme of DAB-based solid-state transformer with extended-phase-shift control. *IEEE Transactions on Industry Applications, 54*(4), 3483–3496.
5. Hebala, O. M., Aboushady, A. A., Ahmed, K. H., & Abdelsalam, I. (2019). Generic closed-loop controller for power regulation in dual active bridge DC–DC converter with current stress minimization. *IEEE Transactions on Industrial Electronics, 66*(6), 4468–4478.
6. Chakraborty, S., & Chattopadhyay, S. (2018). Fully ZVS, Minimum RMS current operation of the dual-active half-bridge converter using closed-loop three-degree-of-freedom control. *IEEE Transactions on Power Electronics, 33*(12), 10188–10199.
7. Tao, H., Kotsopoulos, A., Duarte, J. L., & Hendrix, M. A. M. (2008). Transformer-coupled multiport ZVS bidirectional DC–DC converter with wide input range. *IEEE Transactions on Power Electronics, 23*(2), 771–781.

Chapter 10
An Optimized DPS Control for Dual Active Bridge Converters to Secure Full Load Range ZVS with Low Current Stress

Abstract It is a challenge for dual active bridge (DAB) converters to take good balance between full load range zero-voltage-switching (ZVS) and control complexity with wide voltage range. Dual phase shift (DPS) control can extend ZVS range greatly compared with single phase shift (SPS) control, and it has less control variables with reduced operation modes compared with triple phase shift (TPS) control. However, with conventional DPS control, at least a pair of switches will be operated under hard switching within a large power range, which severely reduces the efficiency especially at high switching frequency. To solve this problem, based on Lagrange multiplier method (LMM), an optimized DPS control utilizing magnetizing current is proposed. In this chapter, the magnetizing inductance is designed to create possibility of full load range ZVS achievement for all power switches. Based on ZVS analysis considering switch junction capacitors and power characteristics analysis, a power control law premised on full load range ZVS achievement is proposed to achieve low current stress. Finally, the effectiveness of the proposed solution is verified by experimental results obtained with a 500 W, 250 kHz GaN-Based prototype.

Keywords Current stress · Dual active bridge (DAB) · Dual-phase shift (DPS) · Magnetizing current · Zero-voltage switching (ZVS)

10.1 ZVS Modeling Considering Junction Capacitors Without Magnetizing Current

10.1.1 Basic Operation Under DPS Control

The topology of DAB dc-dc converter is plotted in Fig. 10.1. The H-bridge of primary side is composed of four MOSFETs S_1-S_4, and its output voltage is u_{AB}. The H-bridge of secondary side is also composed of four MOSFETs Q_1-Q_4, and its output voltage is u_{CD}. The two H-bridges are connected with a leakage inductor L_k and a high

G. Xu and X. Chen, *Topology Deduction and Control for Dual-Active-Bridge Converters*, https://doi.org/10.1007/978-981-97-4856-3_10

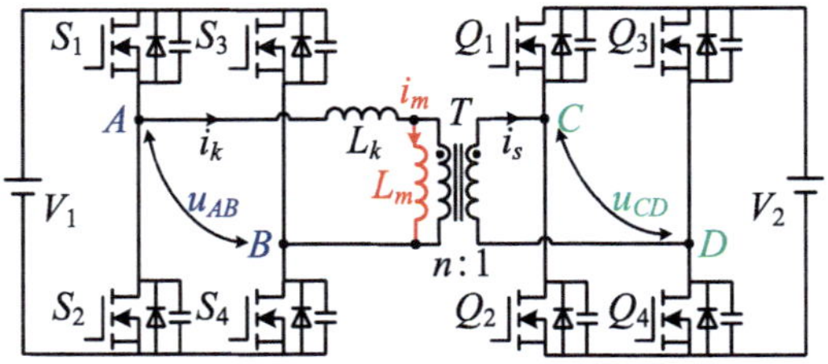

Fig. 10.1 Topology of the DAB dc-dc converter

frequency transformer T. The turns ratio of the transformer is n, and the magnetizing inductance L_m in the previous studies was usually large enough that can be ignored.

Fig. 10.2 shows the possible operation cases in forward working mode for DAB converter under DPS control. D represents the duty cycles of both u_{AB} and u_{CD}, and D_φ is the phase shift angle between u_{AB} and u_{CD}. To simplify the calculation below, ϕ is defined as the corresponding phase shift ratio to D_φ, which is $\phi = D_\varphi/\pi$. T_s is the switching period. Based on the positions of the rising and falling edges of u_{AB} and u_{CD}, these are four possible operation cases can be divided, including case 1 ($0\leq\phi\leq D$&&$0\leq\phi\leq 1$-D), case 2 (1-$D\leq\phi\leq D$), case 3 (1-$D\leq\phi\leq 1$ &&$D\leq\phi\leq 1$) and case 4 ($D\leq\phi\leq 1$-D). Since the topology and control logic are symmetry, the reverse and forward working modes are symmetric, so only forward working mode is studied due to the page limitation.

10.1.2 Basic Mathematical Model

Case 1 ($0\leq\phi\leq D$&& $0\leq\phi\leq 1$-D) is taken as a study case to illustrate the characteristics of DAB converter under DPS control without magnetizing current injection.

Based on Fig. 10.2a, the leakage current i_k is expressed as

$$i_k(t) = \begin{cases} i_k(t_0) + nV_2t/L_k, & t \in [t_0, t_1] \\ i_k(t_1), & t \in [t_1, t_2] \\ i_k(t_2) + V_1t/L_k, & t \in [t_2, t_3] \\ i_k(t_3) + (V_1 - nV_2)t/L_k, & t \in [t_3, t_4] \end{cases} \tag{10.1}$$

Considering that the average leakage current over one switching period should be zero in steady state and the waveform is symmetric during one switching cycle, the leakage currents at the commutating times are expressed as

$$\begin{cases} i_k(t_0) = -2[2\phi + (1/k - 1)D]I_{base} \\ i_k(t_1) = -2(1/k - 1)DI_{base} \\ i_k(t_2) = -2(1/k - 1)DI_{base} \\ i_k(t_3) = 2[2\phi/k - (1/k - 1)D]I_{base} \end{cases} \tag{10.2}$$

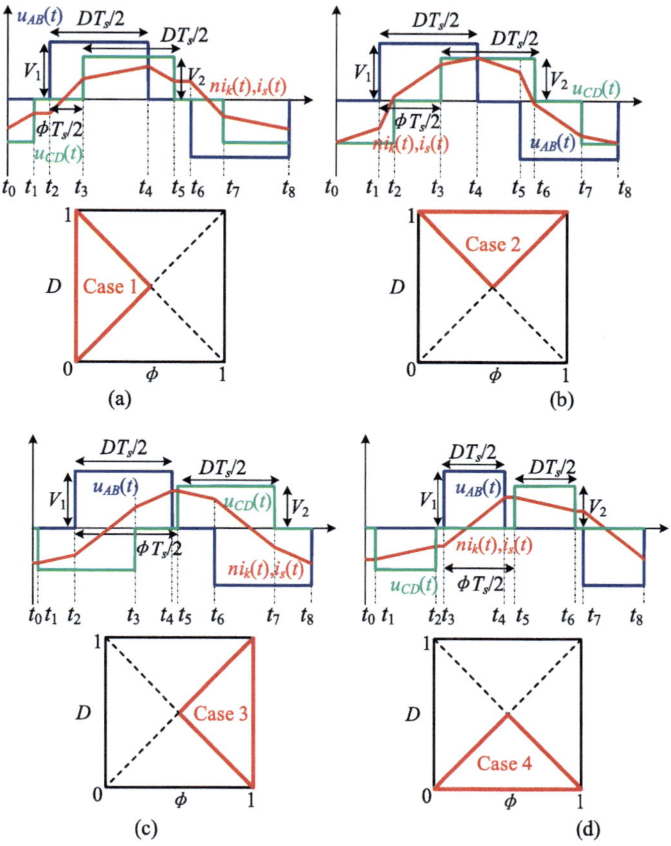

Fig. 10.2 Four operation cases. **a** Case 1: $0 \le \phi \le D$&& $0 \le \phi \le 1\text{-}D$. **b** Case 2: $1\text{-}D \le \phi \le D$. **c** Case 3: $1\text{-}D \le \phi \le 1$&&$D \le \phi \le 1$. **d** Case 4: $D \le \phi \le 1\text{-}D$

where I_{base} is the normalized current, and $I_{base} = nV_2T_s/(8L_k)$. k is the voltage conversion ratio, which is equal to nV_2/V_1.

Since the magnetizing inductance is much larger than the leakage inductance, the effect of magnetizing current can be ignored. Thus, the relationship between leakage current i_k and the current on the secondary side i_s is expressed as

$$i_s(t) = ni_k(t) \tag{10.3}$$

According to Fig. 10.2 and (10.2), the normalized power P_n can be obtained as (10.4), and the base power P_{base} is $P_{base} = (nV_2)^2/(8L_kf_s)$.

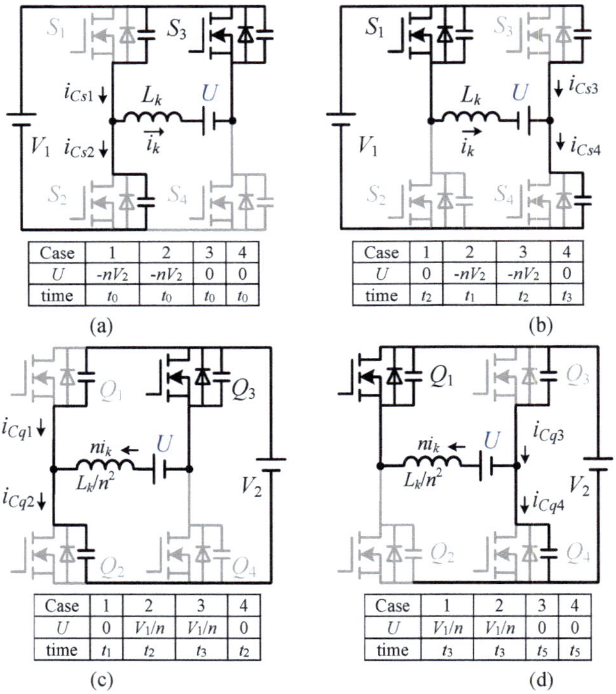

Fig. 10.3 . ZVS models for different cases and different switches. **a** For $S_1\&S_2$. **b** For $S_3\&S_4$. **c** For Q1&Q2. **d** For Q3&Q4

$$P_n = \begin{cases} 2\phi(2D-\phi)/k, & 0 \le \phi \le 1-D \&\& 0 \le \phi \le D \\ [1-2(D-1)^2-(2\phi-1)^2]/k, & 1-D \le \phi \le D \\ 2(1-\phi)(2D+\phi-1)/k, & 1-D \le \phi \le 1 \&\& D \le \phi \le 1 \\ 2D^2/k^2, & D \le \phi \le 1-D \end{cases} \tag{10.4}$$

10.1.3 ZVS Analysis Considering Switch Junction Capacitors

ZVS can be achieved when the currents at the commutating times are large enough. The ZVS conditions are expressed as

$$\begin{cases} S_{1,2}: i_k(t_0) = -2[2\phi + (1/k-1)D]I_{base} \le -I_{\min 1} \\ S_{3,4}: i_k(t_2) = -2(1/k-1)DI_{base} \le -I_{\min 2} \\ Q_{1,2}: i_s(t_1) = -2(1/k-1)DnI_{base} \ge I_{\min 3} \\ Q_{3,4}: i_s(t_3) = 2[2\phi/k-(1/k-1)D]nI_{base} \ge I_{\min 4} \end{cases} \tag{10.5}$$

where I_{minx} ($x = 1, 2, 3, 4$) are the minimum amplitudes of currents for each a pair of complementary switches to charge/discharge the junction capacitors of the switches.

To calculate the minimum currents, the ZVS models are illustrated in Fig. 10.3. The minimum energy on the leakage inductor when the switch is turned off is equal to the energy consumed on the DC sources V_1 and nV_2 during deadtime (Qin et al. in IEEE Transactions on Power Electronics 33:5949–5966, 2018). For switches $S_1\&S_2$ in case 1, based on Fig. 10.3a, the relationship can be calculated as

$$\begin{aligned}\frac{1}{2}L_k I_{\min 1}^2 &= \int_0^{t_m} [-nV_2(C_s\frac{du_{ds1}}{dt} - C_s\frac{du_{ds2}}{dt}) + V_1 C_s\frac{du_{ds1}}{dt}]dt \\ &= (2k-1)C_s(nV_2/k)^2\end{aligned} \tag{10.6}$$

where C_s is the equivalent output capacitance of switches in the primary side, and t_m is the instants when i_k reaches zero during dead time. Based on (10.5) and (10.6), the ZVS condition of $S_1\&S_2$ in case 1 can be deduced as

$$\begin{cases} D \geq -2k\phi/(1-k), & k \leq 0.5 \\ D \geq (a\sqrt{2k-1} - 2k\phi)/(1-k), & k > 0.5 \end{cases} \tag{10.7}$$

where $a = 4f_s(2L_kC_s)^{0.5}$, and f_s is the switching frequency.

For $S_3\&S_4$, with the equivalent circuit and the parameters in Fig. 10.3b, I_{min2} can be calculated by

$$\frac{1}{2}L_k I_{\min 2}^2 = \int_0^{t_m} -V_1 C_s\frac{du_{ds4}}{dt}dt = C_s(\frac{nV_2}{k})^2 \tag{10.8}$$

In terms of (10.5) and (10.8), ZVS condition for switches $S_3\&S_4$ is deduced as

$$\begin{cases} D \geq a/(1-k), & k \leq 1 \\ D \leq 0, & k > 1 \end{cases} \tag{10.9}$$

Also, for switches $Q_1\&Q_2$ and $Q_3\&Q_4$, the relationships with minimum currents are listed in (10.10) and (10.11) respectively.

$$I_{\min 3} = 0 \tag{10.10}$$

$$I_{\min 4} = 0 \tag{10.11}$$

Solving (10.10) and (10.11) with (10.5) leads to

$$k \geq 1 \tag{10.12}$$

$$D \leq 2\phi/(1-k) \tag{10.13}$$

(10.12) is the ZVS condition for switches $Q_1\&Q_2$, and (10.13) is the ZVS condition for switches $Q_3\&Q_4$.

In the same way, utilizing the equivalent circuits and the parameters provided in Fig. 10.3a–d, the ZVS conditions of all switches in all cases can be obtained, which are summarized in Table 10.1, where $b = 4f_s(2L_kC_q)^{0.5}/n$, and C_q is the equivalent switch junction capacitance in the secondary side.

As seen in Table 10.1, if the DAB converter operates in the case when $k \geq 1$, it is impossible for switches $S_3\&S_4$ to achieve ZVS in case 1 and case 4; otherwise, it is impossible for switches $Q_1\&Q_2$ to achieve ZVS in case 1 and case 4 when $k<1$. However, if the DAB converter works with minimum current stress, it needs to operate in case 1 in a large power range.

10.2 ZVS Modeling and Power Control Law Utilizing Magnetizing Current

As seen in (10.5), to achieve ZVS of switches $S_3\&S_4$, $i_k(t_2)$ should be smaller than $-I_{min2}$, so the current should be negative. In the meantime, $i_s(t_1)$ should be larger than I_{min3} to achieve ZVS of switches $Q_1\&Q_2$, and the current needs to be positive. However, in the conventional method, $i_s(t_1)$ is equal to $ni_k(t_2)$, which means the two pairs of switches cannot be achieved ZVS at the same time, as shown in Fig. 10.4. Based on (9), to ensure ZVS achievement for switches $S_3\&S_4$, the voltage conversion ratio k can be limited into the range from 0 to 1. Meanwhile, to create possibility of ZVS achievement for switches $Q_1\&Q_2$ when $k<1$, the magnetizing current i_m can be utilized to make i_s different from ni_k. It is worth noting that the magnetizing inductance can be changed by adjusting the air gap of the transformer without adding auxiliary component.

10.2.1 ZVS Modeling Considering Magnetizing Inductance

Case 1 is taken as an example to show the operation waveforms after utilizing magnetizing current, as shown in Fig. 10.5. As seen, waveform of i_k remains the same as that in Fig. 10.2, but the shape of i_s is changed. Besides, at the rising edge of u_{CD}, i_s becomes larger to the positive direction, which is beneficial to ZVS achievement of switches $Q_1\&Q_2$. The value of i_s is rewritten as

$$i_s(t) = n[i_k(t) - i_m(t)] \tag{10.14}$$

Table 10.1 ZVS Constraints for All Power Switches

Case	$S_1 \& S_2$	$S_3 \& S_4$	$Q_1 \& Q_2$	$Q_3 \& Q_4$
1	$\begin{cases} D \geq \dfrac{-2k\phi}{1-k}, & k \leq \dfrac{1}{2} \\ D \geq \dfrac{a\sqrt{2k-1} - 2k\phi}{1-k}], & k > \dfrac{1}{2} \end{cases}$	$\begin{cases} D \geq \dfrac{a}{1-k}, & k \leq 1 \\ D \leq 0, & k > 1 \end{cases}$	$k \geq 1$	$\begin{cases} D \leq \dfrac{2\phi}{1-k}, & k < 1 \\ D \geq \dfrac{2\phi}{1-k}, & k > 1 \end{cases}$
2	$\begin{cases} D \geq \dfrac{-2k\phi}{1-k}, & k \leq \dfrac{1}{2} \\ D \geq \dfrac{a\sqrt{2k-1} - 2k\phi}{1-k}], & k > \dfrac{1}{2} \end{cases}$	$D \geq \frac{a\sqrt{1+2k}-2k\phi+2k}{1+k}$	$D \geq \frac{2(1-\phi)}{1+k}$	$\begin{cases} D \leq \dfrac{2\phi}{1-k}, & k < 1 \\ D \geq \dfrac{2\phi}{1-k}, & k > 1 \end{cases}$
3	$D \geq 0$	$D \geq \frac{a\sqrt{1+2k}-2k\phi+2k}{1+k}$	$D \geq \frac{2(1-\phi)}{1+k}$	$D \geq \frac{bk}{1+k}$
4	$D \geq 0$	$\begin{cases} D \geq \dfrac{a}{1-k}, & k \leq 1 \\ D \leq 0, & k > 1 \end{cases}$	$k \geq 1$	$D \geq \frac{bk}{1+k}$

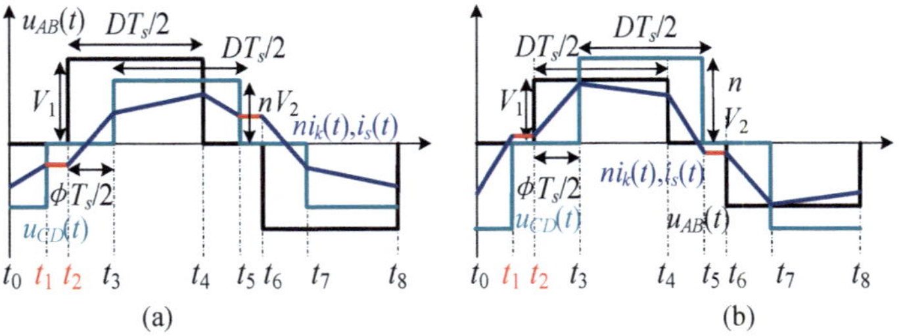

Fig. 10.4 ZVS achievement in case 1. **a** When k<1, is(t1) <0, ZVS cannot be achieved for Q1&Q2. **b** When $k \geq 1$, $i_k(t_2) \geq 0$, ZVS cannot be achieved for S_3&S_4

Fig. 10.5 Operation waveforms utilizing magnetizing current in case 1

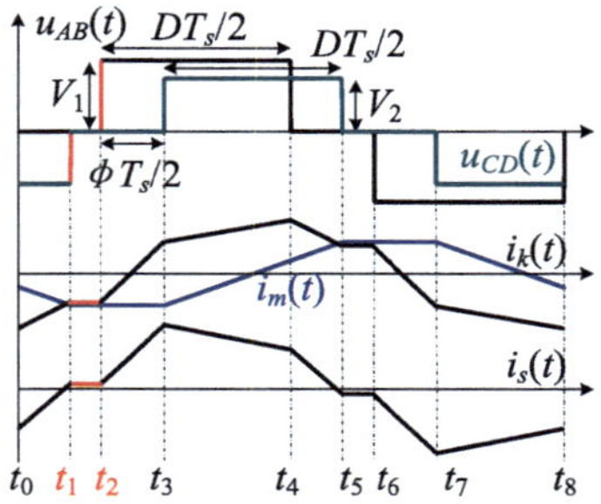

As shown in Fig. 10.5, during the time t_1 to t_3, i_m is constant and is equal to -I_{max}. The following can be obtained as

$$L_m \frac{2I_{\max}}{DT_s/2} = nV_2 \tag{10.15}$$

Rearranging (10.15) yields

$$I_{\max} = nV_2DT_s/(4L_m) \tag{10.16}$$

Based on (10.5) and (10.15), the conditions needing to be satisfied to achieve ZVS are rewritten by

$$\begin{cases} S_{1,2} : i_k(t_0) = -2[2\phi + (1/k - 1)D]I_{base} \leq -I_{\min 1} \\ S_{3,4} : i_k(t_2) = -2(1/k - 1)DI_{base} \leq -I_{\min 2} \\ Q_{1,2} : i_s(t_1) = 2[L_k/L_m - (1/k - 1)]nDI_{base} \geq I_{\min 3} \\ Q_{3,4} : i_s(t_3) = 2[2\phi/k - (1/k - 1 - L_k/L_m)D]nI_{base} \geq I_{\min 4} \end{cases} \tag{10.17}$$

As the magnetizing inductance is clamped by the V_2 side voltage, the ZVS models for switches on primary side is the same as that shown in Fig. 10.3a and b. Thus, the ZVS conditions for switches in primary side are the same with those in

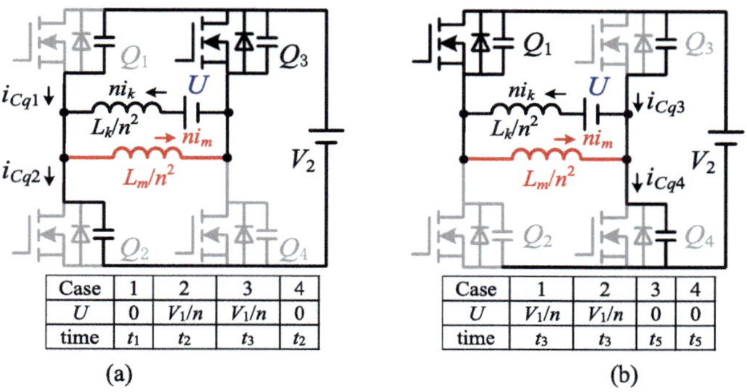

Fig. 10.6 ZVS models utilizing magnetizing current. **a** For switches $Q_1\&Q_2$. **b** For switches $Q_3\&Q_4$

Table 10.1. While, the ZVS models utilizing the magnetizing current for switches on the secondary side are shown in Fig. 10.6. Take case 1 as an example to illustrate the change of ZVS conditions for secondary side switches.

Similarly, the minimum energy on the leakage inductor and the magnetizing inductance when the switch is turned off is equal to the energy consumed on the DC sources V_1 and nV_2 during deadtime. As seen in Fig. 10.6, assuming that the energy loss of circuit is lossless, there is no energy loss in case 1, so the minimum currents are still expressed as (10.10) and (10.11). Based on (10.10) and (10.17), the range of magnetizing inductance to make sure switches $Q_1\&Q_2$ to achieve ZVS is established as

$$L_m \leq k_{\min} L_k/(1 - k_{\min}) \tag{10.18}$$

Also, the ZVS condition for $Q_3\&Q_4$ can be obtained by (10.11) and (10.17).

$$D \geq -2L_m\phi/[kL_k - (1 - k)L_m] \tag{10.19}$$

In the same way, the ZVS conditions in all cases for switches on the secondary side are listed in Table 10.2.

Table 10.2 ZVS Constraints for Switches on Secondary Side

Case	$Q_1\&Q_2$	$Q_3\&Q_4$
1	$L_m \leq kL_k/(1 - k)$	$D \geq -2L_m\phi/[kL_k - (1 - k)L_m]$
2	$D \geq 2(1 - \phi)L_m/[(1 + k)L_m + kL_k]$	$D \geq -2L_m\phi/[kL_k - (1 - k)L_m]$
3	$D \geq 2(1 - \phi)L_m/[(1 + k)L_m + kL_k]$	$D \geq \sqrt{1/[(\frac{1+k}{k\cdot b})^2 + \frac{L_k}{L_m b^2}]}$
4	$L_m \leq kL_k/(1 - k)$	$D \geq \sqrt{1/[(\frac{1+k}{k\cdot b})^2 + \frac{L_k}{L_m b^2}]}$

10.2.2 Power Control Law to Obtain Low Current Stress Based on Lagrange Multiplier Method

High current stress will cause an increase of switching loss and reduction of efficiency. Based on Fig. 10.5, the current stress in secondary side will occur at t_3 or t_4. The current expressions are shown as

$$i_s(t_3) = 2[2\phi/k - (1/k - 1 - L_k/L_m)D]nI_{base} \tag{10.20}$$

$$i_s(t_4) = 2[2(1 + L_k/L_m)\phi + (1/k - 1 - L_k/L_m)D]nI_{base} \tag{10.21}$$

Making comparison between (10.20) and (10.21) leads to

$$i_s(t_3) - i_s(t_4) = 2[2/k - 2(1 + L_k/L_m)](\phi - D) \tag{10.22}$$

Based on (10.18), $[2/k - 2(1+L_k/L_m)]$ is less than zero, and the limitation in case 1 is $\phi \leq D$, so $i_s(t_3)$-$i_s(t_4) \geq 0$. Therefore, current stress i_p in case 1 is equal to $i_s(t_3)$. Likewise, the current stress in other cases can also be obtained and summarized as follows.

$$i_p = \begin{cases} 2[2\phi/k - (1/k - 1 - L_k/L_m)D]nI_{base}, & 0 \leq \phi \leq D \\ 2(1/k + 1 + L_k/L_m)DnI_{base}, & D \leq \phi \leq 1 \end{cases} \tag{10.23}$$

As seen, different combinations of ϕ and D will lead to different current stresses. To optimize the relationship between the two control variables ϕ and D and then get the minimum current stress (MCS) under the given power, Lagrange multiplier method (LMM), one of the popular optimization methods to solve extreme value condition, can be used. By using LMM, the evaluation function is shown as

$$E = i_p + \lambda(P_n - p^*) \tag{10.24}$$

where p^* is the desired and unified transmission power.

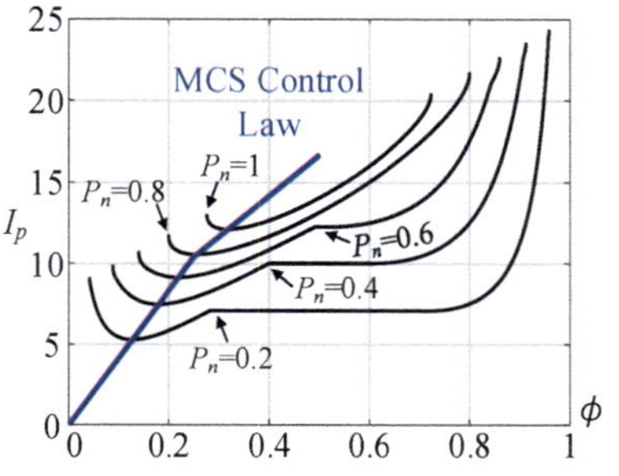

Fig. 10.7 The trajectory of minimum current stress control law

According to (10.4), (10.23) and (10.24), the following constraints can be obtained for the minimum value of i_p.

$$\begin{cases} \partial E/\partial\phi = \partial i_p/\partial\phi + \lambda\partial P_n/\partial\phi = 0 \\ \partial E/\partial D = \partial i_p/\partial D + \lambda\partial P_n/\partial D = 0 \\ \partial E/\partial\lambda = P_n - p^* = 0 \end{cases} \tag{10.25}$$

In terms of the first and the second equations in (10.25), if $(\partial i_p/\partial\phi)/(\partial P_n/\partial\phi) = (\partial i_p/\partial D)/(\partial P_n/\partial D)$, the minimum value of i_p for a desired output power can be obtained. Thus, the control law to get MCS is expressed as (10.25), which is named MCS control law. Besides, the trajectory of the MCS control law is also shown in Fig. 10.7.

$$D_{MCS} = \begin{cases} \dfrac{L_m + kL_m + kL_k}{kL_k - (1-k)L_m}\phi, & 0 \le \phi < \dfrac{kL_k - (1-k)L_m}{2k(L_m + L_k)} \\ \dfrac{kL_k - (1-k)L_m}{L_m}\phi - \dfrac{kL_k - (3-k)L_m}{2L_m}, & \dfrac{kL_k - (1-k)L_m}{2k(L_m + L_k)} \le \phi \le 0.5 \end{cases} \tag{10.26}$$

Based on (10.4), the maximum power occurs at $\phi = 0.5$, so ϕ is limited into [0, 0.5] in forward working mode.

Fig. 10.8 shows the different control laws on ZVS/non-ZVS regions. As seen, the blue long-dotted line is the MCS control law, and it goes through the non ZVS region. To achieve full load range ZVS, the segment from B to C remains the same and the starting point of another segment is proposed to change from origin point (0, 0) to point A, as the red solid line shown. The coordinates of point A and point B are respectively expressed as

$$\mathrm{A}: \begin{cases} \phi_{\mathrm{A}} = 0 \\ D_{\mathrm{A}} = a/(1-k) \end{cases} \tag{10.27}$$

$$\mathrm{B}: \begin{cases} \phi_{\mathrm{B}} = [kL_k - (1-k)L_m]/[2k(L_m + L_k)] \\ D_{\mathrm{B}} = 1 - [kL_k - (1-k)L_m]/[2k(L_m + L_k)] \end{cases} \tag{10.28}$$

Then, the expression of the proposed control law which is from A to B and ends with C is shown as

$$D = \begin{cases} \dfrac{a}{1-k} + [\dfrac{2k(1-k-a)(L_m + L_k)}{(1-k)(kL_k - (1-k)L_m)} - 1]\phi, & 0 \le \phi < \phi_{\mathrm{B}} \\ \dfrac{kL_k - (1-k)L_m}{L_m}\phi - \dfrac{kL_k - (3-k)L_m}{2L_m}, & \phi_{\mathrm{B}} \le \phi \le 0.5 \end{cases} \tag{10.29}$$

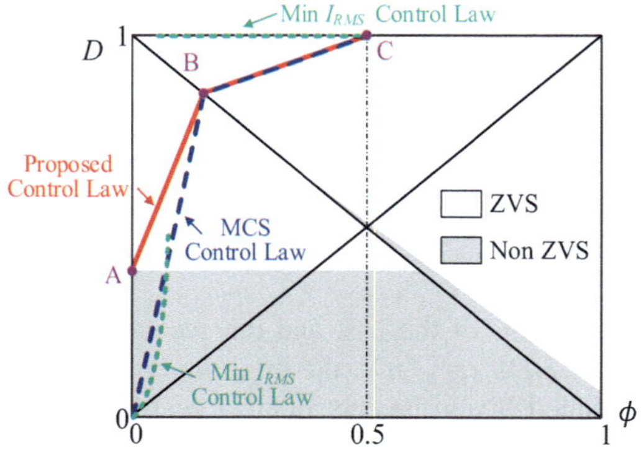

Fig. 10.8 ZVS/non-ZVS regions and power control laws with different optimal goals

By the way, based on LMM, the control law to get minimum RMS current can also be obtained, which is the green short-dotted line (named Min I_{rms} control law) in Fig. 10.8. However, the control law is not unified, and it is also not monotonous with ϕ. So, peak current is set as the optimized target.

Fig. 10.9 is the comparison of current stress between the proposed control law and the MCS control law. As seen, from half load to full load, the current stresses of the two control laws are almost the same under different voltage conversion ratios. Although current stress will increase under the proposed control law within light load range, the difference of i_p between the two control laws becomes smaller as k decreases and finally almost overlaps. What's more, ZVS can be achieved under the proposed control law, which can greatly reduce switching losses.

Fig. 10.10 shows the two control laws on the surface of power curve of D and ϕ with or without magnetizing current. ZVS region and non ZVS region are also indicated on the surface of power curve. The color regions are ZVS regions, and the gray regions are non ZVS regions. Fig. 10.10a is the case where a suitable magnetizing current is utilized. As seen, within some light load ranges, the blue curve corresponding to MCS control law is not in the ZVS region, which means ZVS cannot be achieved under light load. On the contrary, the red curve representing the proposed control law is all in the ZVS region, which means the DAB converter under the proposed control

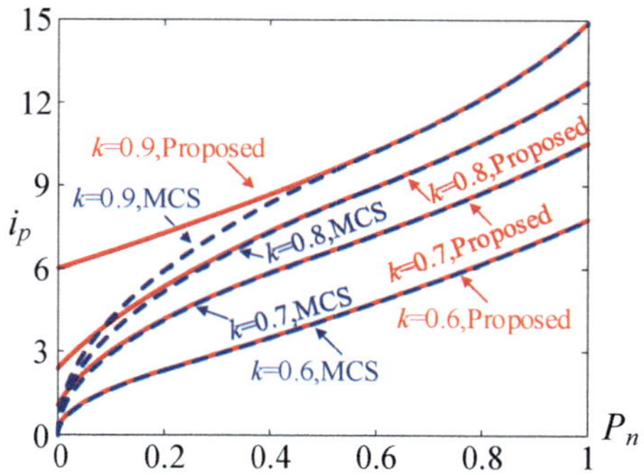

Fig. 10.9 Comparison of current stress between proposed control law and MCS control law under different voltage conversion ratios

law can achieve full load range ZVS. Besides, the power increases monotonically as ϕ increases. As comparison, Fig. 10.10b shows the control laws on the power curve without magnetizing current injection. As seen, under the two control laws, the converter can be operated under ZVS for all switches only at heavy load range, which shows that the magnetizing current is effective to help achieve ZVS.

The control diagram is illustrated in Fig. 10.11. As shown, the instant output voltage v_2 is controlled by a proportional integral (PI) controller. The phase shift ratio ϕ is obtained initially by this controller to ensure that the power is controllable even under special case, and then the duty cycle D is calculated by ϕ with the corresponding expression after judging the turning point (point B in Fig. 10.8). Based on ϕ and D, the driver signals of S_1-S_4 and Q_1-Q_4 are generated. Besides, there is no square root on the expressions, and the expressions are connected without step change. Therefore, it is easy to be implemented even for high frequency applications.

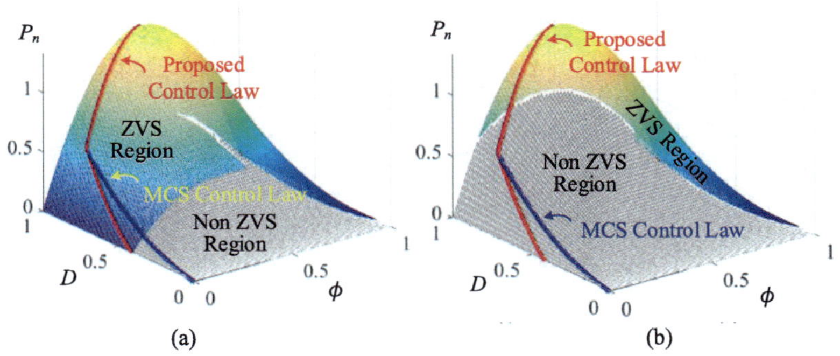

Fig. 10.10 The proposed control law on the surface of power curves. **a** With magnetizing current. **b** Without magnetizing current

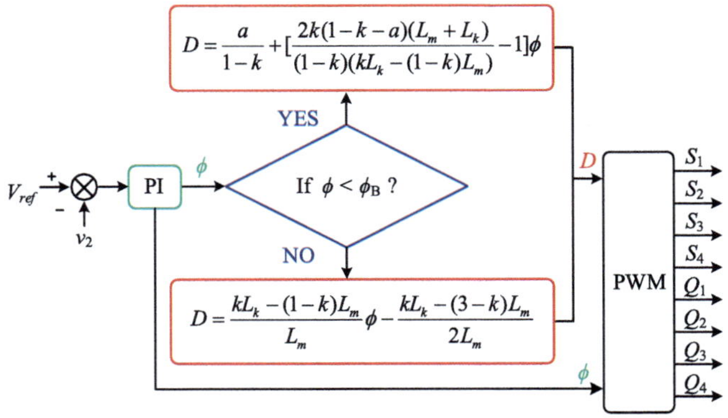

Fig. 10.11 Control diagram with the proposed control law

10.3 Comparative Analysis Among Different Control Methods

The normalized peak currents I_{p_nor} for different modulation schemes under different voltage conversion ratios are shown in Fig. 10.12. The current stress under proposed control law is compared with that under DPS control in [3] and EPS control in [1]. The optimization targets in these three control methods are all current stress. As seen in Fig. 10.12, when $k = 0.6$, the current stress with proposed control law is larger than that with DPS control in [3], but in other cases ($k \neq 0.6$), the current stresses show small deviation between the proposed control law and DPS control in [3]. The current stress under the proposed control law is a bit lower than that under the EPS control in [1] at light load when $k \neq 0.9$ and at heavy load when $k \neq 0.6$.

The normalized rms current I_{rms_nor} for these three control methods under different voltage conversion ratios are shown in Fig. 10.13. Similar with the comparison of current stress, the rms current with the proposed control method is smaller than that with EPS control in [1] at light load when $k \neq 0.9$ and at heavy load when $k \neq 0.6$. At no load, the rms current under the proposed control law is larger than that with DPS control in [3] which is zero. Because the duty cycles of the two H-bridges at

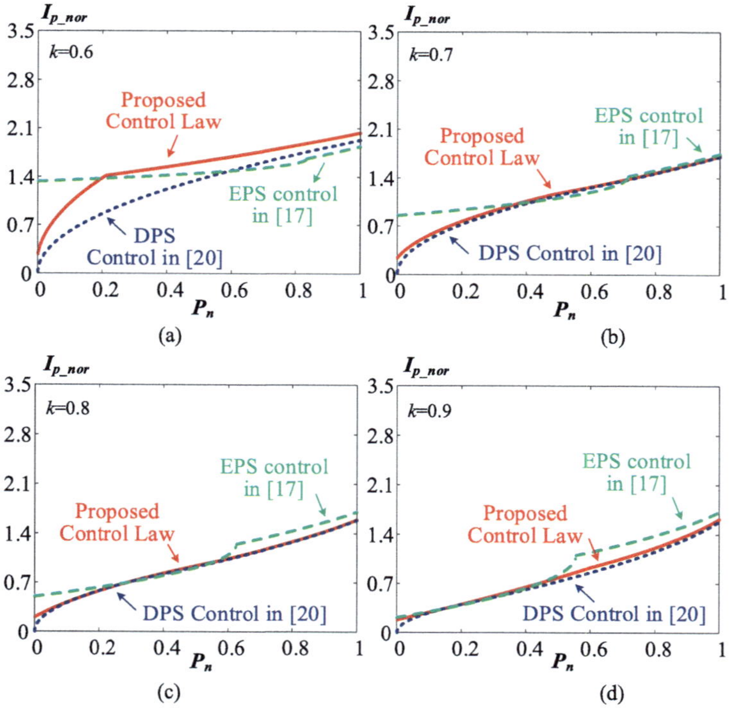

Fig. 10.12 Comparison of current stress among different control methods under different k. **a** k = 0.6. **b** k = 0.7. **c** k = 0.8. **d** $k = 0.9$

no load under the proposed control law are not zero for ZVS achievement, leading a larger rms current compared to DPS control in [3]. However, the difference at low power levels becomes smaller as k increases, and the rms currents are almost equal at heavy power levels. In addition, with DPS control in [3], at lease a pair of switches cannot achieve ZVS within the full load range under some voltage ranges, which will cause some problems including severe switching loss, EMI noise and etc. The maps for showing ZVS regions and non ZVS regions with different loads, voltage gains, and control methods are illustrated in Fig. 10.14. As shown in Fig. 10.14a, with DPS control in [3], ZVS cannot be achieved on all the power ranges within the range of k from about 0.5 to 0.7. Besides, ZVS cannot be achieved under the low power levels within all the voltage gains. With the proposed method shown in Fig. 10.14b, full load range ZVS can be achieved within a wide voltage range, and the range can be changed by designing the magnetizing inductance.

Therefore, comprehensively comparing the current stress and the ZVS achievement, the proposed control law utilizing magnetizing inductance is still preferred especially for high frequency applications with wide voltage range.

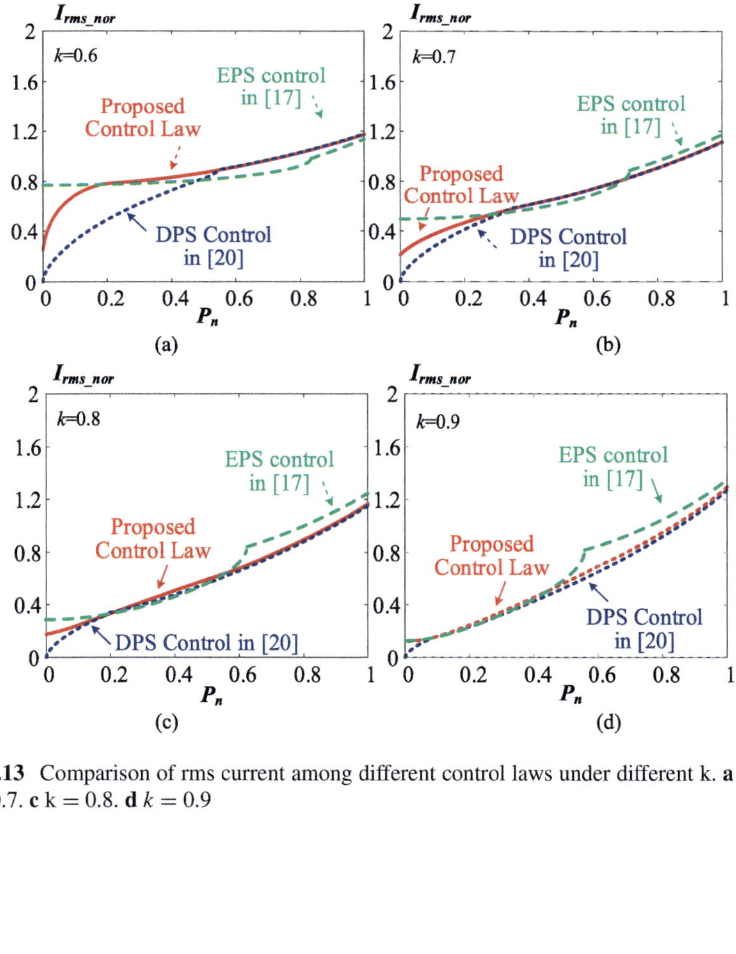

Fig. 10.13 Comparison of rms current among different control laws under different k. **a** k = 0.6. **b** k = 0.7. **c** k = 0.8. **d** $k = 0.9$

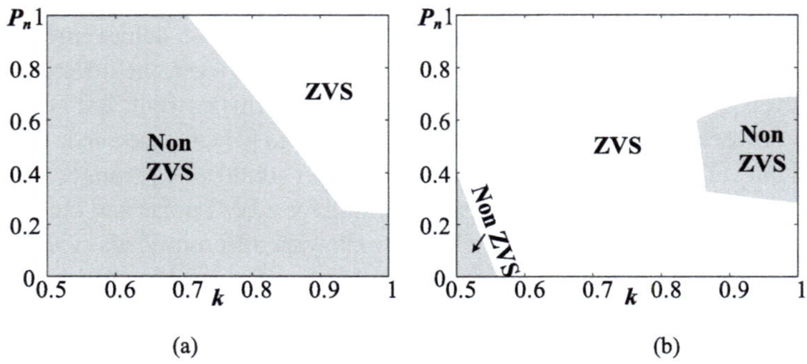

Fig. 10.14 Comparison of ZVS regions. **a** Under DPS control in 3. **b** Under the proposed DPS control solution

10.4 Parameter Design

The process of the parameters design to the DAB converter with the proposed control law is shown in Fig. 10.15. Firstly, determine the voltages V_1 and V_2, rated power P_{rated} and switching frequency f_s according to the application. Then, based on the definition of voltage conversion ratio that $k = nV_2/V_1$, the turns ratio of the transformer n can be determined with input and output voltages as well as the maximum voltage conversion ratio. Power transmission capability is determined by the leakage inductance. The maximum normalized power $P_{n,max}$ can be deduced by (10.4). The maximum power ($P_{n,max} \cdot P_{base}$) should be larger than P_{rated}, and then the leakage inductance L_k can be computed. The design of magnetizing inductance has been analyzed in part A of Section III. According to (10.18), the magnetizing inductance can be sized. Finally, check ZVS range and current stress using the determined parameters. If unsuitable, go back and reselect k_{max} to redesign the parameters.

Fig. 10.15 Parameter design flowchart

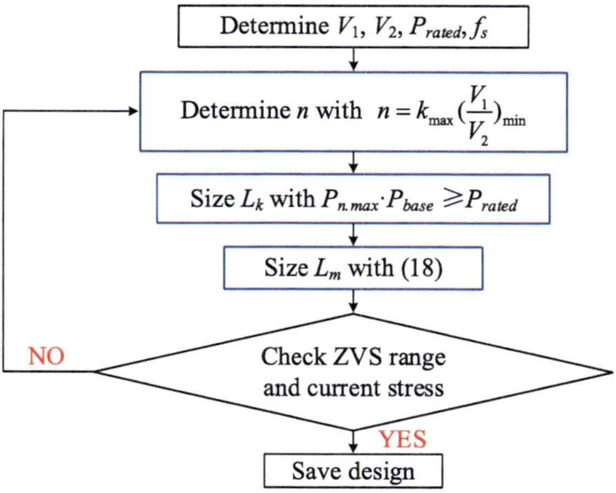

10.5 Experimental Results

A 500 W laboratory prototype of the DAB converter is built to verify the effectiveness of the proposed solution, as shown in Fig. 10.16. Based on the flowchart in Fig. 10.15, the detailed specifications of the prototype can be designed, and they are listed in Table 10.3. Mainly based on voltage and current levels, GaN based devices GS66508T are choiced as the switches on the V_1 side S_1-S_4, and the switches on V_2 side Q_1-Q_4 are used GaN based devices GS61008T.

10.5.1 Steady State Operation

Figures 10.17, 10.18, and 10.19 show the steady state operating waveforms of voltages and currents u_{AB}, u_{CD}, i_k and i_s for 300 V and 450 V under different loads respectively. The loads shown in Figs. 10.17, 10.18, and 10.19 include light load (50 W), half load (250 W) and full load (500 W). As seen in the steady-state waveforms, with different loads and different input voltages, the duty cycle and the phase shift angle are different. The duty cycles of u_{AB} and u_{CD} are the same, which conforms to the definition of DPS control. Compared with the case when $V_1 = 450$ V, the duty cycle has a relatively large change as the load changes under $V_1 = 300$ V. The shape

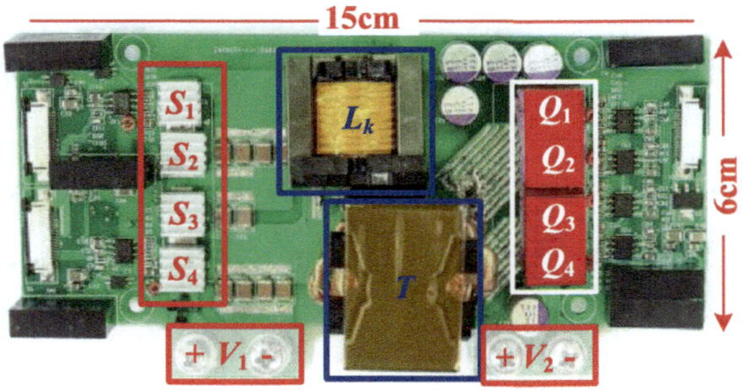

Fig. 10.16 The experimental prototype

Table 10.3 Specification of the Experimental Prototype

Parameter	Symbol	Value
Voltage on V_1 side	V_1	300–450 V
Voltage on V_2 side	V_2	48 V
Switching frequency	f_s	250 kHz
Rated power	P_{rated}	500 W
Turns ratio of the transformer	n	255/48
Leakage inductance	L_k	35 μH
Magnetizing inductance	L_m	44 μH

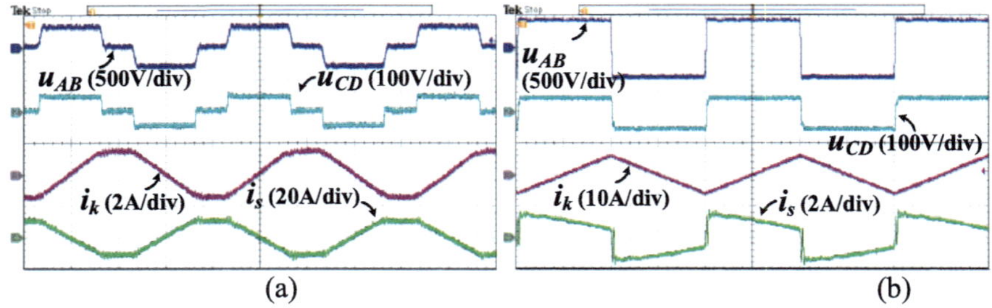

Fig. 10.17 Steady state waveforms under light load (50 W). **a** $V_1 = 300$ V. **b** $V_1 = 450$ V

of secondary current i_s is different from the leakage current i_k by the properly design of magnetizing inductance.

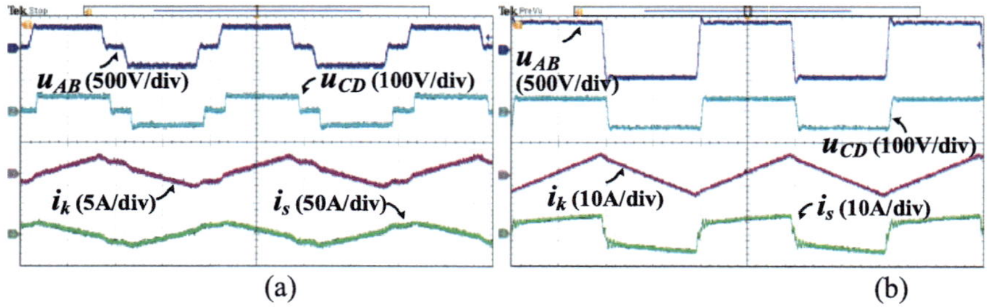

Fig. 10.18 Steady state waveforms under half load (250 W). **a** $V_1 = 300$ V. **b** $V_1 = 450$ V

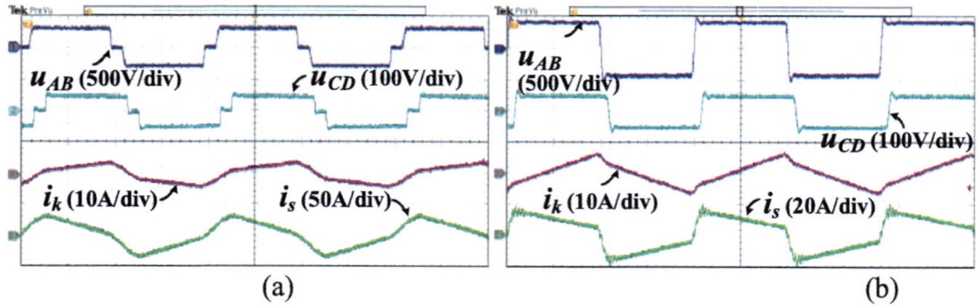

Fig. 10.19 Steady state waveforms under full load (500 W). **a** $V_1 = 300$ V. **b** $V_1 = 450$ V

10.5.2 Steady State Operation

$V_1 = 450$ V, which corresponds to the minimum k, is chosen to show the ZVS results for all the power switches under the proposed method, as shown in Figs. 10.20, 10.21, 10.22, and 10.23.

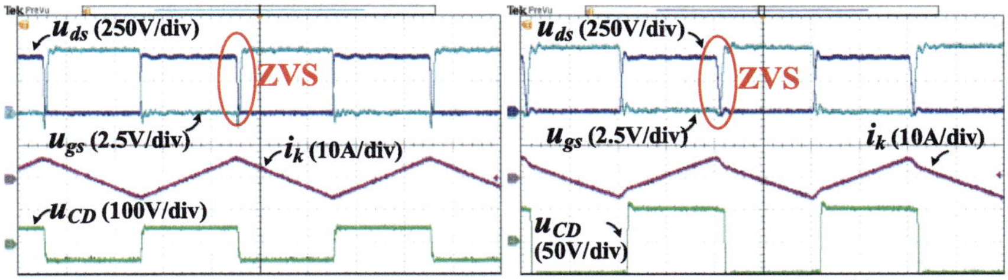

(a) (b)

Fig. 10.20 ZVS waveforms for switch S2 with the proposed method under different loads. **a** Light load (50 W). **b** Full load (500 W)

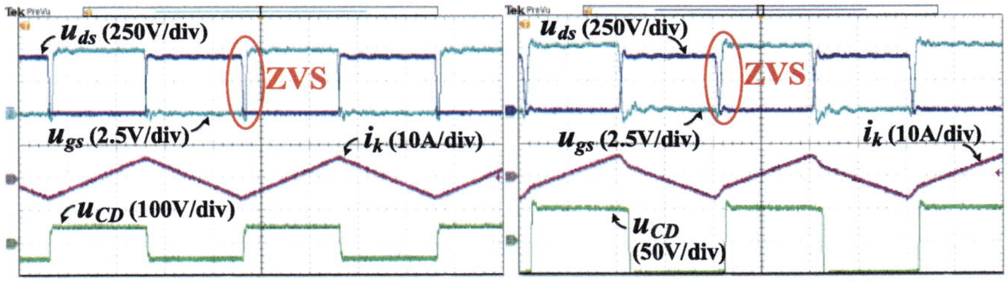

(a) (b)

Fig. 10.21 ZVS waveforms for switch S4 with the proposed method under different loads. **a** Light load (50 W). **b** Full load (500 W)

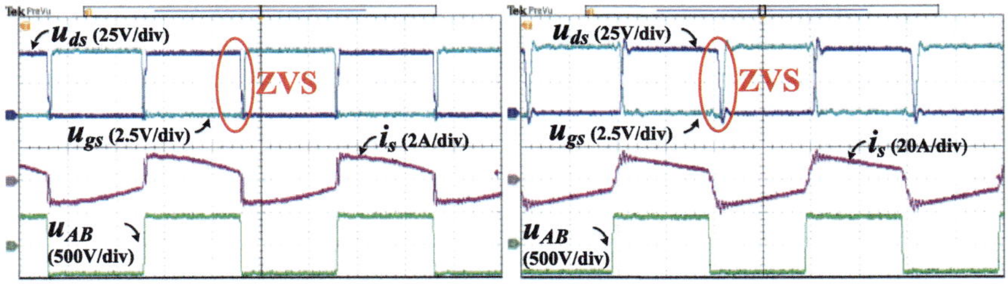

(a) (b)

Fig. 10.22 ZVS waveforms for switch Q_2 with the proposed method under different loads. **a** Light load (50 W). **b** Full load (500 W)

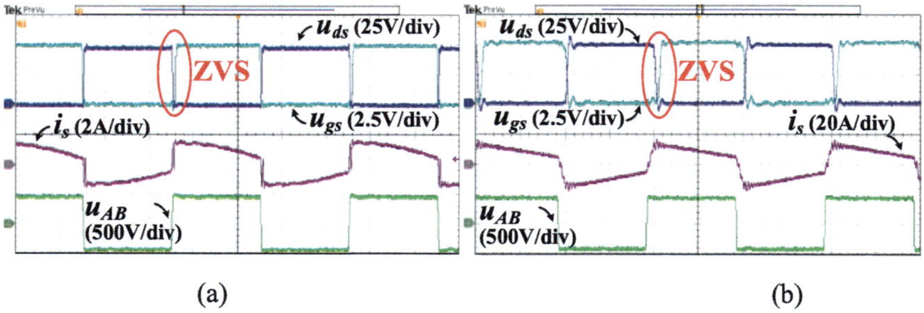

Fig. 10.23 ZVS waveforms for switch Q4 with the proposed method under different loads. **a** Light load (50 W). **b** Full load (500 W)

Fig. 10.20 shows the driver signals vgs and the drain to source voltages vds of switch S2. The leakage current is the peak current when S_2 is turned on, so the energy is sufficient to achieve ZVS.

Fig. 10.21 shows ZVS occurrence of switches S4. The leakage current is large under both light load and full load, which leads to ZVS of S_4.

Fig. 10.22 shows the driver signal and drain-source voltage of switch Q2. Before the driver signal becomes high level, the drain-source voltage reached zero, so ZVS of Q_2 is achieved.

Fig. 10.23 shows ZVS occurrence of switch Q_4. The current on the secondary side is the peak current when Q_4 is turned on, so the energy is sufficient to achieve ZVS.

As comparisons, Figs. 10.24, 10.25, 10.26, and 10.27 show ZVS results under the DPS control in [3]. As seen in Figs. 10.24, 10.25, 10.26, and 10.27, switches $S_1\&S_2$ and $Q_3\&Q_4$ can be achieved ZVS under both light load and full load. In Fig. 25a, the energy in leakage inductor is not large enough to discharge the energy in junction capacitor of switch S_3 to be zero, so ZVS cannot be achieved for switch $S_3\&S_4$ under light load. Fig. 10.26 shows the ZVS waveforms for switch Q_2. To achieve ZVS for Q_2, the current i_s should be negative when the switch Q_1 is turned off. However, Fig. 10.26 shows that the current at the commutating times are positive in both Fig. 10.26a and 10.27b, which indicates that switches $Q_1\&Q_2$ cannot be achieved ZVS under both the light load and full load. These results are all consistent with the analysis.

10.5.3 Loss Breakdown and Efficiency

A power loss breakdown under different voltages and loads is plotted in Fig. 10.28. The main losses include switches conduction loss P_{sw_c}, switches turn-off loss P_{sw_off}, iron loss P_{L_iron} and copper loss P_{L_cop} of inductor respectively, iron loss P_{T_iron} and copper loss P_{T_cop} of transformer respectively. Since all switches are turned ON with ZVS, the switch turn-on loss is zero. As shown, under light load and half load

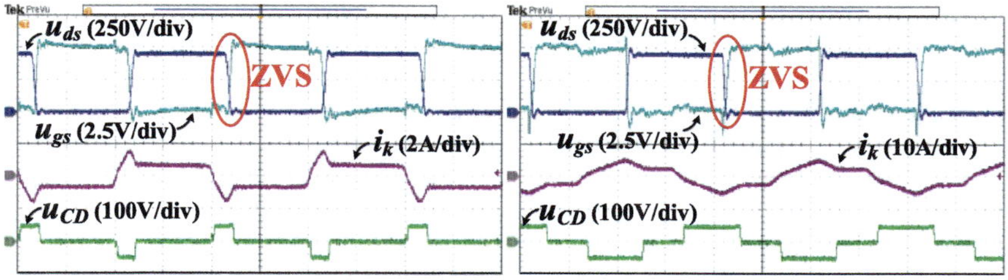

(a) (b)

Fig. 10.24 ZVS waveforms for switch S_2 with DPS control in [3] under different loads. **a** Light load (50 W). **b** Full load (500 W)

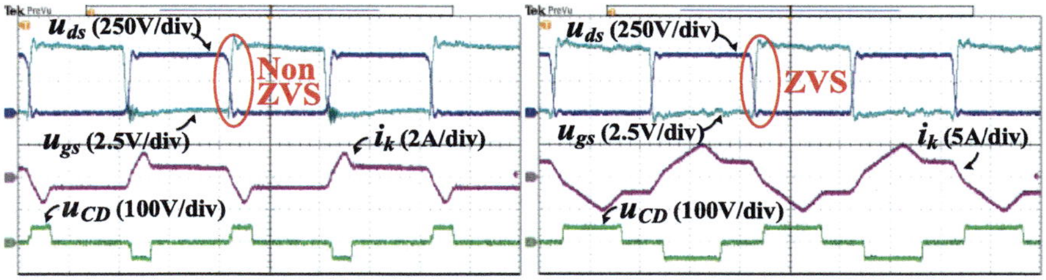

(a) (b)

Fig. 10.25 ZVS waveforms for switch S_3 with DPS control in [3] under different loads. **a** Light load (50 W). **b** Full load (500 W)

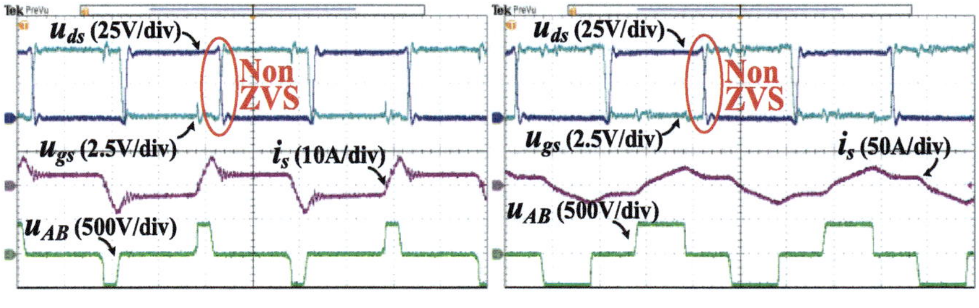

(a) (b)

Fig. 10.26 ZVS waveforms for switch Q_2 with DPS control in [3] under different loads. **a** Light load (50 W). **b** Full load (500 W)

conditions, P_{T_iron} is larger than other losses under different voltages. When $V_1 =$ 300 V, in addition to P_{T_iron}, P_{sw_c} also plays a dominant role under different loads. Different from $V_1 = 300$ V, because RMS current of secondary side is greatly reduced by the magnetizing current, conduction loss is not always dominant when $V_1 = 375$

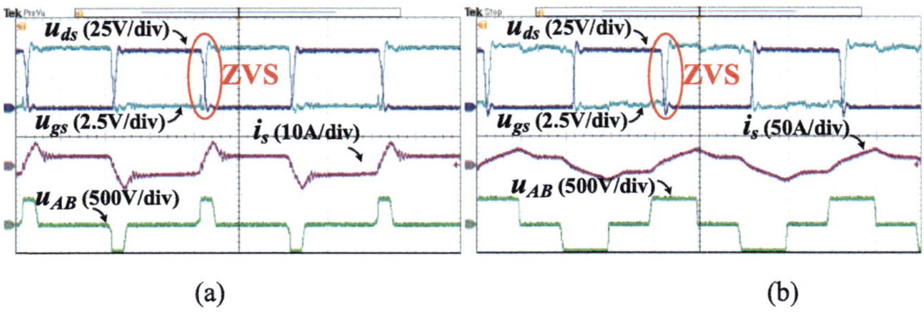

Fig. 10.27 ZVS waveforms for switch Q_4 with DPS control in [3] under different loads. **a** Light load (50 W). **b** Full load (500 W)

Fig. 10.28 The loss breakdown under different loads and voltages

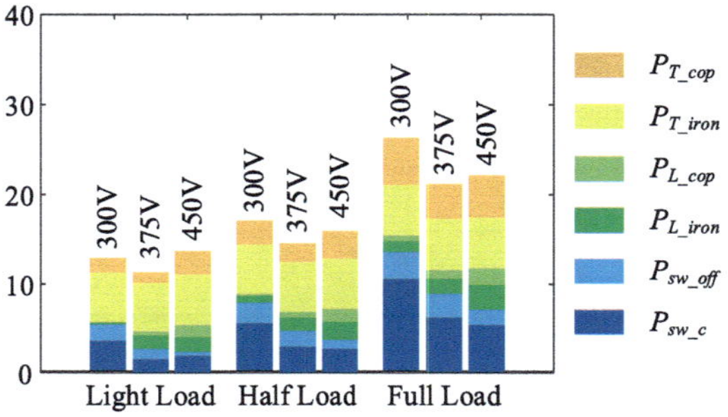

V and 450V. P_{T_cop} is also closely related to RMS current. However, because the turns of the secondary side are much less than that of primary side, the reduction of P_{T_cop} is not so much affected by reduced RMS current of secondary side as P_{sw_c} is by reduced RMS current of secondary side.

In addition, the experimental efficiencies with the proposed method under different operating conditions are measured, as shown in Fig. 10.29. As shown, high efficiency is obtained at $V_1 = 375$ V, mainly because the magnetizing current can reduce the RMS current of secondary side much than that at $V_1 = 300$ V, and the voltages of the two sides are more matched than that at $V_1 = 450$ V. Besides, the maximum efficiency point is close to 96%.

10.5.4 Comparisons

Moreover, a series of comparisons among the proposed solution and some existing relevant solutions are made, and the results are listed in Table 10.4. Compared with the existing solutions, the proposed solution operates at 300–450 V input voltage and 250 kHz switching frequency, which shows that full load range ZVS is achieved with

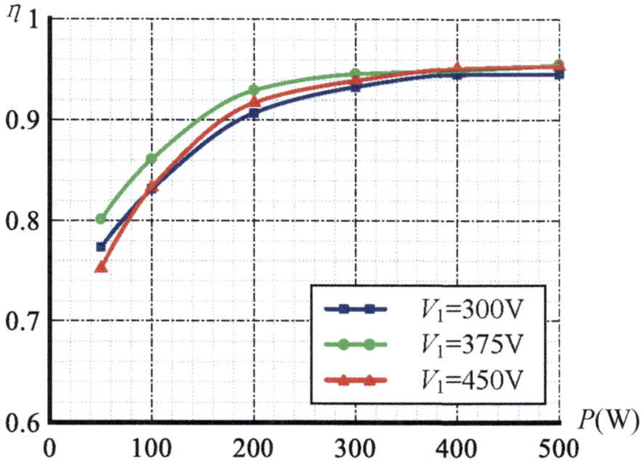

Fig. 10.29 Efficiency curves with different operating conditions

Table 10.4 Comparisons of the Proposed Method and the Existing Methods

Article	Full Load Range ZVS	Input Voltage (V)	Output Voltage (V)	Frequency (kHz)	Efficiency (%)	Control Degrees of Freedom	Modulation
[4]	Yes	400	48	250	97.3	1	SPS
[1]	No	140–300	0–180	10	90.2	2	EPS
[3]	No	140–300	48	10	94	2	DPS
[5]	No	30	30	10	86.4	2	DPS
[6]	Yes	70–110	60	20	97	3	TPS
This chapter	Yes	300–450	48	250	95.6	2	DPS

wide voltage range and high switching frequency. Besides, the maximum efficiency is close to 96% with medium control degrees of freedom.

10.6 Conclusion and Future Work

For a typical DAB converter operated under DPS control, as least a pair of switches cannot achieve full load range ZVS. Thus, high efficiency is not attainable and problems cause by hard switching could be more severe especially for high frequency and wide voltage range applications. Magnetizing inductance, which is large enough to be ignored in the previous DPS control studies, is designed in this chapter to create possibility of full load range ZVS achievement for all switches. When analyzing ZVS performance, the magnetizing inductance and the junction capacitors of switches are taken into consideration. Based on the precondition of full load range ZVS achievement, a power control law is proposed to obtain low current stress utilizing Lagrange multiplier method. Besides, both duty cycle and phase shift ratio vary continuously in all operation cases. The current stress, RMS current and the ZVS range in the DAB converter by using the proposed method is compared with some previous control methods. The results show that the proposed method does not have the minimum current stress and RMS current in some power ranges due to ZVS achievement, but the increments are acceptable. Finally, the feasibility of the proposed control law utilizing magnetizing inductance is confirmed by an experimental prototype.

The proposed solution can also be applied into other power levels. To expand the rated power, multi switches can be connected in parallel to reduce the switch rating and related loss of a single switch. Besides, multi-module DAB converter with different connections structure can also be used. After adopting the multi-module solution, the proposed solution presented in this chapter is also applicable to each sub-module to improve the performance of the converter.

References

1. Qin, Z., Shen, Y., Loh, P. C., Wang, H., & Blaabjerg, F. (2018). A dual active bridge converter with an extended high-efficiency range by DC blocking capacitor voltage control. *IEEE Transactions on Power Electronics*, *33*(7), 5949–5966.
2. Zhao, B., Yu, Q., & Sun, W. (2012). Extended-phase-shift control of isolated bidirectional DC–DC converter for power distribution in microgrid. *IEEE Transactions on Power Electronics*, *27*(11), 4667–4680.
3. Zhao, B., Song, Q., Liu, W., & Sun, W. (2013). Current-stress-optimized switching strategy of isolated bidirectional DC–DC converter with dual-phase-shift control. *IEEE Transactions on Industrial Electronics*, *60*(10), 4458–4467.
4. Chen, X., Xu, G., Han, H., Liu, D., Sun, Y., & Su, M. (2021). Light-load efficiency enhancement of high-frequency dual-active-bridge converter under SPS Control. *IEEE Transactions on Industrial Electronics*, *68*(12), 12941–12946.
5. Liu, X., et al. (2017). Novel dual-phase-shift control with bidirectional inner phase shifts for a dual-active-bridge converter having low surge current and stable power control. *IEEE Transactions on Power Electronics*, *32*(5), 4095–4106.
6. Huang, J., Wang, Y., Li, Z., & Lei, W. (2016). Unified triple-phase-shift control to minimize current stress and achieve full soft-switching of isolated bidirectional DC–DC converter. *IEEE Transactions on Industrial Electronics*, *63*(7), 4169–4179.

Part V
Modular Dual-Active-Bridge Converter

Chapter 11
Single-Stage DAB-LLC Hybrid Bidirectional Converter with Tight Voltage Regulation Under DCX Operation

Abstract Working as a DC-DC transformer (DCX), which operates at the resonant frequency, is a suitable solution to achieve maximum conversion efficiency of LLC converter. However, because the fixed frequency modulation is applied, the voltage regulation performance is limited. To achieve tight voltage regulation and fully utilize the advantages of LLC DCX operation under bidirectional isolated power application, a single-stage DAB-LLC hybrid bidirectional converter based on the sigma converter structure is proposed in this chapter. In the proposed converter, the main power flows through LLC converter, and dual active bridge (DAB) is used for the power flow direction control and output voltage regulation. In addition, a mathematical model of the proposed converter is derived for controller design, which is verified by circuit simulation. Based on the model, the design of the controller and circuit parameters are discussed with the goal of fast dynamics. Finally, the experimental results of a 1200 W prototype verify the effectiveness and advantages of the proposed topology and controller design.

Keywords Bidirectional power flow · Dual active bridge (DAB) · Single-stage DAB-LLC hybrid bidirectional converter · Tight voltage regulation

11.1 System Structure and Converter Modeling

11.1.1 System Structure

The structure diagram of the single-stage DAB-LLC hybrid bidirectional converter proposed in this chapter is shown in Fig. 11.1. The system is composed of an LLC converter and a DAB converter, whose inputs are in series and the outputs are in parallel.

In Fig. 11.1, V_{in} represents the input voltage, V_o represents the output voltage, V_{C1} and V_{C2} are the input voltages of LLC converter and DAB converter, respectively. v_{ab} and v_{cd} are the voltage between the midpoint of the leading and lag bridge arms in primary and secondary side, respectively. C_1, C_2 and C_o are LLC converter input

G. Xu and X. Chen, *Topology Deduction and Control for Dual-Active-Bridge Converters*, https://doi.org/10.1007/978-981-97-4856-3_11

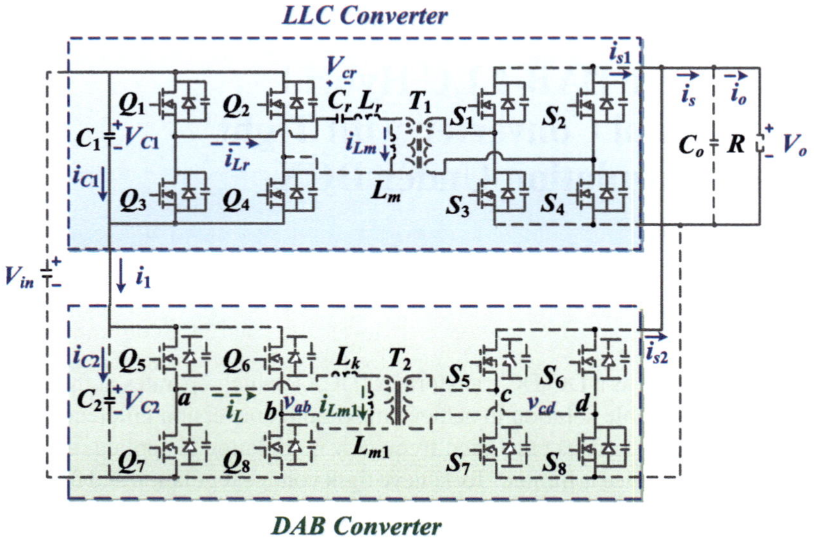

Fig. 11.1 The proposed circuit topology of the single-stage DAB-LLC hybrid bidirectional converter

capacitor, DAB converter input capacitor and output filter capacitor, respectively. R is the load resistor. Transformers T_1 and T_2 are independent. L_r, L_m and C_r are the resonant inductor, magnetizing inductor and resonant capacitor in LLC converter, respectively. L_k and L_{m1} are the leakage inductor and magnetizing inductor of DAB converter. In LLC, Q_1–Q_4 are the primary side switches and S_1–S_4 are the secondary side switches. In DAB converter, Q_5–Q_8 are the primary side switches, and S_5–S_8 are the secondary side switches.

11.1.2 Converter Modeling

Figure 11.2 shows the key waveforms of the proposed converter. The operating frequency of LLC converter is fixed at the resonant frequency, and the duty cycle is 50% for all the power switches. The switch driving signal of Q_i at the primary side is the same with driving signal of S_i at the secondary side (i = 1, 2, 3 and 4), which can easily realize synchronous rectification of LLC converter. Under this modulation strategy, LLC converter is equivalent to a non-regulated DC transformer. DAB converter regulates the output power through adjusting the phase shift angle between the driving signal of the primary side switches and the secondary side switches to control the phase angle between v_{ab} and v_{cd}, thereby changing the voltage on the leakage inductor to realize the adjustment of output power. It is worth noting that the power transfer from left to right is considered to be the positive direction.

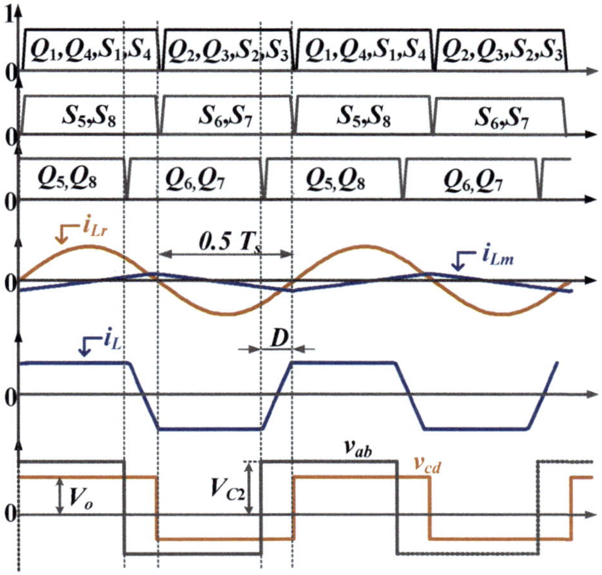

Fig. 11.2 The key waveforms of proposed converter

A DAB converter uses phase shift modulation to achieve energy transfer and the power transmitted from the primary to the secondary side can be expressed by (11.1)

$$P_2 = \frac{K_2 V_{C2} V_o D(1-D)}{2L_k f_s} \tag{11.1}$$

where P_2 is the transfer power of DAB, D is the phase shift angle (value range from −1 to 1), f_s is the switching frequency and K_2 is the transformer ratio of DAB converter from primary side to secondary side.

Neglecting the transmission loss, the power relationship between input and output is shown in (11.2)

$$V_o I_{s2} = V_{C2} I_L \tag{11.2}$$

where I_L is the average inductor current and I_{s2} is the average secondary current of DAB converter.

In addition, the topology is input series and output parallel. The relationships are deduced as follows

$$i_1 = i_{Lr} + i_{C1} = i_L + i_{C2} \tag{11.3}$$

$$i_s = i_{s1} + i_{s2} \tag{11.4}$$

$$V_{in} = V_{C1} + V_{C2} \tag{11.5}$$

where i_1 is the total input current and i_s is the total current in secondary. i_{C1} and i_{C2} are the currents flowing through C_1 and C_2, respectively.

Because LLC converter operates only at the resonant frequency, the impedance is very low. LLC converter in the sigma structure can be regarded as an ideal DC transformer (DCX) 2. The relation between input voltage and output voltage can be expressed by (11.6)

$$V_{C1} = V_o K_1 \tag{11.6}$$

where K_1 is the transformer ratio of LLC converter from primary side to secondary side.

Then, the primary current and secondary current can also be expressed as

$$i_{s1} = K_1 i_{Lr} \tag{11.7}$$

where i_{Lr} is the resonant current and i_{s1} is the corresponding secondary current of LLC converter.

Thus, from (11.2)–(11.7), i_s can be expressed as

$$i_s = K_1 i_{Lr} + \frac{V_{in} - V_o K_1}{V_o} i_L = K_1 i_L + K_1 i_{C2} - K_1 i_{C1} + \frac{V_{in}}{V_o} i_L - K_1 i_L \tag{11.8}$$

Three current expressions in (11.8) are as follows

$$i_L = \frac{P_2}{V_{c2}} = \frac{K_2 V_o D(1-D)}{2L_k f_s} \tag{11.9}$$

$$i_{C1} = C_1 \frac{dV_{C1}}{dt} = K_1 C_1 \frac{dV_o}{dt} \tag{11.10}$$

$$i_{C2} = C_2 \frac{dV_{C2}}{dt} = C_2 \frac{dV_{in}}{dt} - K_1 C_2 \frac{dV_o}{dt} \tag{11.11}$$

Substituting (11.9)–(11.11) into (11.8) can obtain

$$i_s = K_1 C_2 \frac{dV_{in}}{dt} - K_1^2 C_2 \frac{dV_o}{dt} - K_1^2 C_1 \frac{dV_o}{dt} + \frac{K_2 V_{in} D(1-D)}{2L_k f_s} \tag{11.12}$$

Then, the output voltage state equation is expressed as

$$C_o \frac{dV_o}{dt} = i_s - \frac{V_o}{R} \tag{11.13}$$

Substituting i_s into (11.13) yields (11.14)

$$\left(C_o + K_1^2 C_2 + K_1^2 C_1\right)\frac{dV_o}{dt} = \frac{K_2 V_{in} D(1-D)}{2L_k f_s} + K_1 C_2 \frac{dV_{in}}{dt} - \frac{V_o}{R} \tag{11.14}$$

It can be seen from (11.14) that the system output voltage state equation is related to the phase shift angle, input voltage and output voltage.

Setting the differential part to be zero, then the phase shift angle in steady-state is derived as

$$D = \frac{1 \pm \sqrt{1 - 4\left(\frac{2V_o L_k f_s}{K_2 V_{in} R}\right)}}{2} \tag{11.15}$$

In the phase shift modulation, the linear range of D is from −0.5 to 0.5, so D can be determined as

$$D = \frac{1 - \sqrt{1 - 4\left(\frac{2V_o L_k f_s}{K_2 V_{in} R}\right)}}{2} \tag{11.16}$$

11.1.3 Transfer Function

Injecting disturbance near the steady-state solution, the small-signal model can be obtained.

$$\begin{cases} V_o = \overline{V}_o + \tilde{V}_o \\ V_{in} = \overline{V}_{in} + \tilde{V}_{in} \\ D = \overline{D} + \tilde{d} \end{cases} \tag{11.17}$$

where $\overline{V}_o$, $\overline{V}_{in}$ and $\overline{D}$ represent the steady state value, while $\tilde{V}_o$, $\tilde{V}_{in}$ and $\tilde{d}$ represent the injected disturbance.

Substituting (11.17) into (11.14) yields a small signal state equation which is expressed as

$$\left(C_o + K_1^2 C_2 + K_1^2 C_1\right)\frac{d\tilde{V}_o}{dt} = -\frac{\tilde{V}_o}{R} + \frac{K_2 \overline{V}_{in}(1-2\overline{D})}{2L_k f_s}\tilde{d} + \frac{K_2 \overline{D}(1-\overline{D})}{2L_k f_s}\tilde{V}_{in} + K_1 C_2 \frac{d\tilde{V}_{in}}{dt} \tag{11.18}$$

where the high order items are ignored.

The Laplace transform is performed on the small signal model, as shown in (11.19)

$$As\tilde{V}_o = -B\tilde{V}_o + C\tilde{d} + D_1 \tilde{V}_{in} + Es\tilde{V}_{in} \tag{11.19}$$

where the corresponding variables are shown as follows

$$\begin{cases} A = C_o + K_1^2 C_2 + K_1^2 C_1 \ , \ B = \dfrac{1}{R} \\ C = \dfrac{K_2 \overline{V}_{in}(1 - 2\overline{D})}{2L_k f_s} \ , \ D_1 = \dfrac{K_2 \overline{D}(1 - \overline{D})}{2L_k f_s} \ , \ E = K_1 C_2 \end{cases} \tag{11.20}$$

Finally, the transfer function of single-stage DAB-LLC hybrid bidirectional converter is derived as

$$\tilde{V}_o = \frac{C}{As + B}\tilde{d} + \frac{(D_1 + Es)}{As + B}\tilde{V}_{in} = G_d(s)\tilde{d} + G_{in}(s)\tilde{V}_{in} \tag{11.21}$$

where $G_d(s)$ is transfer function from control to output voltage and $G_{in}(s)$ is transfer function from input voltage to output voltage.

11.1.4 Model Verification

In order to verify the correctness of the calculated model, two cases of parameters are selected, as shown in Table 11.1. The proposed topology is simulated in PSIM software to get the amplitude frequency and phase frequency curves by using frequency sweep function. Meanwhile, bode graph of system transfer function $G_d(s)$ is drawn by MATLAB. The comparison results are shown in Fig. 11.3.

It can be seen from Fig. 11.3a and b that within the frequency range from 10 Hz to 10 kHz, which is 1/10 of the switching frequency, the amplitude frequency curves

Table 11.1 System parameters

Symbol	Quantity	Value case 1	Value case 2
V_{in}	Input voltage	400 V	48 V
V_o	Output voltage	48 V	5 V
R	Load	1.92 Ω	0.12 Ω
f_s	Switching frequency	100 kHz	100 kHz
L_k	DAB leakage inductor	25 μH	1.6 μH
D	Steady phase shift angle	0.09	0.409
C_1	LLC converter input capacitor	220 μF	100 μF
C_2	DAB converter input capacitor	220 μF	100 μF
C_o	Output capacitor	2100 μF	2100 μF
K_1	LLC converter transformer ratio	40/6	40/5
K_2	DAB converter transformer ratio	8/6	8/5

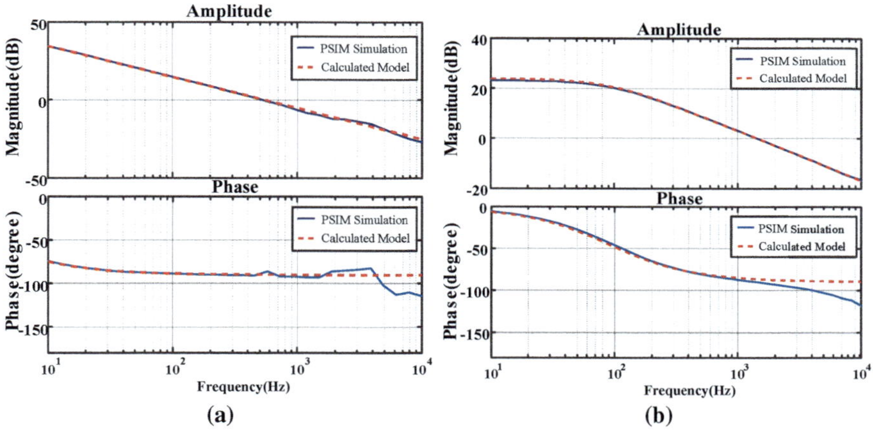

Fig. 11.3 The comparison between calculated model and PSIM simulation results. **a** case 1. **b** case 2

obtained by the mathematical model and the hardware topology simulation are almost overlapped.

11.2 Controller Design

The system transfer function has been obtained in the previous section and the corresponding controller will be designed in this section. The input voltage disturbance is considered to be negligible, and the system model is considered as follows

$$\tilde{V}_o = G_d(s)\tilde{d} = G(s)\tilde{d} \tag{11.22}$$

Ignoring the delay of sensor, the feedback transfer function is a constant value H_v, as follow

$$H(s) = H_v \tag{11.23}$$

The PWM modulation transfer function is shown as

$$G_{PWM}(s) = \frac{1}{K_{PWM}} \tag{11.24}$$

where K_{PWM} is the modulation factor.

Therefore, the system open loop transfer function is shown as follows

$$G_{os}(s) = G(s)G_{PWM}(s)H(s) = G(s)\frac{H_v}{K_{PWM}} \tag{11.25}$$

11.2.1 Controller Design

The proportional integral (PI) controller $G_c(s)$ is shown as follows:

$$G_c(s) = K_p + \frac{K_i}{s} \tag{11.26}$$

The control block diagram is shown in Fig. 11.4. Based on the bode diagram, K_p and K_i can be determined according to the design targets which are crossover frequency w_c and phase margin.

The open-loop transfer function is written by

$$G_o(s) = G_c(s)G_{os}(s) = \frac{K_p CH_v s + K_i CH_v}{AK_{PWM} s^2 + BK_{PWM} s} \tag{11.27}$$

Substituting $s = jw_c$ into (11.27) and calculate $G_o(s)$ as follow

$$G_o(jw_c) = \frac{K_p CH_v w_c j + K_i CH_v}{-AK_{PWM} w_c^2 + BK_{PWM} w_c j} = a + bj \tag{11.28}$$

where $w_c = 2\pi f_c$.

The relationship among the transfer function, phase angle θ and w_c can be expressed as follows

$$G_o(jw_c) = r\cos\theta + jr\sin\theta \tag{11.29}$$

$$r = \sqrt{a^2 + b^2} = |G_o(jw_c)| = 1 \tag{11.30}$$

$$\theta = \gamma - 180^\circ \tag{11.31}$$

where γ is phase margin.

Then the expression of a and b are obtained as

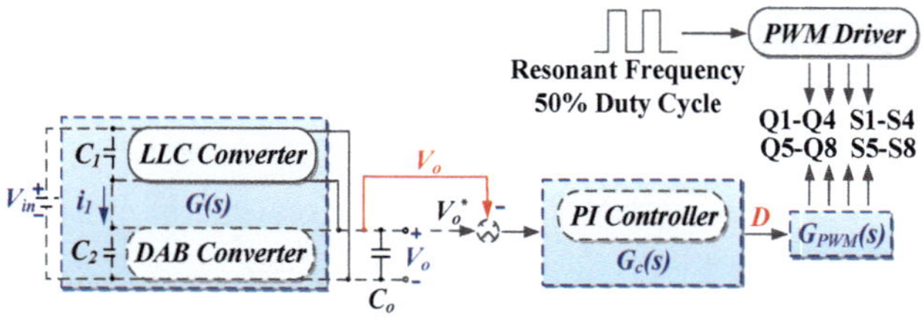

Fig. 11.4 The system control block diagram

$$a = \frac{AK_{PWM}CH_v w_c K_i - BK_{PWM}CH_v w_c K_p}{-(AK_{PWM})^2 w_c^3 - (BK_{PWM})^2 w_c} = \cos\theta \tag{11.32}$$

$$b = \frac{AK_{PWM}CH_v w_c^2 K_p + BK_{PWM}CH_v K_i}{-(AK_{PWM})^2 w_c^3 - (BK_{PWM})^2 w_c} = \sin\theta \tag{11.33}$$

Based on (11.32) and (11.33), K_p and K_i are solved as

$$K_p = \frac{K_{PWM}(-Aw_c \sin\theta + B\cos\theta)}{CH_v} \tag{11.34}$$

$$K_i = -\frac{K_{PWM}w_c(Aw_c \cos\theta + B\sin\theta)}{CH_v} \tag{11.35}$$

11.2.2 Controller Verification

According to (11.34) and (11.35), the parameters of PI controller can be obtained. In order to verify the accuracy of the controller design, parameters of case 2 in the Table 11.1 are substituted into (11.34) and (11.35) to get the values of K_p and K_i. The expected crossover frequency and phase angle margin are set to be 5 kHz and 45 degrees, respectively. The bode graph of open-loop transfer function with and without introducing PI controller are drawn respectively, as shown in Fig. 11.5. It can be seen that the crossover frequency increases to the expected value and the phase margin meets the requirement after controller is designed.

11.3 Parameter Design Consideration

11.3.1 Capacitor

The crossover angle frequency of system is calculated as

$$G(jw_{c1}) = \frac{C}{A(jw_{c1}) + B} = \frac{-BC + ACw_{c1}j}{-A^2 w_{c1}^2 - B^2} \tag{11.36}$$

where $w_{c1} = 2\pi f_{c1}$. According to (11.28)–(11.30), w_{c1} is expressed as

$$w_{c1} = \sqrt{\frac{C^2 - B^2}{A^2}} = \frac{\sqrt{C^2 - B^2}}{C_o + K_1^2 C_2 + K_1^2 C_1} \tag{11.37}$$

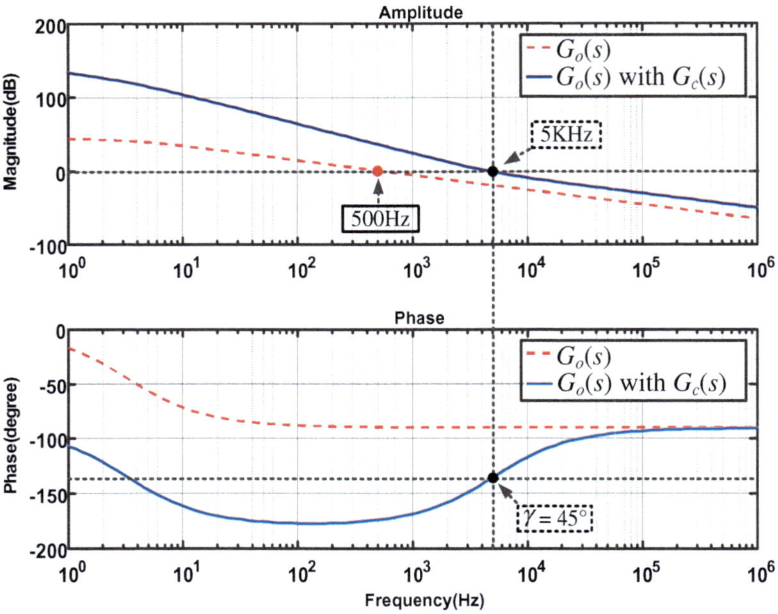

Fig. 11.5 The bode graph of open loop transfer function with and without introducing controller

Combined with (11.20), it can be seen that the larger the sum of capacitors, the smaller the crossover frequency will be, which means the weaker the dynamic performance of the system will be. But when the capacitors are small, the ability of voltage disturbance suppressing will be weaken. Therefore, the maximum capacitor value can be selected in combination with the crossover frequency according to the dynamic requirements, substituting (11.20) into (11.37) can obtain

$$C_o + K_1^2(C_1 + C_2) \leq \frac{\sqrt{\left(K_2 V_{in}\left(1 - 2\overline{D}\right)R\right)^2 - (2L_k f_s)^2}}{2L_k f_s R w_{c\_\min}} \tag{11.38}$$

where $w_{c\_\min}$ is the minimum crossover angle frequency.

11.3.2 DAB Converter Inductor

The DAB converter leakage inductor is a power link between primary and secondary side, and it is related to the maximum power of the converter. The design considerations are shown as follows:

The capability of the transferred power for DAB converter should be greater than or equal to the rated output power P_{o_rated}, as shown in the following:

$$P_{DAB} = \frac{K_2 V_{c2} V_o D(1-D)}{2L_k f_s} \geq \frac{V_{c2}}{V_{in}} P_{o\_rated} = \frac{V_{c2}}{V_{in}} \frac{V_o^2}{R} \tag{11.39}$$

The maximum transfer power occurs at the point when $D = 0.5$. Substituting it into (11.16) leads to the maximum limit for DAB converter leakage inductor, which is expressed as

$$L_k \leq \frac{K_2 R V_{in}}{8 f_s V_o} \tag{11.40}$$

In addition, the following equation can be derived from (11.37)

$$A^2 w_{c1}^2 + B^2 = C^2 \tag{11.41}$$

Substituting C in (11.20) into (11.41) can obtain the relationship between the system open loop crossover frequency and DAB converter leakage inductor.

$$L_k = \frac{K_2 V_{in}(1-2D)}{2 f_s \sqrt{A^2 w_c^2 + B^2}} \tag{11.42}$$

From (11.39), it can be seen that the smaller leakage inductor is, the larger the crossover frequency will be. Unlike analog control, the digital controller introduces a 1.5-step delay in the loop due to the presence of a zero-order holder and one-step delay 3. So, in order to keep the system with sufficient phase margin under digital control, the crossover frequency will be at least lower than half of the switching frequency. In that situation, substituting (11.20) into (11.41), the minimum value of the leakage inductor can be expressed as

$$L_k \geq \frac{K_2 V_{in}(1-2D)R}{2 f_s \sqrt{\left(C_o + K_1^2 (C_1 + C_2)\right)^2 R^2 w_{c\_max}^2 + 1}} \tag{11.43}$$

where, w_{c_max} is the maximum crossover frequency.

11.3.3 Transformer Turns Ratio

The power distribution of the two converters ignoring the conversion loss is shown as

$$\frac{P_{DAB}}{P_o} = \frac{P_{DAB}}{P_{LLC} + P_{DAB}} = \frac{V_{C2} i_1}{V_{C1} i_1 + V_{C2} i_1} = \frac{K_2}{K_1 + K_2} = \lambda \tag{11.44}$$

where λ is power distribution coefficient. It can be seen that λ is related to the transformer turns ratio.

In ideal case, the proportion of the power that flows through DAB converter should be as low as possible to ensure that as much power as possible flows through LLC converter. However, it can be seen from (11.37) that the crossing frequency decreases when the transformer ratio K_1 increases. Therefore, the maximum K_1 can be determined according to the minimum dynamic response requirements.

$$K_1 \leq \sqrt{\frac{\sqrt{C^2 - B^2}}{w_{c1\_\min}(C_1 + C_2)} - \frac{C_o}{(C_1 + C_2)}} \tag{11.45}$$

In addition, ZVS of DAB converter at 10% of the total power can also be considered. The energy relationship between the inductor and the capacitor is shown in (11.46), where I_{L_off} is the turn-off current and C_p is the junction capacitor of primary side.

$$L_k I_{L\_off}^2 \geq C_p V_{C2}^2 \tag{11.46}$$

Then, the boundary of K_2 can be deduced in (11.46)

$$K_2 \geq \frac{8V_o L_k f_s}{5V_{in} R\left(2 - 16C_p f_s^2 L_k\right)} \tag{11.47}$$

So, the boundary of K_1 can be obtained by introducing the total turn ratio of the proposed converter.

$$K_1 = \frac{V_{in}}{V_o} - K_2 \leq \frac{5V_{in}^2 R\left(2 - 16C_p f_s^2 L_k\right) - 8V_o^2 L_k f_s}{5V_{in} V_o R\left(2 - 16C_p f_s^2 L_k\right)} \tag{11.48}$$

Therefore, the maximum value of K_1 should be selected considering both the dynamic response and ZVS of DAB converter.

11.4 Experiments Results

A 1200 W single-stage bidirectional DAB-LLC hybrid converter prototype is built up with 400 V (±30 V) input and 48 V output voltage, as shown in the Fig. 11.6. The upper module is DAB converter and the lower module is LLC converter. MOSFET is used for primary and secondary side switching devices. The system circuit parameters and control parameters are shown in Table 11.2. The PI controller is implemented in a TI digital signal processor TMS320F28069. λ is designed to be 0.2, which determines transformer turns ratios of LLC converter and DAB converter. During the experiment, the load step change from 1.9 to 5.5 Ω, which means the output power step change from 1200 to 420 W.

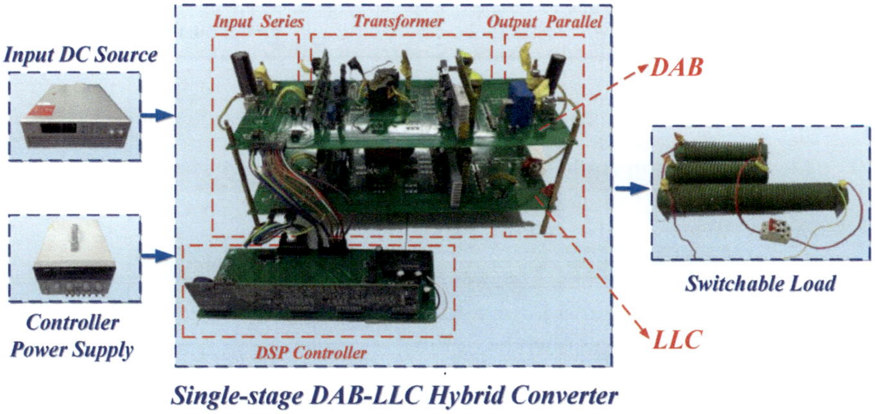

Fig. 11.6 The hardware prototype of single-stage DAB-LLC hybrid bidirectional converter

Table 11.2 System parameters

Symbol	Quantity	Value
V_{in}	Input voltage	370–430 V
V_o	Output voltage	48 V
L_r	LLC converter resonant inductor	25 μH
C_r	LLC converter resonant capacitor	0.1 μF
L_k	DAB converter leakage inductor	25 μH
R	Rated load	1.9 Ω
f_s	Switching frequency	100 kHz
f_c	Crossover frequency	5 kHz
K_p	Proportionality coefficient	0.48
K_i	Integration coefficient	15,125
K_1	LLC converter transformer ratio	40/6
K_2	DAB converter transformer ratio	10/6

The steady state waveforms of proposed converter under 420 W load condition and 1200 W load condition are shown in Fig. 11.7, which include output voltage, output current, resonant current of LLC converter and leakage inductor current of DAB converter, respectively. It can be seen from the experimental results that the output voltage remains unchanged at 48 V under different loads. It also can be seen from Fig. 11.7b and c that LLC converter always operates at the resonant frequency point. In addition, the phase shift angle of DAB converter is changed corresponding to the load, which indicates that the output is regulated through DAB converter.

Figure 11.8 shows the key waveforms of primary side and secondary side under full load condition. It can be seen from Fig. 11.8a that the ratio of input voltage of LLC converter and DAB converter is nearly 4/1. And the amplitude of leakage inductor current is almost the same as the amplitude of the resonant current. From

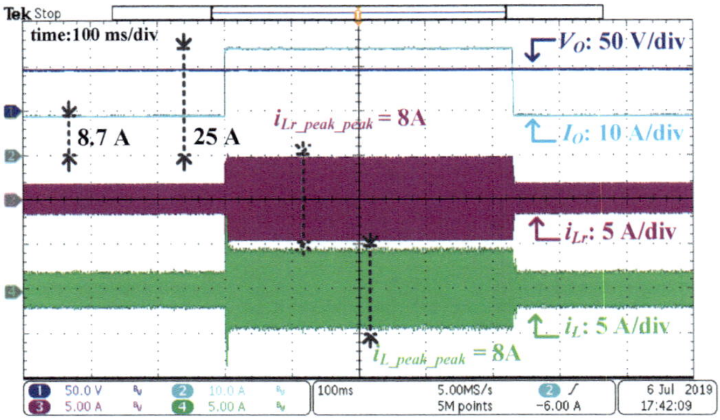

(a)

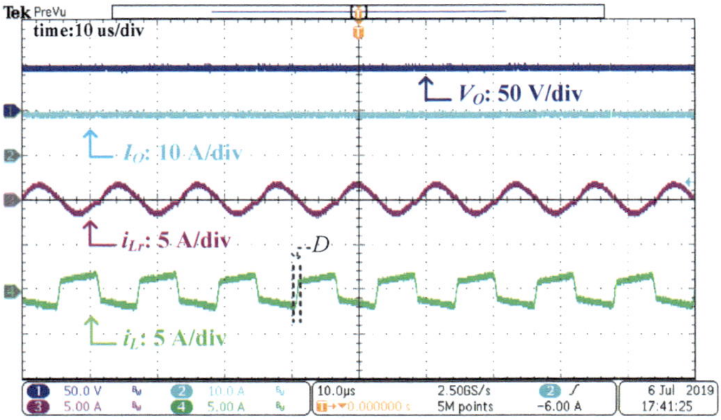

(b)

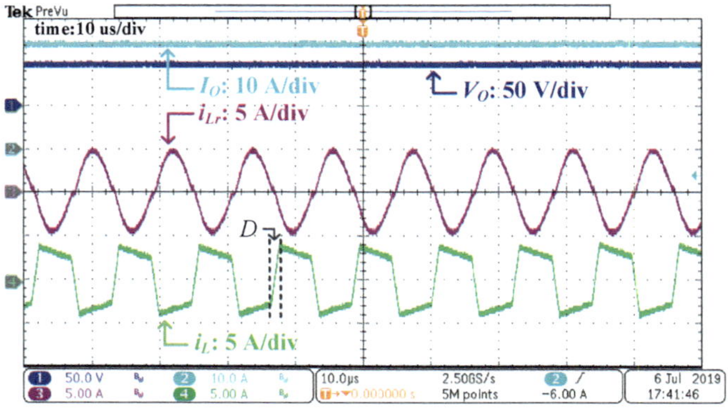

(c)

Fig. 11.7 The key waveforms under different load. **a** Load is step changing between 420 and 1200 W. **b** 420 W load. **c** 1200 W load

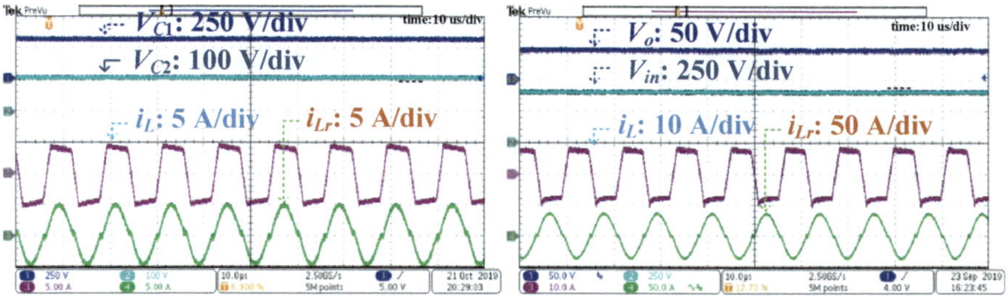

(a) (b)

Fig. 11.8 The key waveforms in different side. **a** primary side. **b** secondary side

Fig. 11.8b, it can be seen that the secondary current of DAB converter is much smaller than the secondary current of LLC converter, which can verify that most power flows through LLC converter.

In order to verify the voltage regulation capability, the experimental results are compared with a conventional full-bridge LLC DCX structure. The conventional LLC DCX is built up with 400 V input and 48 V output. Its transformer turns ratio is 50/8, and the output power is also designed to 1200 W. It operates at the resonant frequency (100 kHz) and is open loop control. During the experiment, the output power is also switched from 420 to 1200 W. The output voltage waveforms of the conventional LLC-DCX and the proposed converter are shown in Fig. 11.9a and Fig. 11.9b, respectively. As seen from Fig. 11.9a, 1 V difference occurs under the two load conditions, which shows that the load regulation of the conventional converter is poor. While for the proposed converter, the output voltage is remained almost constant, as shown in Fig. 11.9b.

Figure 11.10 shows the dynamic response waveforms for proposed converter when the output voltage channel is under AC coupling condition. While the output power jumps between 420 W and full load, the output voltage overshoot is less than 1 V (2%), and the settling time is less than 2 ms.

Figure 11.11 shows the measured open-loop Bode plots of the proposed converter using PSM1700 loop analyzer. It can be seen that after introducing the designed PI controller, the crossing frequency is increased to 3.38 kHz and the phase margin is 46.7°, which is close to the theoretical analysis. In the high frequency region, the difference of phase is mainly caused by the damping, line impedance, sampling delay and other factors.

Since LLC converter is used as a DCX in the system, its output voltage is determined by the input voltage. When the input voltage changes, the proposed converter can stabilize the output voltage by adjusting the phase shift angle in DAB converter. Figure 11.12 shows the key waveforms when the input voltage drops from 400 to 370 V and increases from 400 to 430 V. It can be seen that the output voltage remains constant under different input voltage.

By paralleling a bidirectional current source at the output side of the proposed converter, it can be seen from the experiment that the naturally bidirectional power

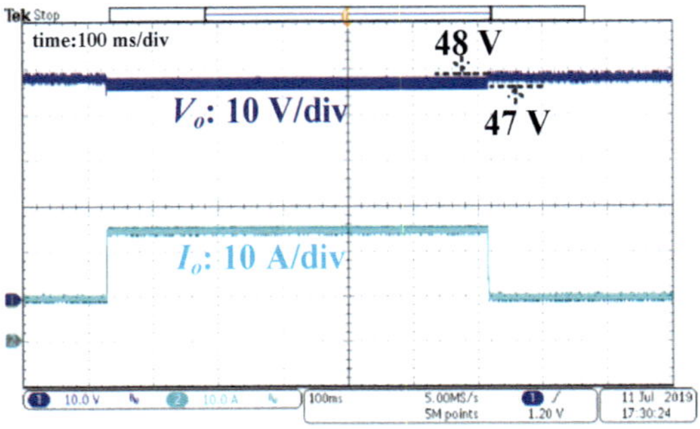

(a)

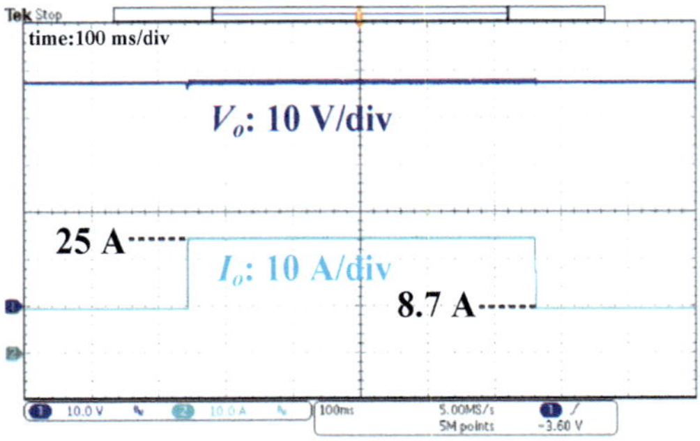

(b)

Fig. 11.9 The output voltage waveforms between the conventional LLC DCX converter and the proposed converter. **a** conventional LLC DCX. **b** proposed converter

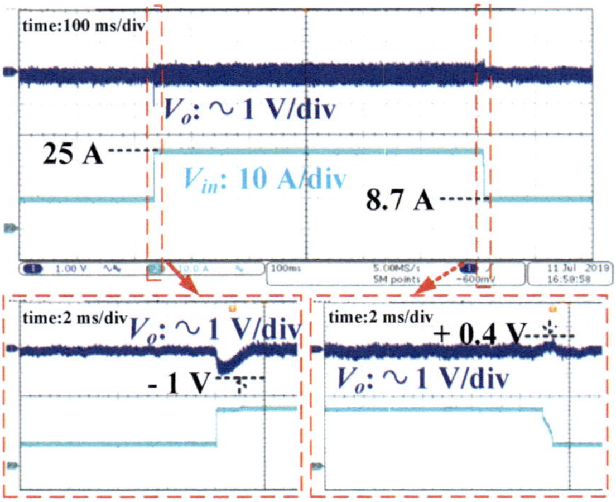

Fig. 11.10 Dynamic response waveforms with load step changing

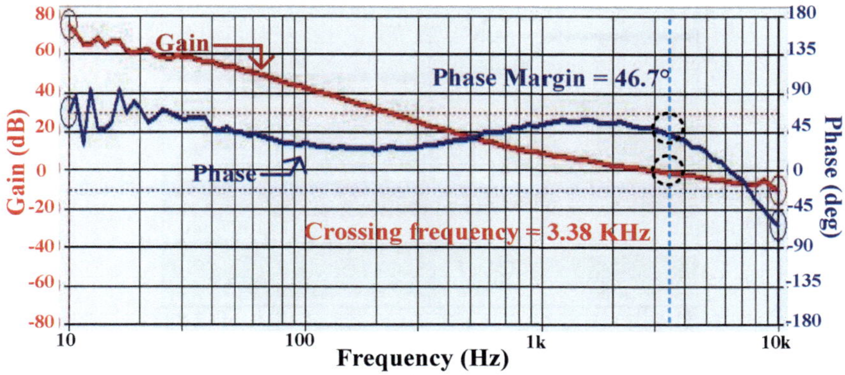

Fig. 11.11 Measured bode plots of the proposed converter

Fig. 11.12 The key waveforms with input voltage variation. a input voltage drops. b input voltage increases

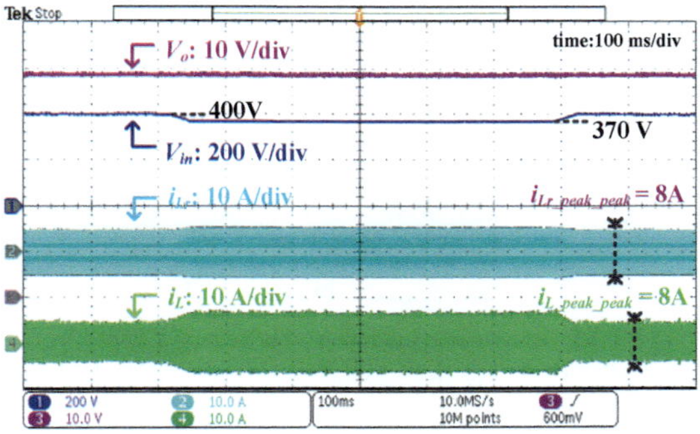

(a)

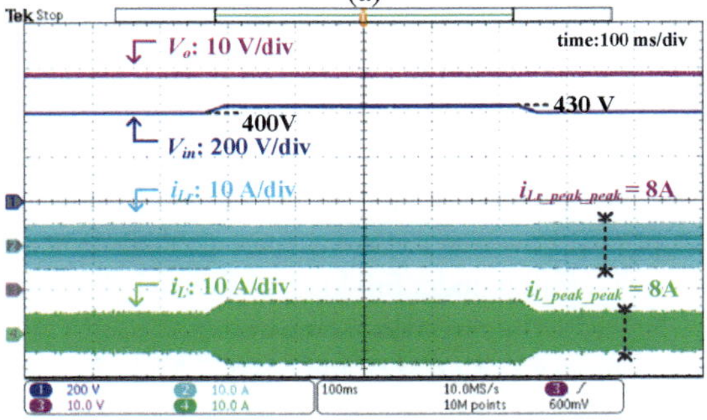

(b)

flows is realized when the power changes between +1200 W and −1200 W, as shown in Fig. 11.13. Figure 11.13a shows the waveforms of forward 1200 W switching to backward 1200 W and Fig. 11.13b shows the waveforms of backward 1200 W switching to forward 1200 W. During the switching process, the output voltage is

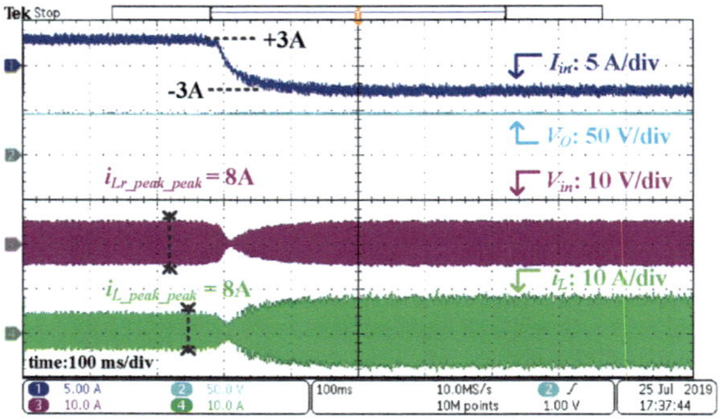

(a)

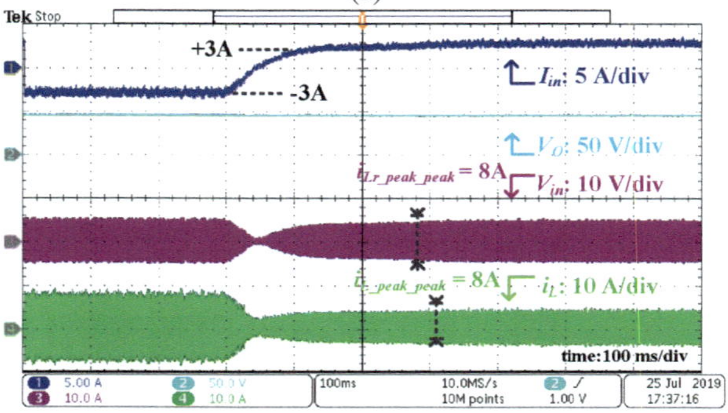

(b)

Fig. 11.13 Dynamic waveforms with power direction switching **a** forward 1200 W switching to backward 1200 W **b** backward 1200 W switching to forward 1200 W

rarely affected, and the resonant current and the inductor current of DAB converter are smoothly changed.

The loss breakdown and cost analysis for the proposed converter and conventional LLC DCX are presented in Fig. 11.14. Although more switches are added in the proposed converter, the losses increased are pretty low because the total device power rating is similar under full load, as shown in Fig. 11.14a. Figure 11.14b shows the unit pricing of switch from the same manufacturer. The switches of the proposed converter are composed of Q_1–Q_8 and S_1–S_8. And, the switches of conventional LLC DCX are composed of Q_1–Q_4 and S_1–S_4. The total price of all switching devices of the proposed converter is just increased by 23.7%. Although the cost of the proposed converter is more expensive, it can take advantage of the natural bidirectional power of DAB converter to achieve tight voltage regulation. The measured efficiencies under different load are shown in Fig. 11.14c. It can be seen that the difference of the peak efficiency between the proposed converter and the conventional LLC DCX is within 0.5%.

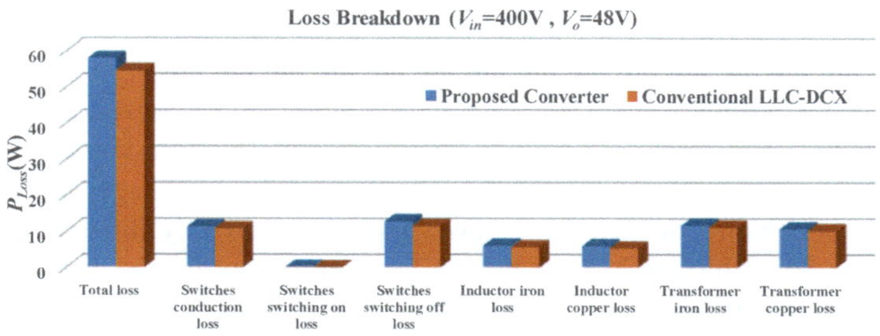

(a)

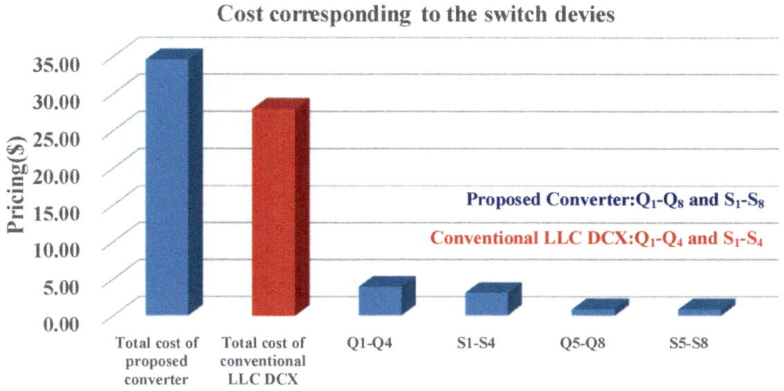

(b)

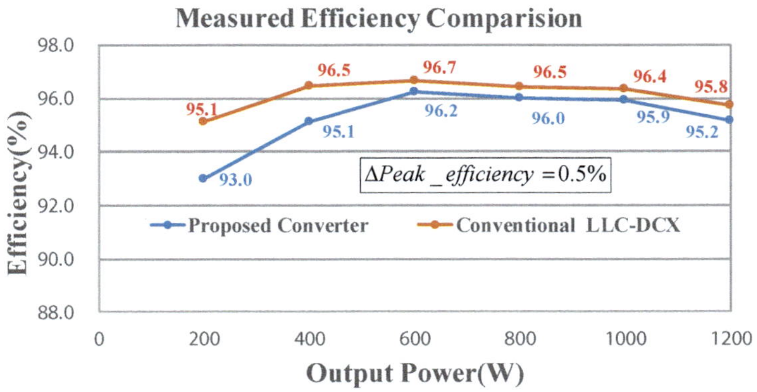

(c)

Fig. 11.14 Comparisons of loss breakdown, cost and efficiency. **a** loss breakdown analysis under full load. **b** cost analysis. **c** measured efficiencies under different load

11.5 Conclusion

In this chapter, a single-stage DAB-LLC hybrid bidirectional converter is proposed, whose inputs are in series and outputs are in parallel. The tight voltage regulation capability is achieved through a DAB converter, and the advantages of LLC DCX operation under bidirectional isolated power application are fully utilized, because the main power flows through LLC converter. In order to improve the dynamic performance, a mathematical model of this new converter is built based on the reduced order modeling method. The accuracy of the model is verified by comparing circuit simulation and calculated results. Then the PI controller and circuit parameters are designed based on the model aiming at the requirements of crossover frequency and phase margin. Finally, the experimental results show that the topology and controller designed in this chapter can achieve tight voltage regulation with good load regulation, fast dynamic response and natural bidirectional power flows. In addition, the proposed converter can maintain the constant output voltage within a certain range of input voltage variation.

Chapter 12
An LLC-DAB Bidirectional DCX Converter with Wide Load Range ZVS and Reduced Switch Count

Abstract An LLC-DAB bidirectional converter with reduced switch count as a DC-DC transformer (DCX) is proposed in this chapter. It can achieve voltage regulation, full load range zero voltage switching (ZVS) for eight main switches and wide load range ZVS for four auxiliary switches, with consideration of the switch junction capacitors under different power direction. A part portion of total power is transferred through an auxiliary DAB circuit to achieve voltage regulation and bidirectional power conversion. The main power is transferred through LLC circuit with an easy PWM scheme achieving high conversion efficiency. Different from the existing LLC-DAB DCX structure, the two secondary side transformer outputs are in parallel, and connected to the same full-bridge circuit. In this way, the auxiliary DAB circuit can not only adjust the output voltage, but also can inject current into the secondary side to help achieving ZVS of switches without synchronous rectifier (SR) circuit. The converter operation principle is analyzed under different power directions, and ZVS is analyzed with the consideration of all the switch junction capacitors. Based on the analysis, the design consideration is introduced. Finally, a 2-kW prototype with 750 V input and 400 V output is built up to verify the effectiveness of the proposed solution.

Keywords Bidirectional power flow · LLC-DAB bidirectional converter · Zero voltage switching

12.1 Proposed Converter and Operation Principles

12.1.1 Converter Topology

The topology of the proposed LLC-DAB bidirectional DCX converter is shown in Fig. 12.2a. Different from the structure shown in Fig. 12.1, the secondary side of the transformer of auxiliary DAB circuit is directly connected in parallel with the secondary side of the transformer of LLC primary circuit. In this way, only four switches are needed in secondary side, which reduces the switch count.

G. Xu and X. Chen, *Topology Deduction and Control for Dual-Active-Bridge Converters*, https://doi.org/10.1007/978-981-97-4856-3_12

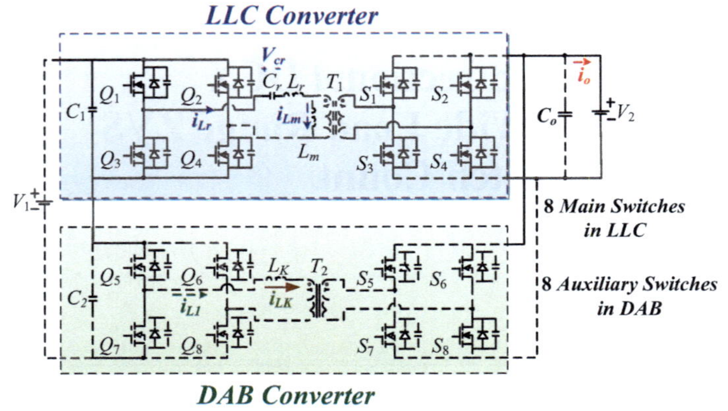

Fig. 12.1 Traditional DAB-LLC DCX converter structure proposed in [1]

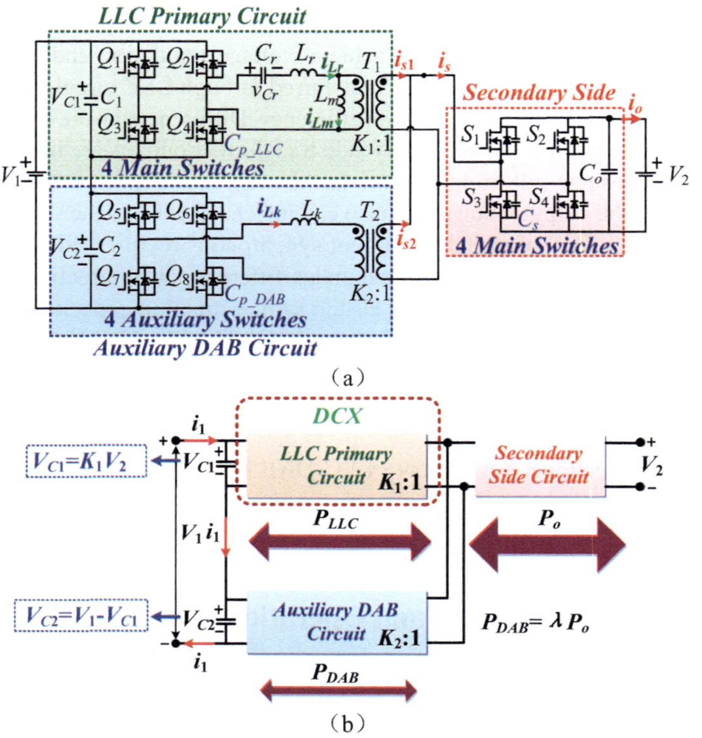

Fig. 12.2 Proposed LLC-DAB bidirectional DCX converter. **a** detailed circuit diagram. **b** pow-er distribution diagram

In Fig. 12.2a, Q1–Q4, Q5–Q8 are the switches of LLC primary circuit and auxiliary DAB, respectively. S_1-S_4 are the secondary side switches. L_r, L_m and L_k are the resonant inductor, magnetizing inductor and leakage inductor, respectively. C_o, C_1, C_2 and C_r are the output filter capacitor, and input capacitor of LLC primary circuit, input capacitor of auxiliary DAB circuit and resonant capacitor, respectively. C_{p_LLC}, C_{p_DAB} and C_s denote the junction capacitors for switches in LLC primary circuit, auxiliary DAB circuit and secondary side respectively, which are equivalently obtained according to 1. V_1, V_2, V_{C1}, V_{C2} and v_{Cr} represent the total input voltage, output voltage, input voltage of LLC primary circuit, input voltage of auxiliary DAB circuit and resonant capacitor voltage, respectively. i_s, i_{Lr}, i_{Lm} and i_{Lk} are the secondary side current, resonant current, magnetizing inductor current and leakage current, respectively. K_1 and K_2 are the turns ratio for transformers T_1 and T_2 respectively. With proper design of these two turns ratios, the distribution of the power for LLC primary circuit and auxiliary DAB circuit can be controlled. It is assumed that $V_{C1}/K_1 = V_2$ under the steady state since LLC circuit works in DCX mode. The main power flows through LLC primary circuit to achieve high conversion efficiency, and the auxiliary DAB circuit utilizes partial power to achieve voltage regulation and bidirectional power flow control, as shown in Fig. 12.2b, where λ is the power distribution coefficient. Under this power distribution, Q_1–Q_4 and S_1–S_4 are selected as main switches and Q_5–Q_8 are selected as auxiliary switches.

The PWM logics are shown in Fig. 12.3, where T_s is the switching period. The proposed converter is operated at a fixed resonant switching frequency with 50% duty cycle. The upper switch and lower switch in any bridge leg are complementary. The driving signals of Q_i ($i = 1, 2, 3$ and 4) are the same as that for S_i ($i = 1, 2, 3$ and 4) at the secondary side. For the auxiliary DAB circuit, Q_5 and Q_8 are turned on and off simultaneously, and has a phase shift angle ϕ with respect to the driving signal of Q_1 and S_1. The voltage regulation and power flow control can be obtained with the controlling of ϕ. It is worth noting that the power transfer from V_1 side to V_2 side is denoted as forward power transfer. Figure 12.3 shows the key waveforms of the proposed converter under forward and backward power transmission mode.

12.1.2 Forward Power Transmission Mode

Figure 12.3a shows the key waveforms under forward power transmission mode. The equivalent circuits during each time interval are shown in Fig. 12.4.

Before t_0, the switches of LLC primary circuit and the secondary side are turned off. And, Q_5 and Q_8 are conducted. The resonant current i_{Lr} flows through the body diodes of Q_1 and Q_4. The secondary side current i_s flows through the body diodes of S_1 and S_4, as shown in Fig. 12.4a.

Stage 1 [t_0, t_1] [see Fig. 12.4b]: At t_0, Q_1, Q_4, S_1 and S_4 are turned on with zero voltage switching. LLC primary circuit enters the resonance process. The resonant current i_{Lr}, magnetizing current i_{Lm}, leakage current i_{Lk} and resonant capacitor voltage v_{Cr} can be expressed as

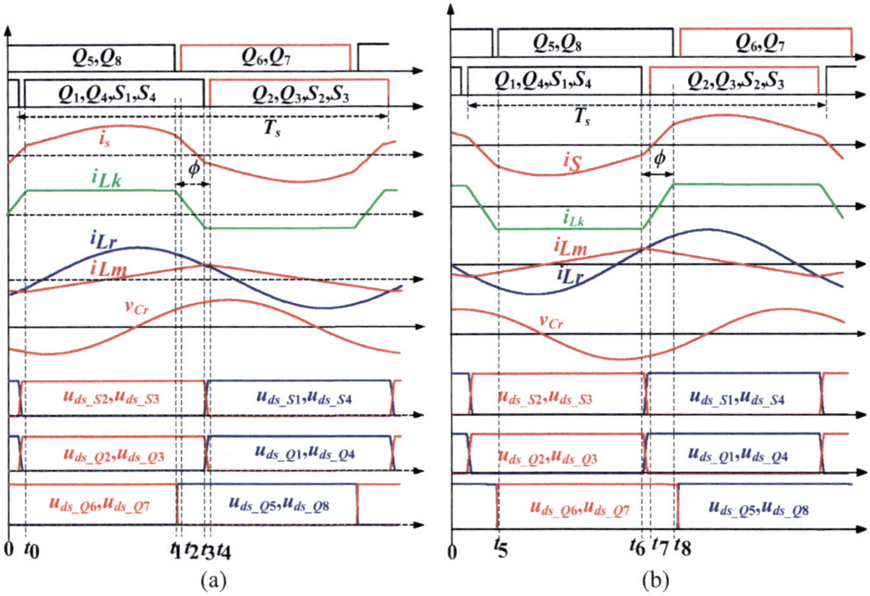

Fig. 12.3 Key waveforms of the proposed converter. **a** forward power transmission mode. **b** backward power transmission mode

$$\begin{cases} i_{Lr}(t) = \sqrt{\left(\dfrac{v_{Cr}(t_0)}{Z_0}\right)^2 + i_{Lr}(t_0)^2}\sin[\omega_r(t - t_0) - \theta] \\ i_{Lm}(t) = \dfrac{K_1 V_2}{L_m}t - \dfrac{K_1 V_2 T_s}{4L_m} \\ i_{Lk}(t) = i_{Lk}(t_0) + \dfrac{V_{C2} - K_2 V_o}{L_k}(t - t_0) \\ v_{Cr}(t) = v_{Cr}(t_0) - \dfrac{1}{C_r}\displaystyle\int_{t_0}^{t} i_{Lr}(t)dt \end{cases} \tag{12.1}$$

where Z_0 is the impedance of resonant tank. The relationship between i_s and the currents of the two transformer secondary sides (i_{s1} and i_{s2}), is shown as follow.

$$i_s = i_{s1} + i_{s2} = K_1(i_{Lr} - i_{Lm}) + K_2 i_{Lk} \tag{12.2}$$

Stage 2 [t_1, t_2] [see Fig. 12.4c]: At t_1, Q_5 and Q_8 are turned off. The leakage current $i_{Lk}(t_1)$ is larger than zero, which is expressed as

$$i_{Lk}(t_1) = \frac{2K_2 V_2 \phi - K_2 V_2 + V_{C2}}{4L_k f_s} = \frac{K_2 V_2 \phi}{2L_k f_s} \tag{12.3}$$

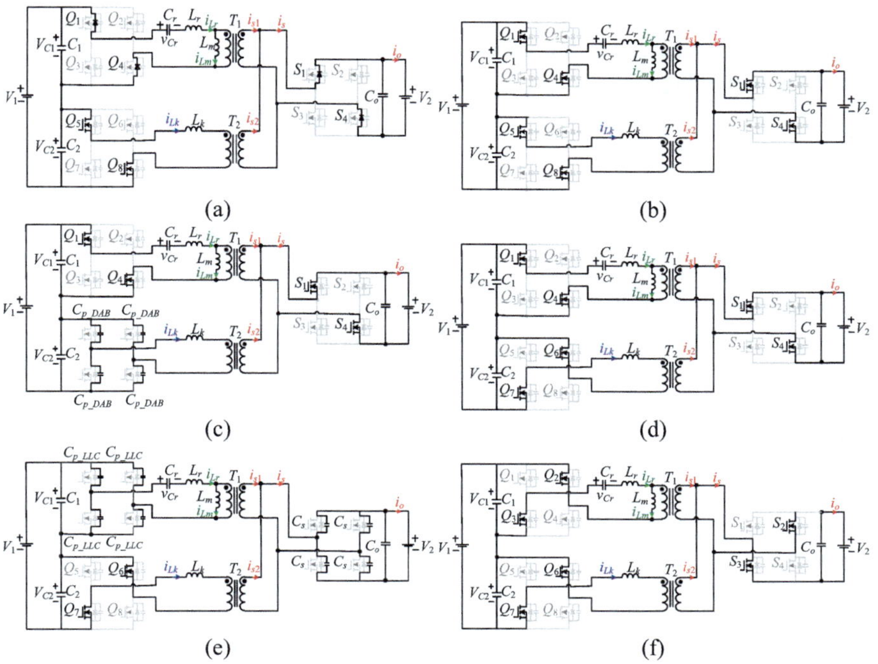

Fig. 12.4 Equivalent circuit of the proposed converter in forward power transmission mode. **a** before t_0. **b** $[t_0, t_1]$. **c** $[t_1, t_2]$. **d** $[t_2, t_3]$. **e** $[t_3, t_4]$. **f** after t_4

where f_s is the switching frequency. The junction capacitors of auxiliary DAB circuit can be discharged/charged through i_{Lk}. At the end of this stage, i_{Lk} flows through the body diodes of Q_6 and Q_7 to create the ZVS turn on condition.

Stage 3 $[t_2, t_3]$ [see Fig. 12.4d]: At t_2, Q_6 and Q_7 are turned on with zero voltage switching. Then i_{Lk} starts to decrease with a constant slop.

Stage 4 $[t_3, t_4]$ [see Fig. 12.4e]: At t_3, Q_1, Q_4, S_1 and S_4 are off. At this time, i_{Lr} is equal to i_{Lm}, which can be expressed as

$$i_{Lr} = i_{Lm} = \frac{K_1 V_2}{4L_m f_s} \tag{12.4}$$

During this stage, the magnetizing inductor current will charge/discharge the junction capacitors of switches in LLC primary circuit. As for the secondary side switches, even though the secondary side current of T_1 will approach zero, the current of T_2 can still provide energy to discharge/charge the junction capacitor of the secondary side switches to achieve ZVS switching. The current of the transformer secondary side can be expressed as (12.5). At the end of this stage, the body diodes of Q_2, Q_3, S_2 and S_3 will be conducted.

$$i_s(t_3) = i_{s2}(t_3) = K_2 i_{Lk} < 0 \tag{12.5}$$

Stage 5 [*After* t_4] [see Fig. 12.4f]: At t_4, Q_2, Q_3, S_2 and S_3 turn on with zero voltage. After t_4, the operation mode is similar to *Stage* 1, and the other half of the switching period begins.

12.1.3 Backward Power Transmission Mode

When the power transfer is reversed, the phase-shift angle ϕ will be negative. The backward operation is shown in Fig. 12.5b.

Stage 1 [t_5, t_6] [see Fig. 12.5a]: At t_5, Q_5 and Q_8 are turn on. During this stage, L_r is resonating with C_r, and i_{Lk} increases linearly since the applied voltage is constant.

Stage 2 [t_6, t_7] [see Fig. 12.5b]: At t_6, Q_1, Q_4, S_1 and S_4 are turned off. And, i_{Lr} equals to i_{Lm}. At this time, $i_s = i_{s2}$, since i_{s1} is approaching zero.

Then, the junction capacitors of switches in LLC primary circuit can be charged/discharged by current i_{Lr}, and those in the secondary side can be charged/discharged by current i_{s2}. At the end of this stage, the body diodes of Q_2, Q_3, S_2 and S_3 will be conducted.

Stage 3 [t_7, t_8] [see Fig. 12.5c]: At t_7, Q_2, Q_3, S_2 and S_3 are on with zero voltage. Then, i_{Lk} starts to increase and the resonant tank enters into the resonance process again.

Stage 4 [*After* t_8] [see Fig. 12.5d]: At t_8, Q_5 and Q_8 are turned off. The discharge/charge process of junction capacitor is similar to the forward operation. After t_8, the operation modes are similar with the previous stage.

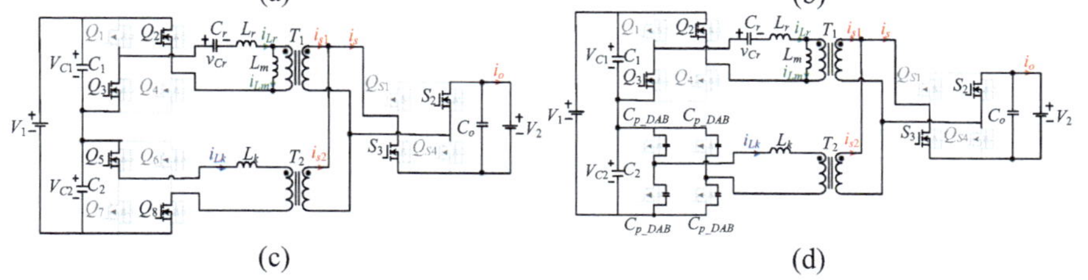

Fig. 12.5 Equivalent circuit of the proposed converter in backward power transmission mode. **a** [t_5, t_6]. **b** [t_6, t_7]. **c** [t_7, t_8]. **d** after t_8

12.2 ZVS Operation Analysis

This section mainly analyzes ZVS conditions of all the switches under the forward power transmission and backward power transmission modes.

12.2.1 ZVS Analysis of the Secondary Side Circuit

(1) **Forward Power Transmission Mode**

As analyzed in the above, i_{Lk} reaches the negative maximum during *Stage* 4 in the forward power transmission mode, which can inject current to the secondary side to assist the discharge/charge process according to (12.5). Moreover, when the junction capacitors of LLC primary circuit complete discharge/charge process, i_{Lr} starts to decrease. The difference between i_{Lr} and i_{Lm} can also assist the secondary side to achieve the ZVS operation. Then, to completely discharge/charge of secondary side junction capacitors before the end of the dead time, the following relationship should be satisfied.

$$2K_2 i_{Lk}(t_3)T_{dead} + \left(T_{dead} - T_{p\_dis}\right)K_1 i_{Lm}(t_3) \geq 4C_s V_2 \tag{12.6}$$

where $i_{Lk}(t_3)$ and $i_{Lm}(t_3)$ can be calculated according to (12.1), and T_{dead} and T_{p_dis} are dead time and discharge/charge time of LLC primary junction capacitors, respectively.

(2) **Backward Power Transmission Mode**

Under the backward power transmission mode, the secondary side current can be deduced from the circuit.

$$i_s = K_1(i_{Lr} - i_{Lm}) + K_2 i_{Lk} \tag{12.7}$$

The difference of the ZVS operation analysis between the backward and forward power transmission mode is that the transformer secondary current of auxiliary DAB circuit starts to decrease toward zero, as shown in Fig. 12.5b. Then, the secondary side current at the end of the discharge/charge process can be expressed as

$$i_{s\_end} = K_1(i_{Lr}(t_6) + \Delta i_{Lr} - i_{Lm}(t_6)) + K_2(i_{Lk}(t_6) + \Delta i_{Lk}) \tag{12.8}$$

where Δi_{Lr} is the variation of resonant current caused by the inconsistent discharge/charge time of C_{p_LLC} and C_s, and Δi_{Lk} means the current injected into the secondary side by auxiliary DAB circuit gradually decreases. According to (12.8), the range of ZVS operation of the secondary side is narrower.

In order to ensure the achievement of ZVS operation of the secondary side, the secondary side current needs to be negative at end of *Stage* 2 $[t_6, t_7]$ in the backward power transmission mode. As the switches are turned off at the resonant point, $i_{Lr}(t_6)$ equals to $i_{Lm}(t_6)$ in (12.8). Thus, the secondary side current at the end of this interval, i_{s_end}, can be expressed as

$$i_{s\_end} = K_1 \Delta i_{Lr} - \frac{K_2^2 V_2 \phi}{2L_k f_s} + K_2 \frac{V_{C2} + K_2 V_2}{L_k}(T_{dead} - T_{dis}) \tag{12.9}$$

where T_{dis} is the total discharge/charge time of primary and secondary. The difference of the discharge/charge time for primary and secondary side junction capacitors will bring a sudden change on the voltage v_{Lr} crossing the resonant inductor L_r, thus changing the sign of Δi_{Lr}. When the discharge/charge time of the secondary side is shorter than LLC primary circuit, $v_{Lr} = V_{C1} + K_1 V_2 + v_{Cr}$ which make Δi_{Lr} to be positive. Conversely, if the discharge/charge time of the secondary side is longer than LLC primary circuit, $v_{Lr} = -V_{C1} - K_1 V_2 + v_{Cr}$ which make Δi_{Lr} to be negative. According to (12.9), the range of ZVS operation can be extended if Δi_{Lr} is negative. Therefore, T_{p_dis} is designed smaller than the discharge/charge time of the secondary side T_{s_dis} to make v_{Lr} smaller than zero in this chapter. The theoretical discharge/charge time can be calculated as

$$\begin{cases} T_{p\_dis} = 8C_{p\_LLC} L_m f_s \\ T_{s\_dis} = \dfrac{4C_s L_k f_s}{K_2^2 \phi} \end{cases} \tag{12.10}$$

An illustration of the discharge/charge process is shown in Fig. 12.6. The equivalent charging capacitor value is used here, so the discharge/charge process can be linear. Then, v_{Lr} at the end of the discharge/charge time of LLC primary circuit, can be calculated as

$$\begin{aligned} v_{Lr}(t_{6a}) &= -2K_1 V_2 + v_{Cr}(t_{6a}) + \frac{T_{p\_dis} K_1 K_2 i_{Lk}(t_6)}{C_s} \\ &= -2K_1 V_2 + \frac{V_2}{4RC_r f_s (K_1 + K_2)} + \frac{T_{p\_dis} K_1 K_2^2 V_2 \phi}{2L_k f_s C_s} < 0 \end{aligned} \tag{12.11}$$

where R denotes the load. Since the discharge/charge process of secondary side is not complete, the remaining discharge/charge time is divided into two intervals from t_{6a} to t_{6c}.

The first duration is from t_{6a} to the time when v_{Lr} becomes zero at t_{6b}, and i_{Lr} will decrease to be less than i_{Lm}, which causes the current of the transformer secondary side of LLC primary circuit i_{s1} less than zero. Then, the discharge/charge process of the secondary side will be accelerated. Denoting that $T_{6ba} = t_{6b} - t_{6a}$, it is obtained that

$$v_{Lr}(t_{6b}) = v_{Lr}(t_{6a}) - \frac{K_1 K_2^2 V_2 \phi}{2C_s L_k f_s} + \frac{K_1^2 T_{6ba}^2 v_{Lr}(t_{6a})}{2C_s L_r} = 0 \tag{12.12}$$

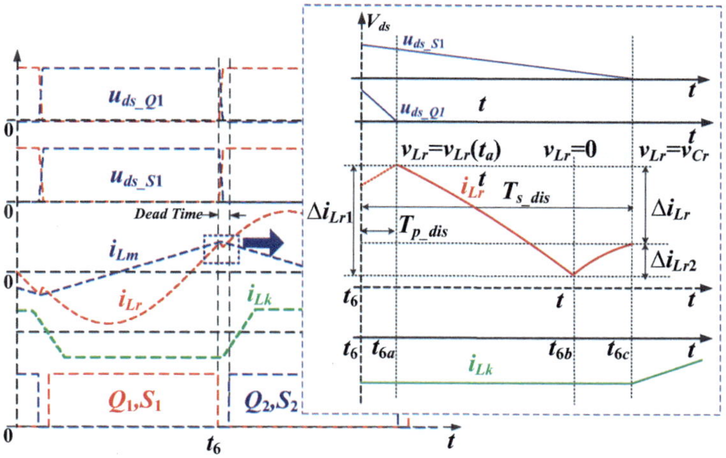

Fig. 12.6 Illustration of the discharge/charge process under backward power transmission mode

At end of this interval, the variation of the resonant current can be calculated as

$$\Delta i_{Lr1} = \frac{K_2^2 V_2 \phi + \sqrt{\left(K_2^2 V_2 \phi\right)^2 + 8 C_s L_k^2 f_s^2 v_{Lr}(t_{6a})^2 / L_r}}{4 K_1 L_k f_s} \tag{12.13}$$

Then, the duration T_{6ba} from t_{6a} to t_{6b} can be expressed as

$$T_{6ba} = t_{6b} - t_{6a} = -\frac{2 L_k f_s C_s v_{Lr}(t_{6\mathrm{a}})}{K_1 \left(K_2^2 V_2 \phi + 2 L_k f_s K_1 \Delta i_{Lr1}\right)} \tag{12.14}$$

The second duration is from t_{6b} to the time when the discharge/charge process of secondary side is complete at t_{6c}, the secondary side junction capacitor voltage can be calculated as

$$u_{ds\_s}(t_{6b}) = \frac{v_{Cr}}{2K_1} = \frac{V_2}{8 R C_r f_s K_1 (K_1 + K_2)} \tag{12.15}$$

When v_{Lr} becomes larger than zero, i_{Lr} starts to increase until the discharge/charge process ends. The variation of the resonant current in this interval can be calculated as

$$\Delta i_{Lr2} = \frac{-K_2^2 V_2 \phi + \sqrt{\left(K_2^2 V_2 \phi\right)^2 + 16 L_k^2 f_s^2 C_s u_{ds\_s}(t_{6b}) v_{Cr} / L_r}}{4 K_1 L_k f_s} \tag{12.16}$$

Then, the duration T_{6cb} from t_{6b} to t_{6c} can be calculated as

$$T_{6cb} = t_{6c} - t_{6b} = -\frac{4L_k f_s C_s u_{ds\_s}(t_{6b})}{K_1\left(-K_2^2 V_2 \phi + 2K_1 L_k f_s \Delta i_{Lr1}\right)} \tag{12.17}$$

Finally, the total variation of the resonant current Δi_{Lr} and the duration from T_{p_dis} to the completion of the discharge/charge process can be obtained as

$$\begin{cases} \Delta i_{Lr} = \Delta i_{Lr2} - \Delta i_{Lr1} \\ T_{6ca} = T_{6ba} + T_{6cb} \end{cases} \tag{12.18}$$

Based on (12.10) and (12.18), i_{s_end} can be calculated as

$$i_{s\_end} = K_1 \Delta i_{Lr} - \frac{K_2^2 V_o \phi}{2L_k f_s} + K_2 \frac{V_{C2} + K_2 V_o}{L_k}\left(T_{dead} - T_{6ca} - T_{p\_dis}\right) \tag{12.19}$$

where Δi_{Lr} can be calculated according to (12.13) and (12.16), and T_{6ca} can be calculated according to (12.10), (12.14) and (12.17). Therefore, through proper parameter design, the current before the switches turned on, i_{s_end} in (12.19), can be negative, thus achieving the ZVS operation of the secondary side.

12.2.2 ZVS Analysis of LLC Primary Circuit

(1) **Forward Power Transmission Mode**

When the switches of LLC primary circuit are turned off, the resonant current equals to the magnetizing inductor current. Then, the ZVS realization depends on the magnetizing inductor current i_{Lm}. To complete the discharge/charge process of the junction capacitor during dead time, according to (12.10), the discharge/charge time can be derived as

$$T_{p\_dis} = 8C_{p\_LLC} L_m f_s \leq T_{dead} \tag{12.20}$$

At the same time, the dead time should not set too large to cause anti-resonance.

(2) **Backward Power Transmission Mode**

As shown in Fig. 12.6b, under the backward power transmission mode, the charging/discharging energy are also provided by the magnetizing current, therefore, (12.20) is still valid.

12.2.3 ZVS Analysis of Auxiliary DAB Circuit

The ZVS operation analysis of auxiliary DAB circuit using the SPS modulation strategy has been discussed extensively in the literature. In the proposed converter, the energy stored in the leakage inductor needs to be higher than the energy stored in the junction capacitor of switches, as shown follow.

$$L_k i_{Lk\_off\ \min}^2 \geq 2C_{p\_DAB} V_{C2}^2 \tag{12.21}$$

where i_{Lk_offmin} is the minimum turn off current.

12.3 Parameters Design

According to the above analysis, to realize the ZVS operation of all the switches under different power transmission directions, the parameters need to be proper designed.

12.3.1 Power Distribution and Transformer Turns Ratios

Since the proposed converter is a sigma structure, the distribution of power flowing through LLC primary circuit and auxiliary DAB circuit is determined by the corresponding transformer turns ratio. Referring the power distribution analysis in [2], this chapter further discusses. It can be known that if the power ratio of auxiliary DAB circuit is very small, it will affect the dynamics of the system. At the same time, the power ratio of LLC primary circuit will increase accordingly, causing high voltage stress of the switches, which will result in the need for higher voltage level switching devices and insulation. In this chapter, the power distribution is designed based on the tradeoff between ZVS region and these aforementioned factors, which will be shown in part C.

12.3.2 Leakage Inductor L_k

The leakage inductor in auxiliary DAB circuit is a power link between the primary and secondary side, and it is related to the maximum transmission power. The maximum steady-state phase shift angle ϕ_{max} is defined as 0.25 considering the maximum power margin and the ratio of the reactive power to active power, as shown in Fig. 12.7.

The maximum value of L_k can be obtained as

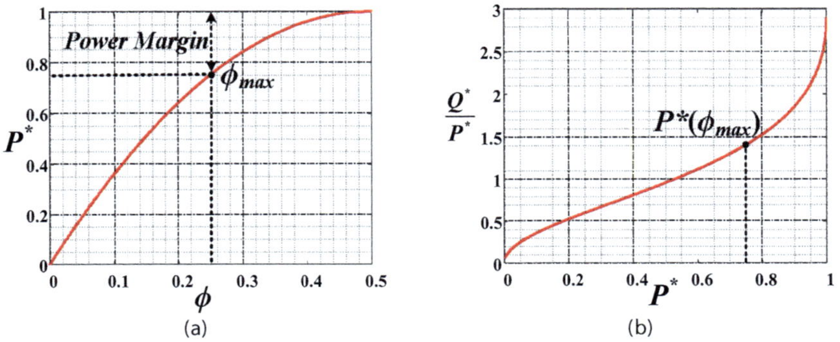

Fig. 12.7 The transferred power characteristics and the ratio of the reactive power for the DAB converter. **a** the transferred power characteristics. **b** the ratio of the reactive power to active power

$$L_k \leq \frac{K_2 R V_{C2} \phi_{\max}(1 - \phi_{\max})}{2 V_2 f_s} \tag{12.22}$$

Then, based on the ZVS operation analysis of auxiliary DAB circuit, the minimum i_{Lk_off} can be deduced according to (12.3) as

$$i_{Lk\_off\ \min} = \frac{K_2 V_2 \phi_{\min}}{2 L_k f_s} = \frac{i_{Lk\_off}}{n} \tag{12.23}$$

where n is the ratio between ϕ_{max} and ϕ_{min}, which can be obtained as

$$n = \frac{\phi_{\max}}{\phi_{\min}} = \frac{2\phi_{\max}}{1 - \sqrt{1 - (0.8)\phi_{\max}(1 - \phi_{\max})}} \tag{12.24}$$

Substituting (12.22) and (12.23) into (12.24) can obtain as

$$L_k \geq \frac{2 C_{p\_DAB} K_2^2 [n R V_{C2}(1 - \phi_{\max})]^2}{V_2^2} \tag{12.25}$$

The minimum leakage inductor corresponding to the minimum turn off current can be obtained, which ensures ZVS operation under light load conditions (10% of rated power). If the value is less than the rated maximum inductor value, a compromise can be made within the interval, given as

$$\frac{2 C_{p\_DAB} K_2^2 [n R V_{C2}(1 - \phi)]^2}{V_2^2} \leq L_k \leq \frac{K_2 R V_{C2} \phi (1 - \phi)}{2 V_2 f_s} \tag{12.26}$$

12.3.3 Magnetizing Inductor L_m

According to the previous ZVS analysis, the secondary side under backward power transmission is the most difficult to achieve ZVS. The discharge/charge time of LLC primary circuit T_{p_dis} is designed to be smaller than the secondary side time T_{s_dis} for extend the ZVS operation range. According to (12.10), L_m should be satisfied the following relationship.

$$L_m \leq \frac{C_s L_k}{2K_2^2 \phi C_{p\_LLC}} \tag{12.27}$$

It can be seen from (12.24) and (12.25), the magnetizing inductor is related to the transformer turns ratio, which means that to determine the above parameters, it is necessary to determine the power distribution. Since the proposed structure, the power distribution of the auxiliary DAB circuit when ignoring the conversion loss is given as

$$\frac{P_{DAB}}{P_o} = \frac{P_{DAB}}{P_{LLC} + P_{DAB}} = \frac{V_{C2} i_1}{V_{C1} i_1 + V_{C2} i_1} = \frac{K_2}{K_1 + K_2} = \lambda \tag{12.28}$$

where i_1 is input current and λ is a power distribution coefficient. In order to determine the power distribution relationship, the basic parameters of the proposed converter are given in Table 12.1, where the value of the junction capacitors can be obtained by referring to the datasheet. To achieve high efficiency, the proportion of the power that flows through auxiliary DAB circuit should be as low as possible to ensure that main power flows through LLC primary circuit. Generally, λ is better smaller than 1/3, thus the leakage inductor L_k and power devices can be initially determined according to (12.26). As the switching frequency selects as 100 kHz, the resonant inductor L_r and the resonant capacitor C_r are selected as shown in Table 12.1 to form a resonant tank, where the relationship can be expressed as

$$f_s = f_r = \frac{1}{2\pi\sqrt{L_r C_r}} \tag{12.29}$$

In order to final design magnetizing inductor $\mathrm{L_m}$, under the premise of satisfying (12.27), substituting system parameters into (12.19) to calculate the secondary side current i_s at the end of dead time under different L_m with different λ, as shown in Fig. 12.8. If the secondary side current i_s can maintain a negative value, the ZVS operation can be achieved. It can be seen from Fig. 12.8, if L_m exceeds the boundary, there will be a case where i_s is greater than zero in the same power range, thus causing ZVS to be lost. It can also be concluded from Fig. 12.8 that when λ is decreased, the ZVS region can be extended.

According to previous analysis, the voltage regulation ability will be weak as λ decreases. Finally, the magnetizing inductor L_m and the power distribution coefficient λ are selected as 500 μH and 4/15 for a compromise according to Fig. 12.8.

Table 12.1 System parameters

Symbol	Parameter	Value
V_1	Input voltage	750 V
V_2	Output voltage	400 V
L_r	Resonant inductor	25 μH
C_r	Resonant capacitor	0.1 μF
L_k	Leakage inductor	55 μH
P_o	Rated power	2 kW
f_s	Switching frequency	100 kHz
C_{p_LLC}	Junction capacitor in LLC primary circuit	200 pF
C_{p_DAB}	Junction capacitor in auxiliary DAB circuit	250 pF
C_s	Junction capacitor in secondary	290 pF

12.4 Experiment Results

A 2 kW LLC-DAB bidirectional DCX converter prototype is built up with 750 V input and 400 V output voltage. The system circuit parameters are previously shown in Table 12.1. The system control block diagram is shown in Fig. 12.9. The LLC circuit is open loop operation, which is operated at fixed resonant switching frequency. A proportional-integral (PI) controller is adopted by auxiliary DAB circuit to regulate the output voltage regulation, which is implemented on a digital signal processor TMS320F28069.

12.4.1 Steady State Performance

The steady state performance of the proposed converter under full load in different power transmission directions is shown in Fig. 12.10, which includes output voltage V_o, leakage inductor current of auxiliary DAB circuit i_{Lk}, resonant current of LLC primary circuit i_{Lr} and output current I_o, respectively.

In order to verify the power distribution relationship, Fig. 12.11 shows the input voltage of LLC primary circuit and auxiliary DAB circuit, V_{C1} and V_{C2}, under light and full load condition. It can be seen that the ratio of input voltage of LLC primary circuit and auxiliary DAB circuit is around 550/200. As the proposed converter is series-connected at the input, the input currents of the two circuits are equal. Hence, it can be calculated that the power flowing through LLC is $550/(550 + 200) = 73.3\%$ of the total power, and the power flowing through DAB accounts for $200/(550 + 200) = 26.7\%$ of the total power. Therefore, the main power is processed by LLC primary circuit, while partial power has been transferred through the auxiliary DAB circuit. This agrees well with the analysis.

Figure 12.12 shows the gate signals of the switches in LLC primary circuit (u_{gs_Q1}) and the secondary side (u_{gs_S1}) under the forward and backward power transmission

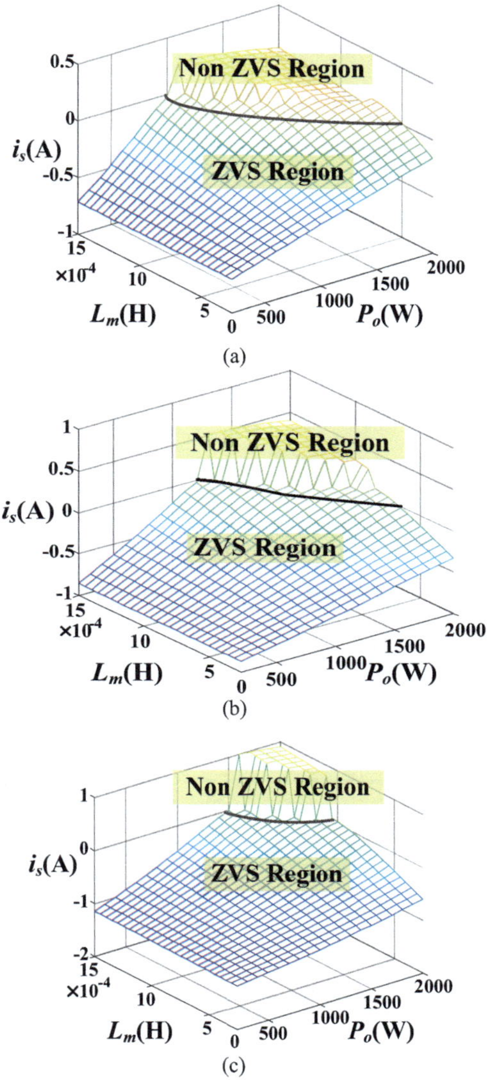

a

Fig. 12.8 ZVS range of the secondary side in backward power transmission mode. **a** λ = 1/3. **b** λ = 4/15. **c** λ = 1/5

modes, which are consistent. It means that the PWM scheme does not need to be changed when the power direction changes, and it is easy to implement.

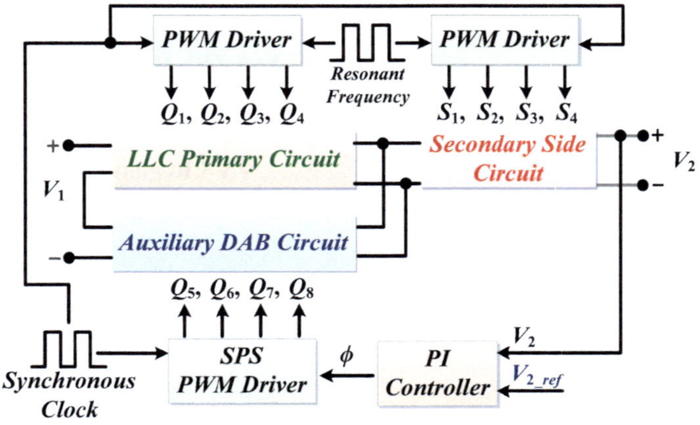

Fig. 12.9 System control block diagram

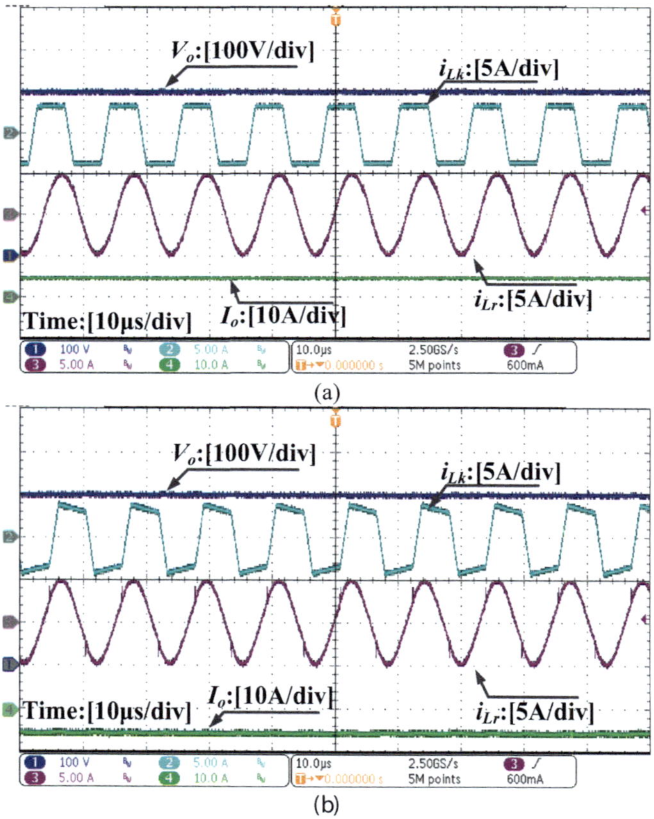

Fig. 12.10 Performance of the proposed converter under different power transmission modes. **a** forward power transmission mode. **b** backward power transmission mode

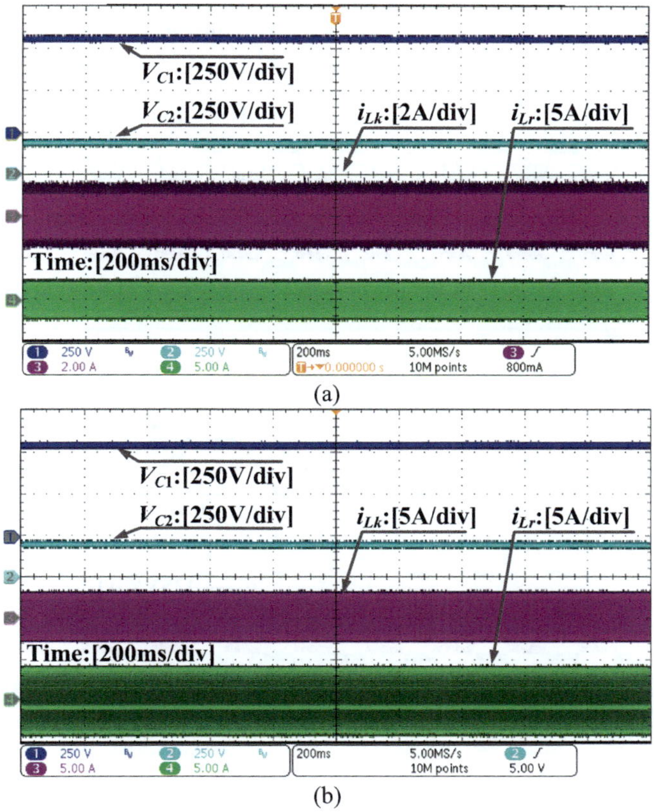

Fig. 12.11 Key waveforms of power distribution of the proposed converter. **a** under light load condition. **b** under full load condition

12.4.2 Soft Switching Performance and Comparison

In order to verify ZVS performance under the forward power transmission mode. The soft switching performance of proposed solution is shown in Fig. 12.13, where Fig. 12.13a, b, d and e show the output voltage V_o, resonant current of LLC primary circuit i_{Lr}, drain to source voltage and drive voltage can be observed. Fig. 12.13c and f show the output voltage V_o, leakage current of auxiliary DAB circuit i_{Lk}, drain to source voltage and drive voltage, respectively. From the measured waveforms, the ZVS operation on all switches of the LLC primary circuit, secondary side and auxiliary DAB circuit can be realized under the full load (2 kW) and light load (200 W).

Similarly, the soft switching performance under backward power transmission mode is shown in Fig. 12.14. It can be seen that the ZVS operation on all switches can also be achieved under full load (2 kW) and light load (200W) condition.

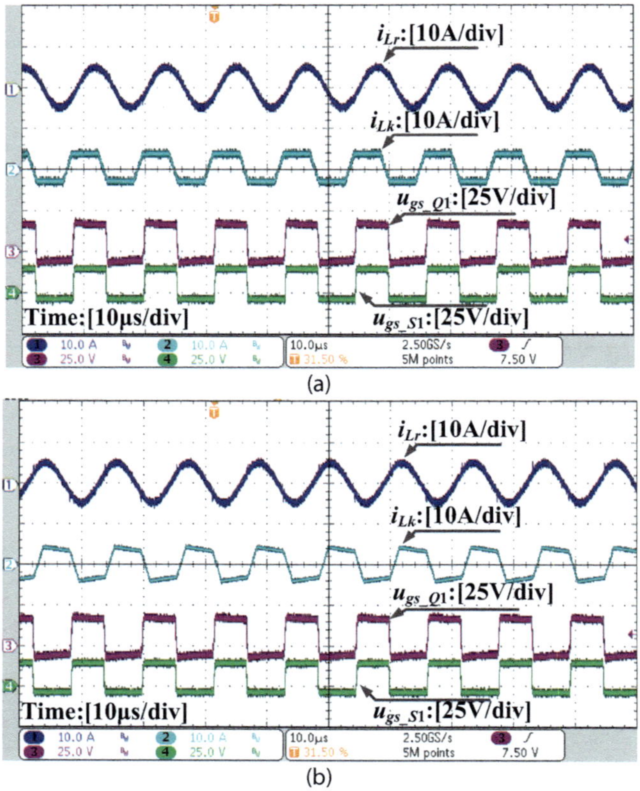

Fig. 12.12 Gate signals of LLC primary circuit and secondary side under the full load condition. **a** forward power transmission mode. **b** backward power transmission mode

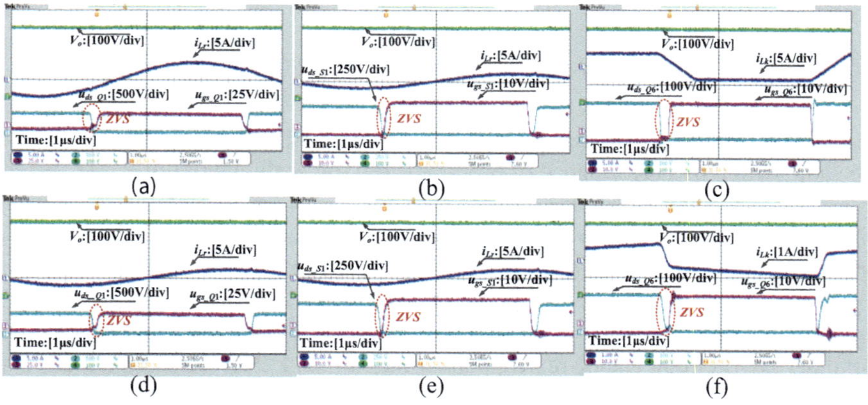

Fig. 12.13 Soft switching performance of the proposed converter under forward power transmission mode. **a** and **d** are for LLC primary circuit in full and light load conditions. **b** and **e** are for the secondary side in full and light load conditions. **c** and **f** are for auxiliary DAB circuit in full and light load conditions

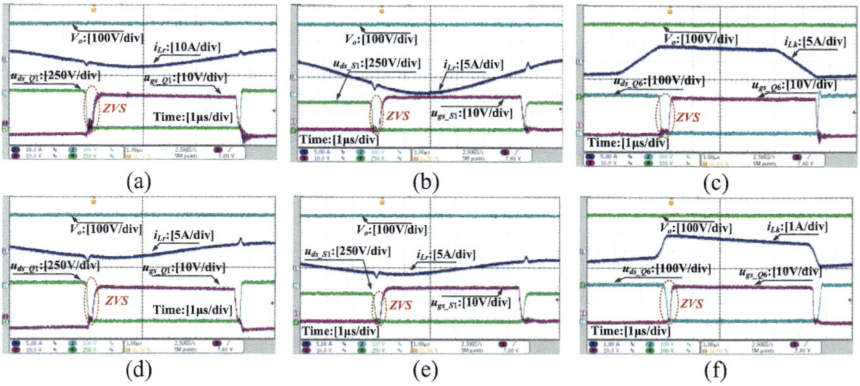

Fig. 12.14 Soft switching performance of the proposed converter under backward power transmission mode. **a** and **d** are for LLC primary circuit in full and light load conditions. **b** and **e** are for the secondary side in full and light load conditions. **c** and **f** are for auxiliary DAB circuit in full and light load conditions

Table 12.2 Switch types for the converters

Item	Proposed converter	Converter in [2]
Q_1–Q_4	NTHL190N65S3HF	
S_1–S_4	FCH072N60	IPW60R170CFD7
Q_5–Q_8	IRFP4137PBF	
S_5–S_8	Not used	IPW60R125P6

To further demonstrate the superiority of the proposed converter in terms of ZVS operation, comparison experiments are carried out on the traditional DAB-LLC DCX converter proposed in [2], where the system parameters are the same as those in Table 12.1, and the switching devices on the secondary side are chosen to be the same, as shown in Table 12.2. Figure 12.15 shows its soft switching performance. It can be seen that the ZVS operation is lost at half load and full load conditions when the power transfers backward. Therefore, it can also be verified that under the same modulation strategy, the traditional DAB-LLC DCX converter is difficult to completely realize the ZVS operation in full load range under different power transmission modes. While the proposed solution is effective which has been shown in Figs. 12.13 and 12.14, respectively.

12.4.3 Dynamic Performance and Efficiency

In order to verify the voltage regulation capability, the load is changed from 400 to 1800 W. The key waveforms are shown in Fig. 12.16. It can be seen that the output voltage remains stable at 400 V under the load step change.

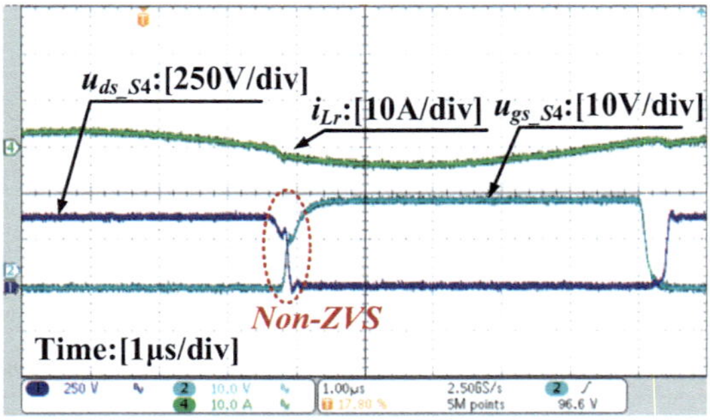

(a)

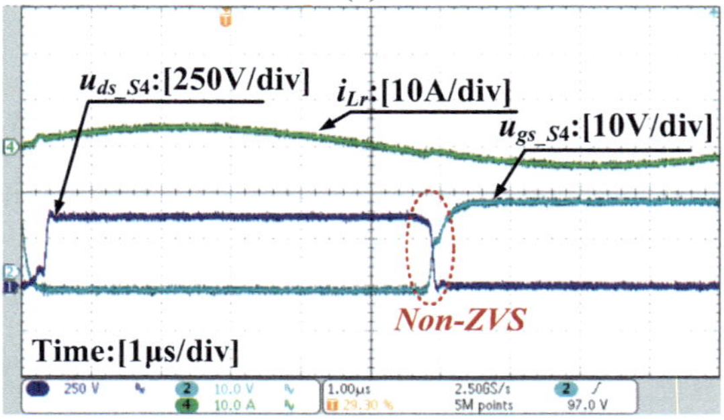

(b)

Fig. 12.15 Soft switching performance of secondary side in the traditional DAB-LLC DCX converter under backward power transfer. **a** half load condition. **b** full load condition

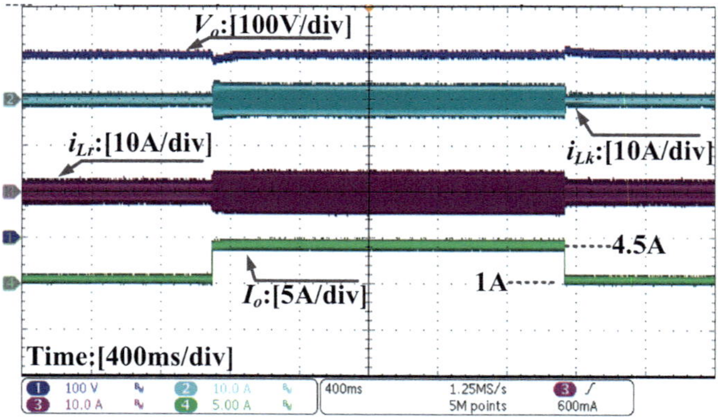

Fig. 12.16 Dynamic performance of the proposed converter with a load step change

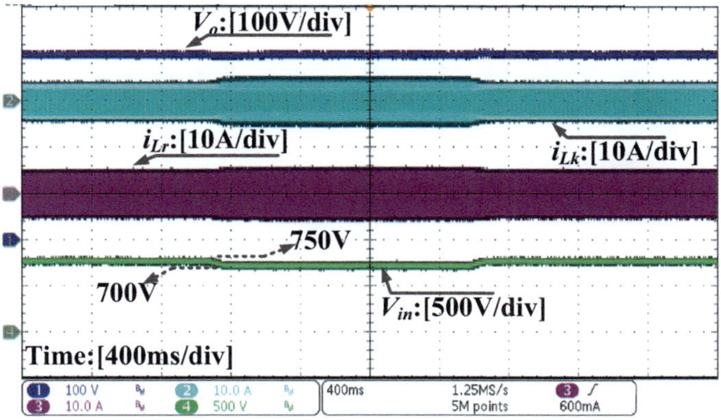

(a)

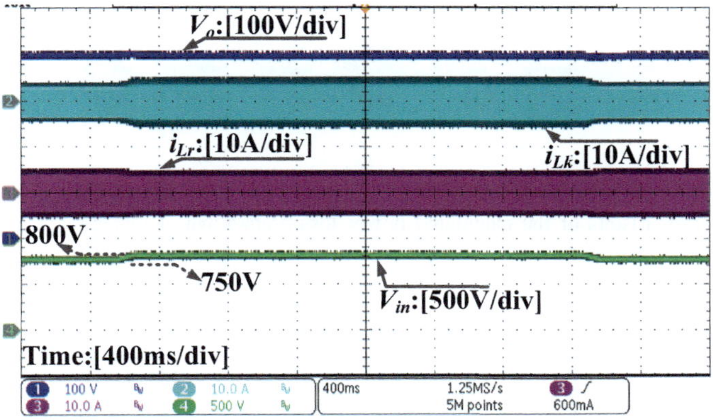

(b)

Fig. 12.17 Performance of the proposed converter with input voltage variation. **a** input voltage drops. **b** input voltage increases

Since LLC primary circuit and the secondary side is used as a DCX in the proposed converter, its output voltage is determined by the input voltage. When the input voltage changes, the proposed converter can stabilize the output voltage by adjusting the phase shift angle in auxiliary DAB circuit. Figure 12.17 shows the key waveforms when the input voltage drops from 750 to 700 V and increases from 750 to 800 V. It can be seen that the output voltage remains constant under different input voltage.

12.4.4 Efficiency Comparison

In order to verify the advantages of choosing LLC circuit as the main power processing unit, Fig. 12.18 shows the loss comparison between the proposed converter with LLC as the main power processing unit and the proposed converter

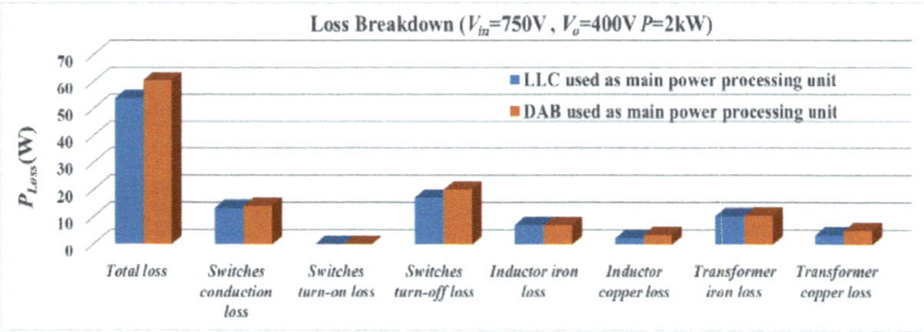

(a)

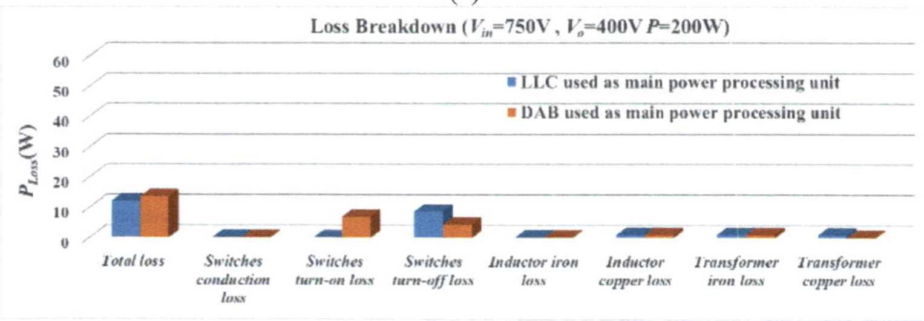

(b)

Fig. 12.18 Comparisons of the calculated power loss breakdown. **a** under rated power (2 kW) **b** under light load power (200 W)

with DAB as the main power processing unit, under rated power (2 kW) and light load power (200 W), respectively. It can be seen that compared with DAB as the main power processing unit under rated power and light power, the loss reduction of the proposed converter is 6.8W and 1.6W, respectively. Therefore, the proposed converter with LLC as the main power processing unit is more efficiency.

For better illustrating the efficiency performance of the proposed converter within a certain input voltage range, the comparison of loss breakdown results are presented in Fig. 12.19 when the input voltage are 700 V, 750 V and 800 V, respectively. It can be seen that when the input voltage is increased, the total loss are increased. That is because the power flowing through the DAB circuit is increased, resulting in increased loss. However, the increased loss is relatively small and acceptable.

The comparisons of the measured efficiency under different load conditions between the proposed converter and the traditional DAB-LLC DCX converter are shown in Fig. 12.20. It can be seen that the peak efficiency of the proposed converter is 97.3%. The efficiency of the traditional converter under light load is much smaller, and the efficiency under other loads is also smaller than the proposed converter. In addition, the proposed converter also has less switch number compared with the traditional converter.

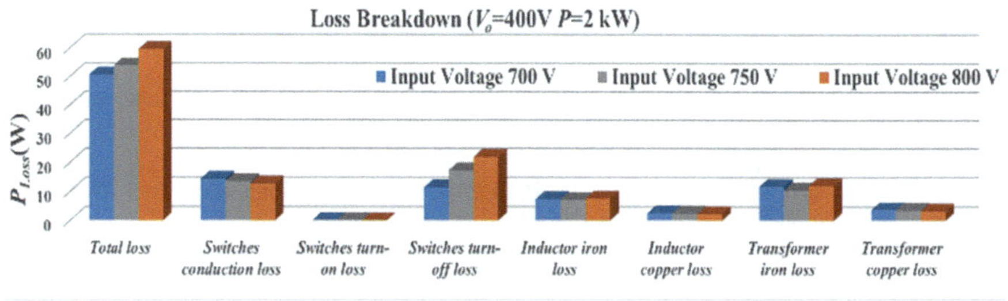

Fig. 12.19 The loss breakdown results of the proposed converter under the different input voltage

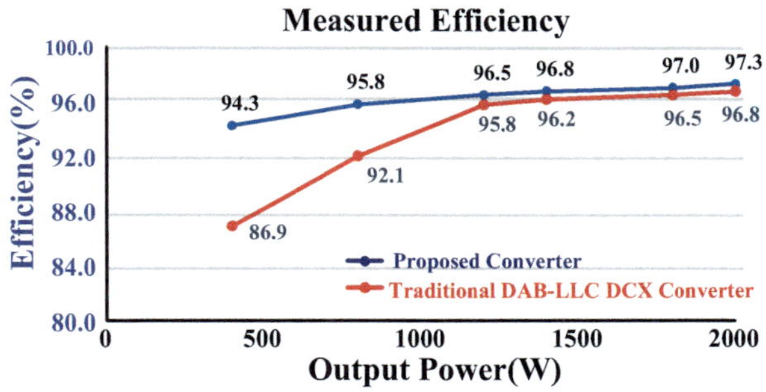

Fig. 12.20 Measured efficiency under different load conditions

12.5 Conclusion

To achieve flexible voltage regulation, and a wide load range of ZVS operation of all switches under the forward and backward power transmission, this chapter proposed an LLC-DAB bidirectional DCX converter. The proposed converter adopts a sigma structure. By connecting to the same full-bridge circuit in the secondary side, the proposed converter requires four fewer switches than the conventional solution. In this way, DAB is used as an auxiliary power regulation unit that only processes partial power, and LLC primary circuit transfers the main power with a simple PWM scheme, achieving a high efficiency. The operation modes, the ZVS operation range considering the switch junction capacitors, and the design are discussed in detail. Finally, the experimental results have verified the effectiveness of the proposed converter by comparing its performance with that of the conventional solution. As for the future work, since the outputs of the two transformers are connected together, the modeling of this structure to analyze the dynamic performance is worth of study.

References

1. Sun, Y., Gao, Z., Fu, C., Wu, C., & Chen, Z. (2020). "A hybrid modular DC solid-state transformer combining high efficiency and control flexibility," in *IEEE Transactions on Power Electronics*, *35*(4), 3434–3449, April 2020. https://doi.org/10.1109/TPEL.2019.2935029.
2. Liao, Y. et al. (2021). "Single-Stage DAB-LLC Hybrid Bidirectional Converter With Tight Voltage Regulation Under DCX Operation," in *IEEE Transactions on Industrial Electronics*, *68*(1), 293–303, Jan 2021. https://doi.org/10.1109/TIE.2020.2965495.

Chapter 13
Stability Analysis and Design of Common Phase Shift Control for Input Series Output Parallel Dual Active Bridge with Consideration of Dead-Time Effect

Abstract Input-series output-parallel (ISOP) connected Dual Active Bridge (DAB) converters allow the use for occasions requiring high input voltage and galvanic isolation with bidirectional power capability. However, when DAB converters are ISOP connected, the traditional analysis without consideration of the dead-time cannot effectively explain the reason why the ISOP converter system is stable under common phase shift control (CPSC) when parameters are mismatched. In this chapter, it is firstly revealed that the stability of the ISOP DAB converter under common phase ratio control can be ensured by consideration of the dead-time. Based on the analysis of the DAB power transmission characteristics with consideration of dead time, a new model is derived for the ISOP DAB converter, which can reveal the reason why the converter can operate stably under common phase shift control when the design parameters of the individual module are mismatched. Then based on the derived model, the stability analysis and the parameters design are further performed. Finally, the effectiveness of the working principle and analysis are verified through results from a two-module ISOP experimental prototype.

Keywords Common duty ratio control · Common phase shift control (CPSC) · Input-series output-parallel (ISOP) dual active bridge converter · Stability analysis

13.1 Existence of Steady Operation Point of ISOP DAB System Under CPSC

Existing dedicated voltage sharing control loop strategies [1, 2] and natural power equalization strategies [3, 4] have been extended to ISOP DAB to realize power equalization. For these power sharing strategies, the additional control loop must be added. And even in some work, it is pointed out that the ISOP DAB system using common phase shift ratio control (it should be noted that it is named as "common duty ration control" for a converter with conventional PWM control, and it is named as "common phase shift control (CPSC)" for DAB converter with phase shift ratio control) will be unstable [1].

G. Xu and X. Chen, *Topology Deduction and Control for Dual-Active-Bridge Converters*, https://doi.org/10.1007/978-981-97-4856-3_13

The conclusion that the ISOP system is stable under CDRC in [5] does not apply to ISOP DAB. The reason is mainly that the power sharing performance of ISOP DAB under common phase ratio control is different from traditional PWM converters under CDRC due to the characteristics of the current source. As Ref. [6] mentioned that the ISOP DAB system is unstable, but a steady state is achieved in practice. It holds the opinion that the ISOP DAB system with parameter mismatch can achieve the steady state because of its damping of line impedance in practice, which is not the main reason according to the analysis in this chapter. Moreover, Ref. [5] did not discuss this issue in detail either.

13.1.1 CPSC Strategy for ISOP DAB System

Figure 13.2 shows the CPSC scheme applied for a two-module ISOP DAB system. v_L and i_L are the voltage and current of leakage inductor of modules, respectively. Figure 13.2 shows the gate signal of CPSC and the typical waveforms of module #1. v_{gs} is the gate signal. T_s is the switching period; d_φ is the phase shift ratio; M is the ratio of dead time to half cycle time.

Single-phase-shift (SPS) modulation and fixed 50% duty ratio is applied for each DAB in CPSC. As shown in Fig. 13.3, the phase shift between two bridges, which is acquired from the common output voltage control loop, is used to adjust the power. Under the CPSC strategy, d_φ of the different modules is equal, and the value is changed under different voltages and loads.

13.1.2 Conditions for the Existence of Steady-State of ISOP DAB System Under CPSC

To clarify the stability of the system, it is necessary to determine how the power is distributed among submodules. To simplify the analysis, the ISOP system in this chapter consists of two DAB converter modules without loss of generality. Assuming that the energy of the circuit is lossless, the transmission power P of DAB can be expressed as

$$P = \frac{NV_{in}V_oD_\varphi(1 - D_\varphi)}{2Lf} \tag{13.1}$$

where f is switching frequency, N represents the transformer turns ratio, L is the leakage inductance, V_{in} and V_o stand for the voltages of the high-voltage-side and low-voltage-side,

Based on $P = \frac{NV_{in}V_oD_\varphi(1-D_\varphi)}{2Lf}$ (13.1), the power distribution relationship between the two modules in Fig. 13.1 holds

$$\frac{P_1}{P_2} = \frac{N_1 V_{in1} V_o D_\varphi (1 - D_\varphi) 2L_2 f}{N_2 V_{in2} V_o D_\varphi (1 - D_\varphi) 2L_1 f} = \frac{N_1 L_2 V_{in1}}{N_2 L_1 V_{in2}} \tag{13.2}$$

where P_1 and P_2 represent the power of two modules; V_{in1} and V_{in2} are the corresponding high-voltage-side voltage of two modules; N_1 and N_2 are the turns ratios

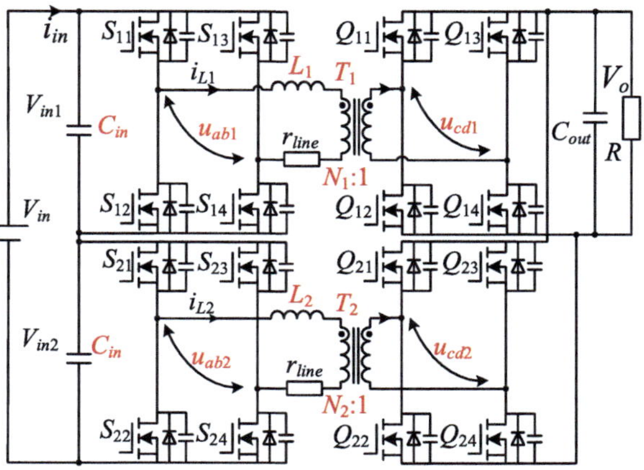

Fig. 13.1 Topology of the high-voltage-side-series low-voltage-side-parallel DAB

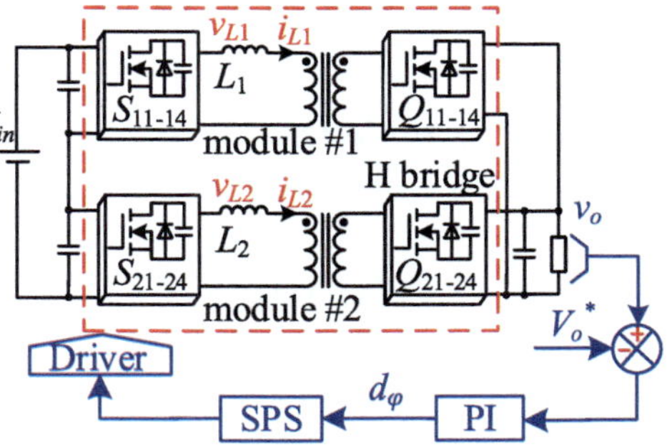

Fig. 13.2 CPSC scheme applied for high-voltage-side-series low-voltage-side-parallel DAB system

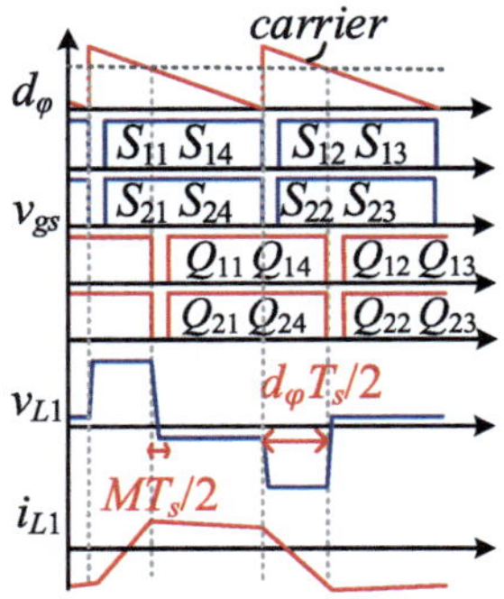

Fig. 13.3 Drive signal of CPSC scheme

of the high-frequency transformers; L_1 and L_2 are the leakage inductance of two modules.

According to the series connection on the high-voltage-side, the power distribution relationship can be expressed as

$$\frac{P_1}{P_2}=\frac{V_{in1}I_{in}}{V_{in2}I_{in}}=\frac{V_{in1}}{V_{in2}} \tag{13.3}$$

where I_{in} represents the total input current of the system.

Based on $\frac{P_1}{P_2}=\frac{N_1V_{in1}V_oD_\varphi(1-D_\varphi)2L_2f}{N_2V_{in2}V_oD_\varphi(1-D_\varphi)2L_1f}=\frac{N_1L_2V_{in1}}{N_2L_1V_{in2}}$ (13.2) and $\frac{P_1}{P_2}=\frac{V_{in1}I_{in}}{V_{in2}I_{in}}=\frac{V_{in1}}{V_{in2}}$ (13.3), the power distribution of the ISOP DAB is

$$\frac{P_1}{P_2}=\frac{N_1L_2V_{in1}}{N_2L_1V_{in2}}=\frac{V_{in1}}{V_{in2}} \tag{13.4}$$

Formula $\frac{P_1}{P_2}=\frac{N_1L_2V_{in1}}{N_2L_1V_{in2}}=\frac{V_{in1}}{V_{in2}}$ (13.4) can be satisfied only when $N_1L_2=N_2L_1$. Therefore, the ISOP DAB system with parameter mismatch, that is $N_1L_2\neq N_2L_1$, is not stable without a steady state. $N_1L_2=N_2L_1$ is a difficult condition to satisfy for a real ISOP system. However, an unstable steady state means only one module will support the whole high-voltage-side voltage, and the other module will have zero input voltage.

13.1.3 Unstable Operation Simulation Under Parameter Mismatch when Line Damping is Considered

In practice, when two DAB converters are ISOP connected, the high-voltage-side voltage is distributed in all modules, and a steady state is achieved [1, 2]. It has been mentioned in a previous study that the loss caused by line damping may lead to the stable operation of the system. However, the work of [2] is focused on the input voltage sharing by means of an active control of voltages, and detailed analysis and discussion of the conditions to make the system under CPSC stable is not carried out. In our work, the stability of the system is analyzed from the point of view of dead time affecting stability. Since it is difficult to build a DAB model that considers both dead time effect and damping, the simulation experiment is adopted to exclude the influence of losses on the conclusions about our work.

The effect of line resistance and deadtime on stability are compared through simulation experiments. Figure 13.4 shows the high-voltage-side voltage of the DAB module and low-voltage-side voltage of the system when line resistance and dead time vary respectively. Values for line resistance and deadtime are indicated in corresponding simulation results. M equals 0, 0.01, 0.05 corresponds to a dead time of 0, 0.05 and 0.1 μs. Other simulation parameters are detailed in Table 13.1.

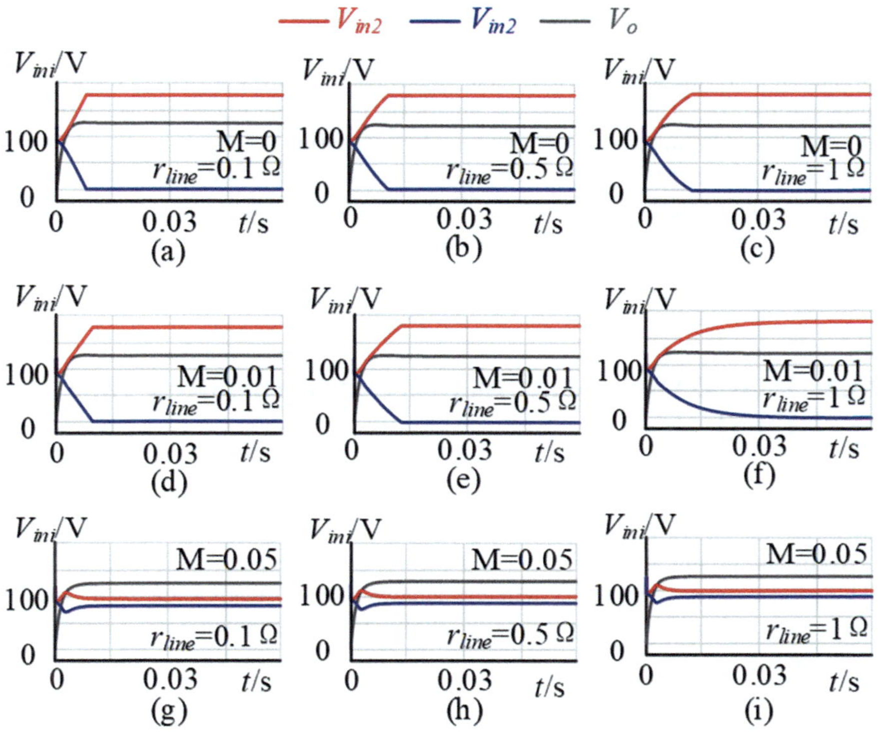

Fig. 13.4 Simulation results with consideration of the damping of line impedance. **a** $r_{line} = 0.1\ \Omega$, $M = 0$. **b** $r_{line} = 0.5\ \Omega$, $M = 0$. **c** $r_{line} = 1\ \Omega$, $M = 0$. **d** $r_{line} = 0.1\ \Omega$, $M = 0.01$. **e** $r_{line} = 0.5\ \Omega$, $M = 0.01$. **f** $r_{line} = 1\ \Omega$, $M = 0.01$. **g** $r_{line} = 0.1\ \Omega$, $M = 0.05$. **h** $r_{line} = 0.5\ \Omega$, $M = 0.05$. **i** $r_{line} = 1\ \Omega$, $M = 0.05$

Table 13.1 Parameter of simulation

Symbol	Parameter	Value
V_{in}	High-voltage-side voltage	180 V
N_1	Transformer turns ratio of module #1	1
N_2	Transformer turns ratio of module #2	1.1
L_1/L_2	Leakage inductance of submodules	44.5 μH
R	Load resistance	33.3 Ω
f	Switching frequency	100 kHz
C_{in}	Input capacitance of submodules	10 μF
C_{out}	Output capacitance at low voltage side	50 μF
D_{φ}	Phase shift ratio value	0.25

By comparing the results in Fig. 13.4a, b, and c, it can be seen that when M equals 0, the system is still unstable even though r_{line} goes from 0.1 to 1 Ω and only one module supports the whole high-voltage-side voltage. It can be seen by comparing the results in Fig. 13.4d, e and f, when M equals 0.01, different line resistance still does not make the system stable. It can be seen by comparing the results in Fig. 13.4g–i, when M increases to 0.05, the system is stable under different line resistance. Hence, it can be concluded that the change of line resistance will rarely affect the stability of the system, and the change of the dead time significantly affects the stability of the analyzed system.

13.2 ISOP DAB Stability Analysis Considering Dead-Time Effect

In this section, different modes of the ISOP DAB system considering the dead-time effect are analyzed. Based on the large-signal model, the steady state analysis is derived. Unlike the conclusion derived from the ideal model, the proposed ISOP DAB model reveals why the system can achieve a stable steady state. At the end of this section, a small-signal model of the ISOP DAB system under dead-time effect is also derived.

13.2.1 Model of DAB with Consideration of Dead-Time Effect

The transmission power characterization of a DAB is different when the impact of the dead time is considered. Figure 13.5 shows three typical theoretical waveforms for the DAB converter under the dead-time effect in boost state ($V_{in}/NV_o < 1$). The significant difference between different modes is that the reversal of voltage u_{ab} polarity is different during dead time. The detailed analysis of the operation characteristics of each mode is as follows.

Characteristics of DAB converter under MODE I are consistent with the ideal DAB. The polarity of u_{ab} is changed exactly when the switch is turned off. In Fig. 13.5a, the current is positive after S_1 and S_4 are turned off and flows through the parasitic diodes of S_2 and S_3. Therefore, the polarity of u_{ab} is reversed immediately at the beginning of the dead time. The state after S_2 and S_3 are turned off is the same. The average power expression is

$$P = NV_{in}V_o f_1(D_\varphi) \tag{13.5}$$

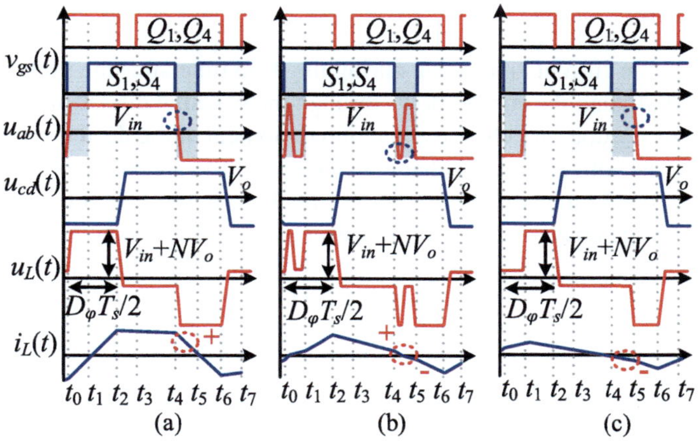

Fig. 13.5 Key waveforms of the single DAB with consideration of dead time effect. **a** MODE I. **b** MODE II. **c** MODE III

where

$$f_1(D_\varphi) = D_\varphi(1 - D_\varphi)\big/(2Lf).$$

Constraint condition for MODE I can be derived as follows:

$$M_0k + M + 0.5 \le D_\varphi \le 1, \tag{13.6}$$

where $k = V_{in}/NV_o$; $M_0 = M$-0.5.

(1) MODE II:

In Fig. 13.5b, after switches S_1 and S_4 are turned off, the current changes from positive to negative during the dead time. Positive current flows through the parasitic diodes of S_2 and S_3, and negative current flows through the parasitic diodes of S_1 and S_4. Therefore, the polarity of u_{ab} is reversed at the beginning of the dead time, and it changes again during the dead time. The polarity of u_{ab} is overturned during dead time. Therefore the average power is changed. The average transmission power is

$$P = NV_{in}V_of_2(D_\varphi) + V_{in}^3M_0^2/(2LfNV_o), \tag{13.7}$$

where

$$f_2(D_\varphi) = [4D_\varphi(1 - D_\varphi) - (2M - 2D_\varphi + 1)^2]\big/(8Lf).$$

Constraint condition for MODE II can be derived as follow:

$$M_0k + 0.5 \le D_\varphi < M_0k + M + 0.5. \tag{13.8}$$

(2) MODE III:

In Fig. 13.5c, the current is negative during the dead time after the switches S_1 and S_4 are turned off, and the current flows through the parasitic diodes of S_1 and S_4, so the polarity of u_{ab} does not change at the initial moment of the dead time. At the end of dead time, the voltage polarity changes when S_2 and S_3 are turned on. The polarity of u_{ab} during dead-time is the same as the state before the switches are turned off. Therefore average power is changed. The average transmission power is

$$P = NV_{in}V_o f_3(D_\varphi), \tag{13.9}$$

where

$$f_3(D_\varphi) = [D_\varphi(1 - D_\varphi) + M(2D - M - 1)]/(2Lf).$$

Constraint condition for MODE III can be expressed as:

$$M \le D_\varphi < M_0 k + 0.5. \tag{13.10}$$

13.2.2 *Large Signal Model of ISOP DAB System with Consideration of the Dead-Time Effect*

The stability of ISOP DAB based on the new DAB model is analyzed in this section. Since the DAB converter is symmetrical both in topology structure and modulation under different power transfer direction (only the polarity of the phase shift angle is changed), the power transfer characteristics and model for ISOP converter under forward and backward power is similar. So, here only the analysis under forward power transfer is conducted. Modules in an high-voltage-side-series low-voltage-side-parallel system may work in the same or different modes, and the system can work in different combination modes. Since MODE I*I (one module works in MODE I and the other module also works in MODE I, and the presentation for other combination modes are the same) is unstable as previously discussed in Sect. 13.2.2, take MODE II*II as an example for analysis. And the difference of transformer turns ratios of the submodules is used to characterize parameter mismatch. The other parameters are exactly the same.

The transmission power of two modules working in MODE II can be expressed as follows according to $P = NV_{in}V_o f_2(D_\varphi) + V_{in}^3 M_0^2/(2LfNV_o)$, (13.7).

$$P_1 = N_1 V_{in1} V_o f_2(D_\varphi) + V_{in1}^3 M_0^2/(2LfN_1 V_o) \tag{13.11}$$

$$P_2 = N_2 V_{in2} V_o f_2(D_\varphi) + V_{in2}^3 M_0^2/(2LfN_2 V_o). \tag{13.12}$$

Due to series connection on the high-voltage-side, the voltage and power at high-voltage-side hold

$$V_{in1} + V_{in2} = V_{in} \tag{13.13}$$

$$P_1 + P_2 = P. \tag{13.14}$$

Solving $\frac{P_1}{P_2} = \frac{V_{in1}I_{in}}{V_{in2}I_{in}} = \frac{V_{in1}}{V_{in2}}$ (13.3), $P_1 = N_1 V_{in1} V_o f_2(D_\varphi) + V_{in1}^3 M_0^2/(2LfN_1 V_o)$ (13.11), $P_2 = N_2 V_{in2} V_o f_2(D_\varphi) + V_{in2}^3 M_0^2/(2LfN_2 V_o)$. (13.12), $V_{in1} + V_{in2} = V_{in}$ (13.13), $P_1 + P_2 = P$. (13.14) leads to the solutions of V_{in1}, V_{in2} and D_φ

$$V_{in2}\left(\sqrt{\begin{matrix} N^2 V_{in}^2 + (N - 1/N)* \\ [NV_{in}^2/4 + 2V_o AQ] \end{matrix}} - NV_{in}\right)/(N - 1/N) \tag{13.15}$$

$$V_{in1} = V_{in} - V_{in2} \tag{13.16}$$

$$D_\varphi = (M+1)/2 - \sqrt{\frac{V_{in2}^2 M_0^2}{2V_o^2 N_2^2} - \frac{A}{V_o N_2} + \frac{1 - 2M^2}{8}}. \tag{13.17}$$

where $A = PLf/V_{in}$; $N = N_1/N_2$; $Q = (N_1 - N_2)/(2M_0)^2$. As analyzed above, the ISOP DAB working under MODE II*II has a steady state operating point. By substituting the power expressions in other cases, the existence of the steady state of all combined modes can be analyzed, and the results are as follow:

(1) MODE I*II

$$V_{in1} = V_{in} + \sqrt{\begin{matrix} (N_2 V_o)^2/4 + 2AN_2 V_o - 4AN_2^2 V_o/N_1 \\ +M\,(N_2 V_o)^2 (M + \sqrt{1 - 8A/(N_1 V_o)} \end{matrix}}\,/M_0 \tag{13.18}$$

$$V_{in2} = V_{in} - V_{in1} \tag{13.19}$$

$$D_\varphi = 0.5 - \sqrt{0.25 - 2A/N_1 V_o}. \tag{13.20}$$

(2) MODE II*III

$$V_{in1} = -\sqrt{\begin{matrix} (N_1 V_o)^2/4 + 2AN_1 V_o - 4AN_1^2 V_o/N_2 \\ +M\,(N_1 V_o)^2 (M - \sqrt{1 - 8A/(N_2 V_o)}) \end{matrix}}\,/M_0 \tag{13.21}$$

$$V_{in2} = V_{in} - V_{in1} \tag{13.22}$$

$$D_\varphi = 0.5 + M - \sqrt{0.25 - 2A/(N_2 V_o)}. \tag{13.23}$$

(3) MODE I*III

Without a steady state.

(4) MODE III*III

Without a steady state.

13.2.3 Small Signal Model of ISOP DAB System with Consideration of the Dead-Time Effect

To perform stability analysis, it is necessary to derive the small-signal model based on the above-mentioned large-signal model of the system It is worth noting that, The small-signal stability analysis is established on the premise that the system has an equilibrium point. Therefore, only the small-signal model of the system under Mode I*II, II*II and III*II is established.

For a DAB converter, due to one of the state variables, the leakage inductor current i_L, is purely ac with an average value of 0, it is difficult to establish a switching cycle average model. One way to model DAB is simply ignoring the dynamics of i_L, which is called the reduced-order model [7]. Hence, the reduced-order model of DAB is adopted in our work which is widely used due to its simplicity and correctness.

The relationship between voltage and current and average power can be expressed as

$$P_{in} = P_o = V_{in}I_{in} = V_oI_o. \tag{13.24}$$

Based on $P = NV_{in}V_of_2(D_\varphi) + V_{in}^3M_0^2/(2LfNV_o)$, (13.7) and $P_{in} = P_o = V_{in}I_{in} = V_oI_o$. (13.24), input and output average current can be expressed as follows, respectively.

$$I_{ini} = N_iV_of_2(D_\varphi) + V_{ini}^2M_0^2\big/\,(2LfN_iV_o) \tag{13.25}$$

$$I_{oi} = N_iV_{ini}f_2(D_\varphi) + V_{ini}^3M_0^2/(2LfN_iV_o^2). \tag{13.26}$$

The small-signal model obtained perturbing is derived as follows:

$$\begin{aligned} \hat{i}_{ini} &= a_i\hat{d}_\varphi + b_i\hat{v}_{in} + c_i\hat{v}_o \\ \hat{i}_{oi} &= l_i\hat{d}_\varphi + m_i\hat{v}_{in} - h_i\hat{v}_o, \end{aligned} \tag{13.27}$$

where

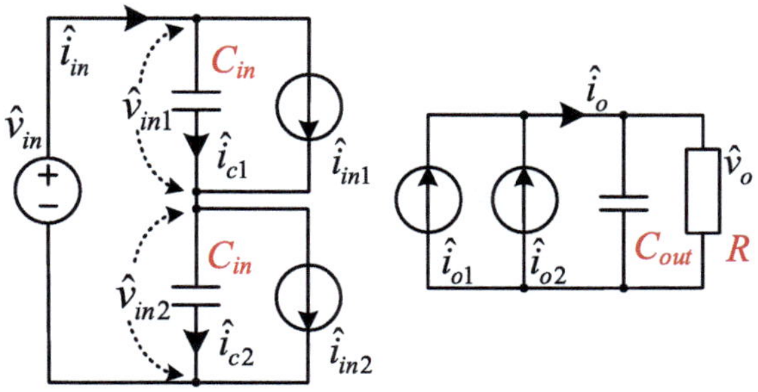

Fig. 13.6 Small signal model of two module ISOP DAB

$$\begin{cases} a_i = N_i V_o[M+1-2D_\varphi)]/(Lf) \\ b_i = V_{ini} M_0^2/(N_i V_o L f) \\ c_i = N_i f_2(D_\varphi) - V_{in}^2 M_0^2/(2N_i V_o^2 Lf) \\ l_i = N_i V_{ini}[M+1-2D_\varphi)]/(Lf) \\ m_i = N_i f_2(D_\varphi) + 3V_{ini}^2 M_0^2/(2N_i V_o^2 Lf) \\ h_i = V_{ini}^3 M_0^2/(N_i V_o^3 Lf) \end{cases}.$$

The corresponding small-signal circuit model is shown in Fig. 13.6. And for the convenience of presentation, define the parameters as follow:

$$x_1 + x_2 = x^+, x_1 - x_2 = x^-, x = a, b, c, l, m, h. \tag{13.28}$$

Based on the small-signal circuit, the following expressions for the system can be obtained as:

$$\begin{cases} \hat{i}_{c1} = \hat{i}_{in} - \hat{i}_{in1}, \hat{i}_{c2} = \hat{i}_{in} - \hat{i}_{in2} \\ \hat{v}_{in1} = \hat{i}_{c1}/(sC_{in}), \hat{v}_{in2} = \hat{i}_{c2}/(sC_{in}) \\ Z_{out}(\hat{i}_{o1} + \hat{i}_{o2}) = \hat{v}_o \\ \hat{v}_{in1} + \hat{v}_{in2} = \hat{v}_{in} \end{cases}, \tag{13.29}$$

where

$$Z_{out} = sRC_{out}/(sRC_{out}+1).$$

The ISOP DAB system can meet the output voltage control requirements with a single-loop PI controller, therefore, the relationship between control signal disturbance and low-voltage-side voltage disturbance satisfies $\hat{d}_\varphi = -(k_p + k_i \frac{1}{s})\hat{v}_o$. (13.30),

$$\hat{d}_\varphi = -\left(k_p + k_i \frac{1}{s}\right)\hat{v}_o. \tag{13.30}$$

Substituting $\hat{d}_{\varphi} = -(k_p + k_i \frac{1}{s})\hat{v}_o$. (13.30) into
$$\begin{cases} \hat{i}_{c1} = \hat{i}_{in} - \hat{i}_{in1}, \hat{i}_{c2} = \hat{i}_{in} - \hat{i}_{in2} \\ \hat{v}_{in1} = \hat{i}_{c1}/(sC_{in}), \hat{v}_{in2} = \hat{i}_{c2}/(sC_{in}) \\ Z_{out}(\hat{i}_{o1} + \hat{i}_{o2}) = \hat{v}_o \\ \hat{v}_{in1} + \hat{v}_{in2} = \hat{v}_{in} \end{cases}$$
, (13.29) can obtain the transfer function from the individual input voltages to perturbation in the total input voltage. The input-voltage transfer function G(s) can be derived as follows:

$$G(s) = \frac{v_{in2}(s) - v_{in1}(s)}{v_{in}(s)} = \frac{\alpha_2 s^2 + \alpha_1 s - \alpha_0}{\beta_3 s^3 + \beta_2 s^2 + \beta_1 s + \beta_0}, \tag{13.31}$$

where

$$\begin{cases} \alpha_0 = C_{out} R b^- \\ \alpha_1 = b^- - Rhb^+ + Rc^- m^+ + Rk_p(b^- l^+ - a^- m^+) \\ \alpha_2 = -Ra^- k_i m^+ + Rb^- k_i l^+ \\ \beta_0 = -m^- a^- + b^+ l^+, \beta_3 = 2C_{in} C_{out} R \\ \beta_1 = b^+ + 2C_{in} R k_i l^+ + R(-b^+ h^+ + c^- m^- + k_p b^+ l^+ - k_p a^- m^-) \\ \beta_2 = C_{out} R b^+ + 2C_{in}(1 - Rh^+) + 2C_{in} R k_p l^+ \end{cases}$$

And the transfer function of the ISOP DAB system under MODE I*II and II*III are presented in appendix.

It is worth noting that this analysis will be used to draw the root distribution in Sect. 13.4.

13.3 Analysis of Parameter Influence for Steady State and Stability

Combination modes of the system and the influence of circuit parameters are analyzed in this section. The stability analysis of the typical system obtained through deadtime parameter design is carried out then. The specifications of the system are listed as follows: (1) high-voltage-side voltage: $V_{in} = 180$ V; (2) low-voltage-side voltage: $V_o = 100$ V; (3) rated power of system: 500 W; (4) switching frequency: 100 kHz; (5) input capacitance: $C_{in1} = C_{in2} = 100\ \mu$F; (6) $L_1 = L_2 = 44.5\ \mu$H; (7) $M = 0.05$–0.2; (8) $N_1 = 1$, $N_2 = 1.01$–$1.1(\Delta N = 10$–$20\%)$. The turns ratios of two transformers are purposely designed to be different to characterize the parameter mismatch.

13.3.1 Analysis of the Combination Modes Corresponding to the Transmission Power

The boundary of a certain combination mode should be derived first. In MODE II*II, the condition of each module is different due to different input voltages. The boundaries of module #1 and #2 working in MODE II can be expressed as follows according to (8).

$$\begin{aligned} D_{iup} &= M_0 V_{ini} / (N_i V_o) + M + 0.5 \\ D_{idown} &= M_0 V_{ini} / (N_i V_o) + 0.5. \end{aligned} \tag{13.32}$$

where D_{iup} and D_{idown} represent upper and lower boundaries of module #i (i = 1, 2). The condition of MODE II*II can be expressed as

$$D_{1down} < D_{\varphi} < D_{1up} \text{ and } D_{2down} < D_{\varphi} < D_{2up}, \tag{13.33}$$

where D_{φ} is the common phase shift of two modules. Therefore, the boundaries are as follows

$$D_{up} = \min\{D_{1up}, D_{2up}\} \tag{13.34}$$

$$D_{down} = \max\{D_{1down}, D_{2down}\}, \tag{13.35}$$

where D_{up} and D_{down} represent upper and lower boundaries of MODE II*II. When D_{φ} is close to D_{up} or D_{down}, the boundary power of the system under combination mode can be obtained.

By calculating the power boundary of different combined modes, the corresponding modes of the system over full power range can be determined. Figure 13.7 shows the power boundary of a certain system ($M = 0.05$, $\Delta N = 10\%$) under different power range. Figure 13.7a–c shows the middle power range, low power range, and high power range, respectively.

It can be seen from Fig. 13.7a, the system operates in MODE II*II under middle power range. D_{φ} is close to D_{1up} at the intersection of the red line and blue line, and the upper boundary power (266.5 W) is determined. It means module #1 will work in MODE I (According to (6) and (8), MODE I and MODE II are adjacent modes), and the system will work in MODE I*II when the power further increases. Similarly, since D_{φ} is close to D_{2down} at lower boundary power (141.4 W), the system will work in MODE II*III when the power further decrease. The blue area identifies the power range in which the system works in MODE II*II. Figure 13.7b and c show that the system operates in MODE II*III under low power range and operates in MODE I*II under high power range. When the power is greater than the upper boundary in Fig. 13.7c, the system will work in MODE I*III, which is unstable for the system. Figure 13.7d shows the distribution of different combined modes of the system over

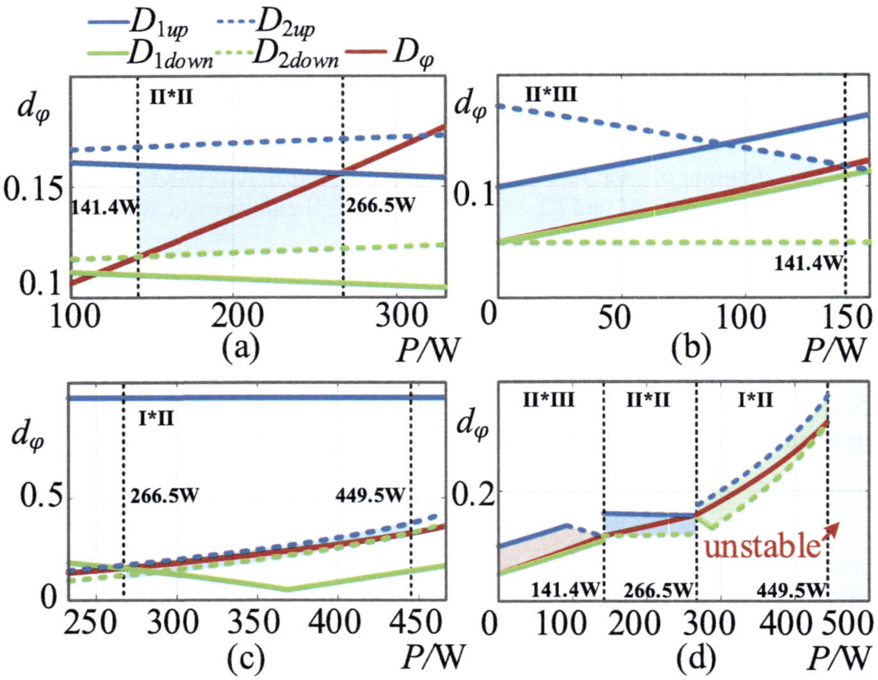

Fig. 13.7 Combined modes of system with M = 0.05 and ΔN = 10%. **a** MODE II*II. **b** MODE II*III. **c** MODE I*II. **d** stable range

the full power range. The red area, blue area, green area, and a gray area identify the power range under MODE II*III, II*II, I*II, and unstable power range, respectively.

As can be seen from Fig. 13.7c, when the power is increased, module #2 will enter MODE III, and the system will go from MODE I*II to I*III, which is unstable without a steady-state. Hence, the condition of stability can be expressed as:

$$D_\varphi > D_{2down} \tag{13.36}$$

Substitute formula (13.19), (13.20) and (13.32) into formula (13.36), the condition for the system to have a steady-state working solution can be obtained

$$\sqrt{\frac{1}{4}-\frac{2A}{N_1V_o}} < \sqrt{\begin{matrix}(N_2V_o)^2/4+2AN_2V_o-4AN_2^2V_o/N_1\\ +M(N_2V_o)^2\left(M+\sqrt{1-8A/(N_1V_o)}\right)\end{matrix}}\Big/(N_2V_o) \tag{13.37}$$

where $A = PLf/V_{in}$.

Therefore, the stability of the system can be guaranteed by adjusting the proportion of dead time M. The larger M is, the more favorable it is to the stability.

13.3.2 Analysis of Parameter Influence for Steady State Tability

(1) Influence of dead time

Dead-time is a key factor affecting the stability of the system. For an ideal ISOP DAB system without dead time, there is no steady-state solution and is unstable over the full power range under parameter mismatching.

Figure 13.8a shows the stable range of a system with $M = 0.001$; $\Delta N = 10\%$. D_φ is close to D_{up} at the intersection of the red line and blue line, and the upper boundary power (21.8 W) is determined. It can be seen that the stable range of the system is very small with tiny dead time. Figure 13.8b shows the stable range of the system with $M = 0.2$. Compared to $M = 0.05$ (Fig. 13.8d), a longer dead time can obtain a larger stable range and wider power range of MODE II*II.

(2) Influence of parameter mismatch

Parameter mismatch will also affect the voltage distribution and stable range of the system. Figure 13.9 shows the distribution of combined modes of system with different ΔN respectively. When other parameters are the same, the system with $\Delta N = 1\%$ has a steady-state operating point over the full power range, as shown in Fig. 13.9a, while the system with $\Delta N = 20\%$ becomes unstable under heavy load. When ΔN increases, the power range of the system under MODE II*II is smaller. In an actual combined converter, the parameter difference between the modules is limited and generally does not exceed 10%.

Fig. 13.8 Combined modes of system with ΔN = 10% and different M. **a** M = 0.001. **b** M = 0.2

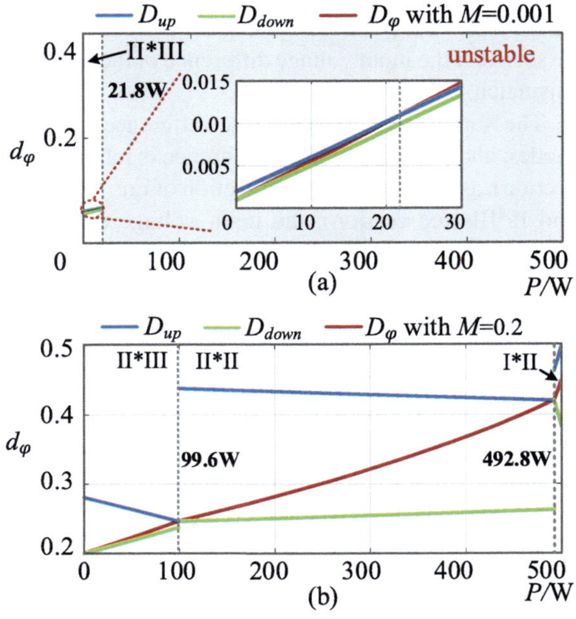

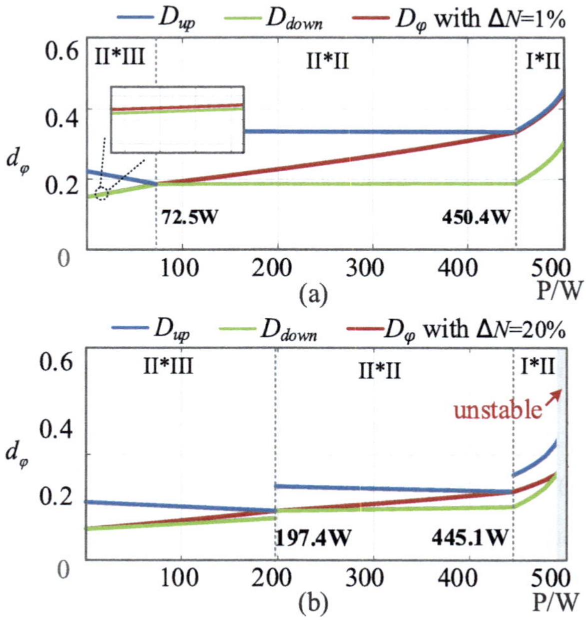

Fig. 13.9 Combined modes of system with M = 0.15 and different ΔN. **a** ΔN = 1%. **b** ΔN = 20%

(3) Input voltage distribution

Figure 13.10 shows the input voltage difference of submodules corresponds to different power where $\Delta V_{in} = V_{in1} - V_{in2}$. Figure 13.10a shows the effect of different M. It can be seen that the input voltage difference is smaller under the longer dead time over a wide power range. Figure 13.10b shows the effect of different ΔN. It can be seen that the input voltage difference of the system is smaller under tiny parameter mismatch.

The variation of input voltage difference is different under different combined modes, and the input voltage difference is relatively small under MODE II*II (green section), as shown in the green section of Fig. 13.10. Under MODE I*II (blue section) and II*III (red section), the input voltage difference changes a lot as the power changes.

13.3.3 Stability Verification of ISOP DAB System

For a certain system with $M = 0.2$ and $\Delta N = 10\%$, according to the analysis method in Sect. 13.3, the transfer functions under the three combined modes are obtained. By substituting the parameters, the transfer function at different steady-state operating points can be obtained.

Figure 13.11 shows the distribution of root when the power of the system changes over the full power range. Different power corresponding to the three combination

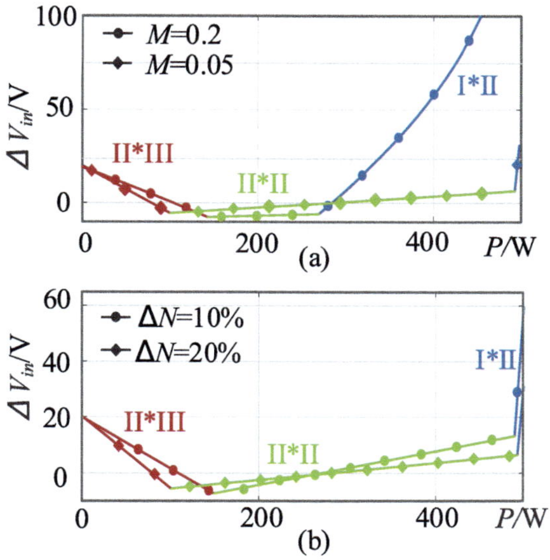

Fig. 13.10 Input voltage distribution of system with different M and different ΔN. **a** different M. **b** different ΔN

modes has different transfer functions. Figure 13.11a–c shows the distribution of the root under the corresponding power and combination modes. Since the system is a third-order system, the distribution of the three poles is shown in the figure. As shown in Fig. 13.11, the roots of the system with typical parameters are distributed in the left half-plane, and the system is stable.

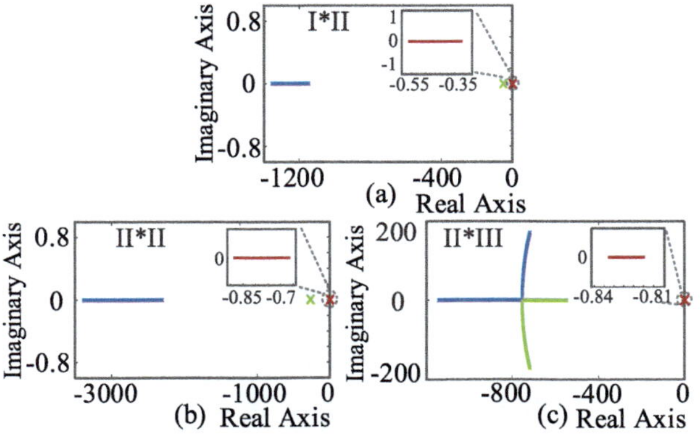

Fig. 13.11 The distribution of typical system roots over full power range. **a** MODE I*II, **b** MODE II*II, **c** MODE II*III

13.4 Experimental Verification

To validate the theoretical analysis in the chapter, a two-module ISOP DAB prototype was designed and implemented according to the parameters given in Table 13.2. The converter is controlled by a Texas Instruments TMS320F28377S digital signal processor (DSP). The input voltage is provided by a DC switching power supply. The experimental circuit is shown in Fig. 13.12.

(1) Dead-time effect on stability

The comparative experimental results showing the influence of the dead time on two module ISOP converter system stability are shown in Figs. 13.13 and 13.14.

Figure 13.13a shows that under heavy load, the system is in an unstable steady state with $M = 0.02$, only module #1 supports the input voltage. When the system's

Table 13.2 Parameter of experimental circuit

Symbol	Parameter	Value
V_{in}	High-voltage-side voltage	180 V
V_o	Low-voltage-side voltage	100 V
N_1	Transformer turns ratio of module #1	1
N_2	Transformer turns ratio of module #2	1.1
M	Dead time ratio	0.2
L_1/L_2	Leakage inductance of submodules	44.5 μH
P	Maximum power	500 W
f	Switching frequency	100 kHz
C_{in}	Input capacitance of submodules	100 μF
C_{out}	Output capacitance at low voltage side	100 μF
r_{line}	Line damping resistance	0.185 Ω

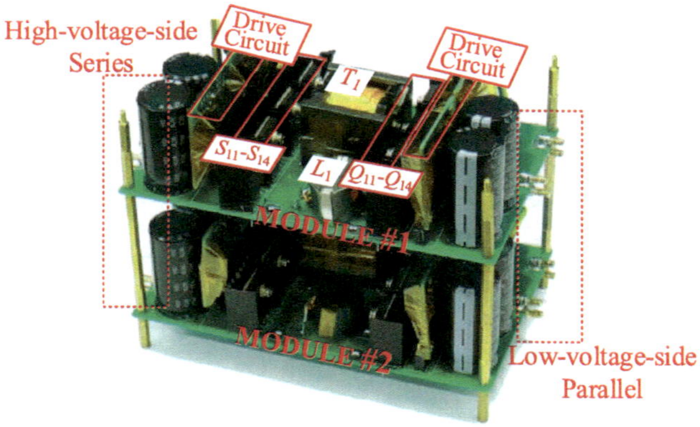

Fig. 13.12 Experimental hardware of the ISOP system

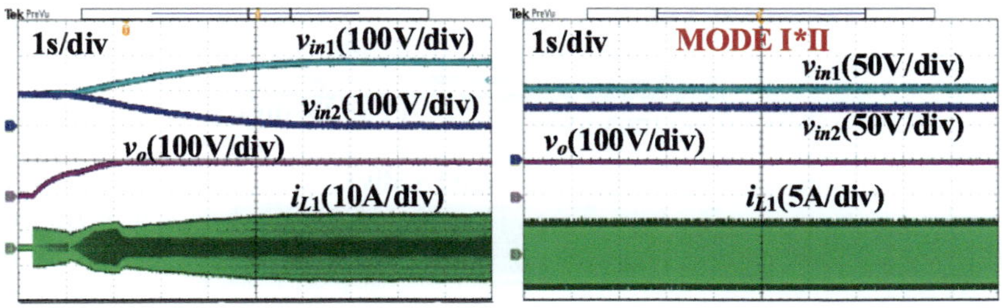

Fig. 13.13 Heavy load waveforms for system. **a** M = 0.02. **b** M = 0.2

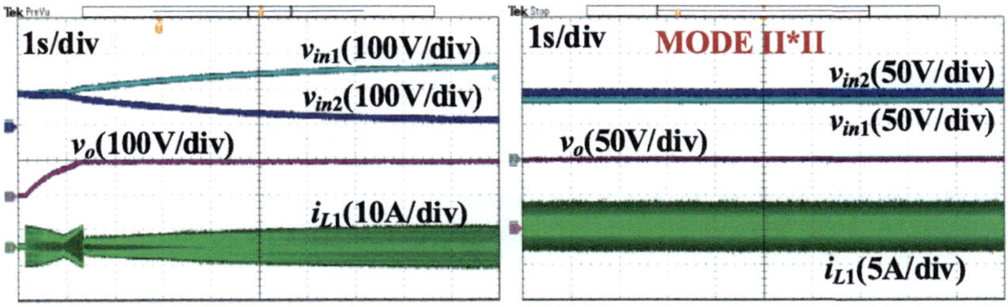

Fig. 13.14 Half load waveforms for system. **a** M = 0.02. **b** M = 0.2

dead time ratio is 0.2, the input voltage is distributed in two modules, and the system is stable, as shown in Fig. 13.13b. Similarly, under half load, Fig. 13.14a, b show that system with $M = 0.02$ is unstable, and when $M = 0.2$, the system can achieve a stable steay state.

(2) Verification of working mode

The analysis in Sect. 13.4 shows that the ISOP DAB system can enter different combined modes when the transmission power changes. To demonstrate the different characteristics under different combined modes, a set of comparative experiments was designed based on the analysis.

Figure 13.15 shows the steady state of the system under heavy load. Figure 13.15a shows the individual input voltage and the output voltage of the system. The input voltage is shared by two modules stably, and $(v_{in1} - v_{in2})/v_{in}$ is 8% which is moderate to the theoretical analysis shown in Fig. 13.15b. In Fig. 13.15b, c, v_{gs1} is the gate signal of S_{11} and S_{21}, u_{ab1} and u_{ab2} are the middle point voltage of the primary H bridge, and i_{L1} and i_{L2} are the currents of leakage inductance. As seen from Fig. 13.15b, i_{L1} is positive during dead time, the polarity of u_{ab1} is reversed immediately when the switch S_1 is off, and module #1 is working in MODE I. In Fig. 13.15c, during dead time, i_{L2} changes from positive to negative, u_{ab2} is reversed when the switch is off, and it changes again because of the negative current. Therefore, module #2 is working in MODE II. This validates that the system is working in MODE I*II under heavy load.

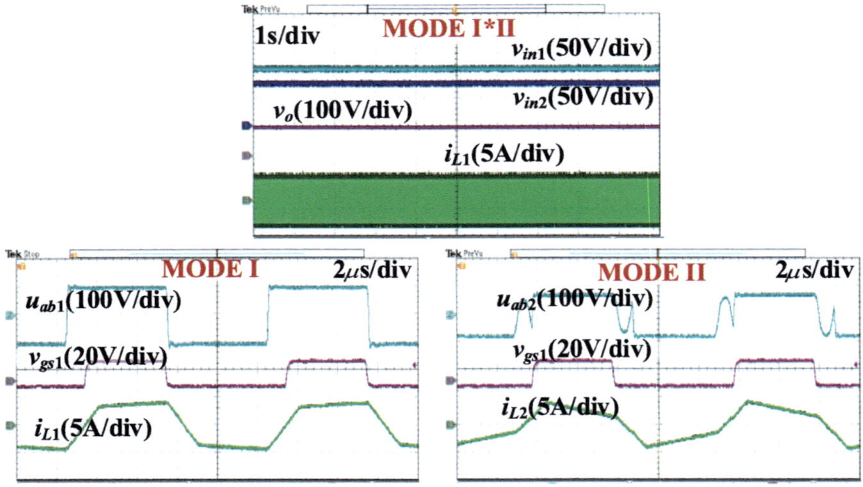

Fig. 13.15 Heavy load waveforms for system (Mode I∗II). **a** Overall. **b** Module #1. **c** Module #2

Figure 13.16 shows the steady state waveforms of the system under half load. It can be seen from Fig. 13.16a that the output voltage of the system is stable, and the input voltages of the two modules are distributed stably, and $(v_{in1} - v_{in2})/v_{in}$ is -2%, which is moderate to the theoretical analysis values shown in Fig. 13.16b. In Fig. 13.16b, c, during dead time, i_{L1} and i_{L2} change from positive to negative, u_{ab1} and u_{ab2} are reversed when the switch is off and change again because of the negative current. Therefore, module #1 and module #2 are both working in MODE II, and the system is working in MODE II*II under half load.

Figure 13.17 shows the steady state waveforms of the system under light load. Figure 13.17a shows that the individual input voltage of the modules and the output voltage of the system. $(v_{in1} - v_{in2})/v_{in}$ is 3% which is moderate to the theoretical analysis values shown in Fig. 13.17b. In Fig. 13.17b, during dead time, i_{L1} changes from positive to negative, u_{ab1} is reversed when the switch is off, and it changes again because of the negative current. Therefore, module #1 is working in MODE II. As seen from Fig. 13.17c, i_{L2} is negative during dead time, the polarity of u_{ab2} is reversed when the dead time is off, and module #2 is working in MODE III. This validates that the system is working in MODE II*III under light load, which matches the theoretical analysis.

(3) Load switching

Figure 13.18 shows the results of the load switching experiment for the two-module ISOP converter system. As shown in Fig. 13.18a, when the system is switched between heavy load and half load, the system can still run stably. In Fig. 13.18b, c, the system is working in MODE I*II under heavy load and in MODE II*II under light load, respectively.

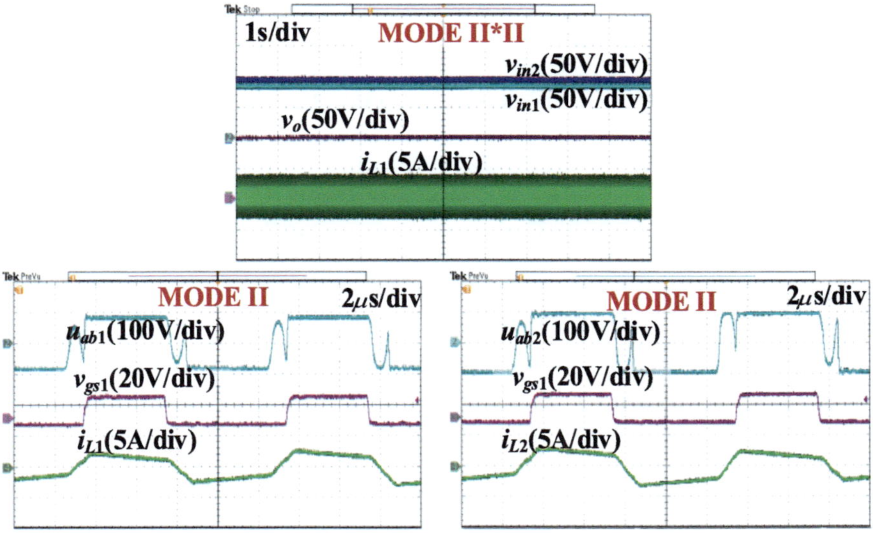

Fig. 13.16 Half load waveforms for system (Mode II*II). **a** Overall. **b** module #1. **c** module #2

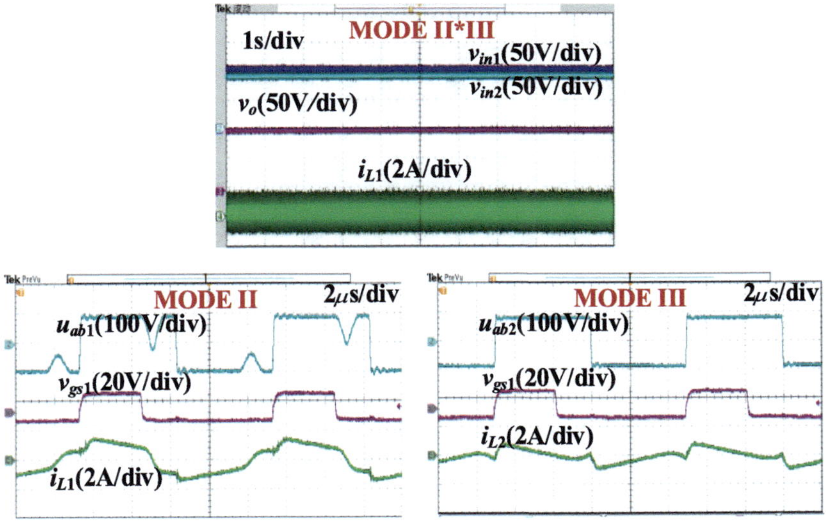

Fig. 13.17 Light load waveforms for system (Mode II*III). **a** Overall. **b** module #1. **c** module #2

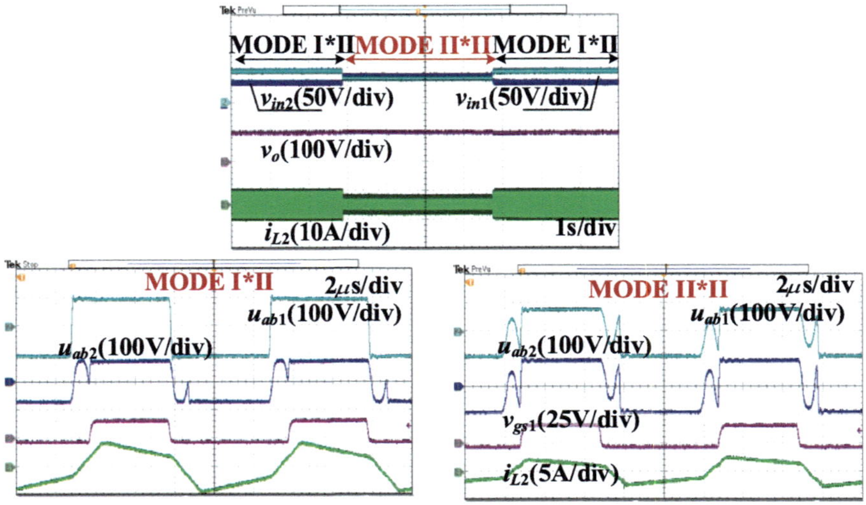

Fig. 13.18 Load switch waveforms for the system. **a** Overall. **b** Heavy Load (before load switch). **c** Half load (after load switch)

13.5 Conclusion

In this Chapter, the stability of the ISOP DAB converter under CPSC is analyzed. It is revealed that the dead-time effect will affect the stable steady state of the ISOP DAB system. Based on the single DAB model with consideration of the dead-time effect, the model of the ISOP DAB system is derived. The stable steady state existence of ISOP DAB is analyzed by deriving the new combined system model. The stability analysis and dead time design of the system are also carried out based on the new model. It can be known from the analysis that the dead time will significantly affect the stable power range of the system under different combination modes. In addition, parameter analysis and dead time design guidelines are also discussed in this chapter to make the system to be stable under the full power range. The experimental results have shown that the dead time will affect the stability of the converter and also verified the correctness of the mode combination analysis, power range analysis, and stability design with consideration of dead-time.

As a future work, the voltage level of the experimental platform can be designed as a standard voltage level to be more tightly connected to the real industrial application.

References

1. Zumel, P., Ortega, L., Lázaro, A., Fernández, C., & Barrado, A. (2013). Control strategy for modular Dual Active Bridge input series output parallel. In *Control and modeling for power electronics* (pp. 1, 7, 23–26). IEEE.
2. Zumel, P., Ortega, L., Lázaro, A., Fernández, C., Barrado, A., Rodríguez, A., et al. (2016). Modular dual-active bridge converter architecture. *IEEE Transactions on Industry Applications, 52*(3), 2444–2455.
3. Ding, R., Wang, F., & Zhang, N., et al. (2020) A decentralized control strategy with output voltage deviation-correction for input-series-output-parallel DC transformer based on dual-active-bridge. In *9th international power electronics and motion control conference* (pp. 2458–2462). IEEE.
4. Chen, X., & Xu, G. (2022). Modulated coupled inductor for input-serial- output-parallel dual-active-bridge converter. *IEEE Transactions on Industrial Electronics, 69*(6), 6450–6455.
5. Giri, R., Choudhary, V., Ayyanar, R., & Mohan, N. (2006). Common-duty-ratio control of input-series connected modular DC-DC converters with active input voltage and load-current sharing. *IEEE Transactions on Industry Applications, 42*(4), 1101–1111.
6. Zumel, P., Ortega, L., Lazaro, A., Fernandez, C., & Barrado, A. (2013). Control strategy for modular Dual Active Bridge input series output parallel. IEEE 14th Workshop on Control and Modeling for Power Electronics (COMPEL), Salt Lake City, UT, USA, 2013, pp. 1–7. https://doi.org/10.1109/COMPEL.2013.6626418.
7. Shao, S., Chen, L., et al. (2022). Modeling and advanced control of dual active bridge DC-DC converters: A review. *IEEE Transactions on Power Electronics, 37*(2), 1524–1547.

GPSR Compliance

The European Union's (EU) General Product Safety Regulation (GPSR) is a set of rules that requires consumer products to be safe and our obligations to ensure this.

If you have any concerns about our products, you can contact us on ProductSafety@springernature.com

In case Publisher is established outside the EU, the EU authorized representative is:

Springer Nature Customer Service Center GmbH
Europaplatz 3
69115 Heidelberg, Germany

Batch number: 10726135

Printed by Printforce, the Netherlands